MEDICAL
INTELLIGENCE
UNIT 21

Transplant-Associated Coronary Artery Vasculopathy

Marlene L. Rose, Ph.D., MRCPath

National Heart and Lung Institute
Imperial College School of Medicine
Royal Brompton and Harefield Hospital NHS Trust
Harefield, Middlesex, U.K.

LANDES BIOSCIENCE
GEORGETOWN, TEXAS
U.S.A.

EUREKHAH.COM
AUSTIN, TEXAS
U.S.A.

TRANSPLANT-ASSOCIATED CORONARY ARTERY VASCULOPATHY
Medical Intelligence Unit
EUREKAH.COM
Landes Bioscience
Designed by Judith Kemper

Copyright ©2001 EUREKAH.COM
All rights reserved.
No part of this book may be reproduced or transmitted in any form or by any means, electronic or mechanical, including photocopy, recording, or any information storage and retrieval system, without permission in writing from the publisher.
Printed in the U.S.A.

Please address all inquiries to the Publishers:
Eurekah.com / Landes Bioscience
810 South Church Street, Georgetown, Texas, U.S.A. 78626
Phone: 512/ 863 7762; FAX: 512/ 863 0081
www.landesbioscience.com / www.EUREKAH.COM

ISBN: 1-58706-005-1 hard cover version
ISBN: 1-58706-078-7 soft cover version

While the authors, editors and publisher believe that drug selection and dosage and the specifications and usage of equipment and devices, as set forth in this book, are in accord with current recommendations and practice at the time of publication, they make no warranty, expressed or implied, with respect to material described in this book. In view of the ongoing research, equipment development, changes in governmental regulations and the rapid accumulation of information relating to the biomedical sciences, the reader is urged to carefully review and evaluate the information provided herein.

Library of Congress Cataloging-in-Publication Data

Transplant-associated coronary artery vasculopathy / [edited by] Marlene L. Rose.
    p. ; cm. -- (Medical intelligence unit)
    Includes bibliographical references and index.
    ISBN 1-58706-005-1 (alk. paper)
    1. Heart--Transplantation--Complications. 2. Heart--Transplantation--
Immunological aspects. 3. Coronary artery disease--Pathophysiology. I. Rose, Marlene
L. II. Series.
    [DNLM: 1. Heart Transplantation--adverse effects. 2. Coronary Disease--
etiology. 3. Coronary Disease--immunology. 4. Postoperative Complications. WG 169
T7718 2001]
RD598 .T645 2001
617.4'120592--dc21   00-021706

CONTENTS

1. **Coronary Arterial Disease in the Cardiac Allograft** 1
   Nicholas R. Banner
   Terminology 1
   Characteristics of Transplant-Related Coronary Disease 2
   Incidence and Prognosis 7
   Methods of Detecting Transplant-related Coronary Disease
       for Clinical Diagnosis and Research 9
   Therapeutic Approaches to Transplant-Related Coronary Disease 16
   Conclusions 18

2. **Atherosclerosis: Response to Injury** 26
   Nick Gall and Robin Poston
   Introduction 26
   General Pathology 26
   Important Molecules
       in Atherogenesis 30
   Mechanisms in Atherogenesis 41
   Risk Factors for Atherosclerosis 57
   The Positive Feedback Hypothesis 60

3. **The Immune Response to Endothelial Cells** 71
   Marlene L. Rose
   Basic Mechanism of Rejection 71
   Endothelial Cells as Antigen Presenting Cells 77

4. **Role of Hemostasis, Anticoagulation, Fibrinolysis
   and Endothelial Activation** 90
   Carlos A. Labarrere and David R. Nelson
   Patients and Methods 91
   Myocardial Fibrin and Transplant-Associated CAD 93
   Vascular Antithrombin and Transplant-Associated CAD 98
   Arterial and Arteriolar Endothelial Activation
       and Transplant-Associated CAD 102
   Arteriolar Tissue Plasminogen Activator
       and Transplant-Associated CAD 106
   Relationship of all Immuno-histochemical Measures
       with Transplant-Associated CAD 108
   Conclusions and Future Directions 112

5. **Nonimmune Factors in the Etiology
   of Graft Coronary Arteriosclerosis** 118
   Hiroaki Nagano and Nicholas L. Tilney
   The Cadaver Donor and Early Organ Injury 119
   Late Influences on Graft Arteriosclerosis 121
   The Common Pathway 123
   Conclusions 125

6. Tissue Remodeling and Extracellular Matrix Proteins 129
   Marlene Rabinovitch
   Fibronectin and Smooth Muscle Cell Migration 129
   Conclusions 138
   Abbreviations 140

7. Insights Regarding the Pathogenesis
   of Transplant Arteriopathy
   from Experiments with Animals 141
   Paul S. Russell, Catharine M. Chase and Robert B. Colvin
   Experimental Systems 142
   Quantification of Vascular Changes 145
   Some Findings from Animal Experiments 146
   Involvement of Cytokines, Adhesion Molecules
        and Growth Factors 154
   Pathogenesis of Transplant Arteriopathy 158
   Some Approaches to the Control of Transplant Vasculopathy 159
   Summary 161

8. Cytomegalovirus as a Contributing Factor
   in Transplant Vascular Sclerosis 166
   W. James Waldman, Deborah A. Knight and Adriana Zeevi
   Overview 166
   Biology of Human Cytomegalovirus 166
   CMV as a Contributing Factor in TVS: Clinical Observations
        and Animal Studies 168
   Graft Endothelial Activation
        and Host Inflammatory Responses 171
   CMV as a Catalyst of TVS Development: Potential Mechanisms 172
   Epilogue 175

9. Lipids and Lipid Lowering Therapy 181
   Jon A. Kobashigawa
   Potential Causes
        of Hyperlipidemia 181
   Clinical Course 182
   Treatment of Hyperlipidemia 183
   Summary 186

10. Immunosuppressive Drugs for the Prevention and Treatment
    of Transplant Coronary Artery Vasculopathy 190
    Norman P. Briffa, C.R. Gregory and Randall E. Morris
    Biology 190
    Studies of Chronic Rejection 191
    Mechanisms 191
    Azathioprine 194

Cyclosporine 194
In Vitro Studies 194
Preclinical Studies 195
Tacrolimus 196
Mycophenolate Mofetil 198
Pharmacokinetics
    and Mechanism of Action 198
Sirolimus (Rapamycin) 199
Leflunomide/Malononitrilamides 201
Future 201
Conclusion 202

Index 209

EDITORS

Marlene L. Rose
National Heart and Lung Institute
Imperial College School of Medicine
Royal Brompton and Harefield Hospital NHS Trust
Harefield, Middlesex, U. K.
Chapter 3

CONTRIBUTORS

Nicholas R.Banner
Royal Brompton and Harefield Hospital
   NHS Trust
Harefield Hospital
Harefield, Middlesex, U.K.
Chapter 1

N.P.Briffa
Transplantation Immunology
Department of Cardiothoracic Surgery
Stanford University School of Medicine
Stanford, California, U.S.A.
Chapter 10

Catharine M. Chase
Transplantation Unit
   of the Department of Surgery
   Department of Pathology
Massachusetts General Hospital
Harvard Medical School
Boston, Massachusetts, U.S.A.
Chapter 7

Robert B. Colvin
Transplantation Unit
   of the Department of Surgery
   Department of Pathology
Massachusetts General Hospital
Harvard Medical School
Boston, Massachusetts, U.S.A.
Chapter 7

Nick Gall
Kings College Hospital
Denmark Hill, London
Chapter 2

C.R. Gregory
Transplantation Immunology
Department of Cardiothoracic Surgery
Stanford University School of Medicine
Stanford, California, U.S.A.
Chapter 10

Deborah A. Knight
Department of Pathology
The Ohio State University College
   of Medicine and Public Health
Columbus, Ohio, U.S.A.
Chapter 8

Jon A. Kobashigawa
UCLA Heart Transplant Program
University of California
Los Angeles Medical Center
Los Angeles, California, U.S.A.
Chapter 9

Carlos A. Labarrere
Division of Experimental Pathology
Methodist Research Institute
Clarian Health
Methodist Indiana University
Riley Hospitals
Indianapolis, Indiana, U.S.A.
Chapter 4

Randall E. Morris
Transplantation Immunology
Department of Cardiothoracic Surgery
Stanford University School of Medicine
Stanford, California, U.S.A.
Chapter 10

Hiroaki Nagano
Department of Surgery
Brigham and Women's Hospital
Surgical Research Laboratory
Harvard Medical School
Boston, Massachusetts, U.S.A.
Chapter 5

David R. Nelson
Department of Biostatistics
  and Epidemiology
Cleveland Clinic Foundation
Cleveland, Ohio, U.S.A.
Chapter 4

Robin Poston
Department of Cardiology
Centre for Cardiovascular Biology,
  Medicine
New Hunt's House
King's College
Guy's Hospital
London, England, U.K.
Chapter 2

Marlene Rabinovitch
Division of Cardiovascular Research
The Hospital for Sick Children
Toronto, Ontario, Canada
Chapter 6

Paul S. Russell
Transplantation Unit
  of the Department of Surgery
  Department of Pathology
Massachusetts General Hospital
Harvard Medical School
Boston, Massachusetts, U.S.A.
Chapter 7

Nicholas L. Tilney
Department of Surgery
Brigham and Women's Hospital
Surgical Research Laboratory
Harvard Medical School
Boston, Massachusetts, U.S.A.
Chapter 5

W. James Waldman
Department of Pathology
The Ohio State University College
  of Medicine and Public Health
Columbus, Ohio, U.S.A.
Chapter 8

Adriana Zeevi
Thomas E. Starzl Transplantation
Institute and Department of Pathology
University of Pittsburgh Medical Center
Pittsburgh, Pennsylvania, U.S.A.
Chapter 8

FOREWORD

Cardiac transplantation has had a major impact on the quality of life and longevity of an ever-increasing number of patients. This benefit is significantly eroded by the development of an accelerated form of coronary arterial disease, which shows some, but not all, of the characteristics of native coronary artery disease, which in itself is one of the major indications for transplantation.

If cardiac transplantation is to realize its potential, it is essential to prevent transplant related coronary disease. This can only be done by thorough understanding of the basic mechanisms involved. In addition this could help in the fight against native atherosclerosis, which have a major impact on the community and in preventing vascular damage after other solid organ transplantation. To date there is no agreement or good guidelines about the management of chronic rejection.

Transplant associated coronary disease is a multifactorial disease contributed to by genetic factors in the donor and recipients, events occurring during brain death, harvesting and implantation and most importantly events after transplantation. These latter events can be conveniently divided into antigen dependent and antigen independent with immunological causes playing a part in both. Recent work has resulted in major and significant accumulation of knowledge in this field, particularly in the molecular mechanisms and, to some extent, management of the disease. This knowledge is extensively and methodically reviewed in this volume by a group of experts in the field. I am confident that this will help in the management and, most importantly, further understanding of this important disease and help future research with the ultimate aim of reducing or abolishing the impact of the disease.

Sir Magdi Yacoub, FRCS

ACKNOWLEDGMENT

This book was published with the assistance of an educational grant from Roche Products, LTD., Welyn Garden City, Herts., U.K.

CHAPTER 1

Coronary Arterial Disease in the Cardiac Allograft

Nicholas R. Banner

Cardiac transplantation has become established as the treatment of choice for selected patients with advanced heart failure. The survival rate achieved with transplantation greatly exceeds that achieved with optimum medical therapy for this condition. Nevertheless, the long-term results of heart transplantation are far from perfect. The Registry of the International Society of Heart and Lung Transplantation (ISHLT) indicates that the overall patient half-time (time to 50% survival after transplantation) is only 8.7 years and the conditional half-life in those surviving beyond the first year is only 11.4 years.[1] After the first year there is a constant mortality rate of 4% a year. Although there was a substantial improvement in survival in the mid-1980s, further progress has been slow.[1] The leading causes of death more than 1 year after transplantation are coronary artery disease, infection and malignancy.[2] In one series, coronary artery disease accounted for about a quarter of deaths occurring 1-3 years after transplantation and the mortality rate had not improved with time;[2] in another series, coronary disease caused the majority of late deaths.[3] The angiographic incidence of coronary abnormalities increases steadily with time and, once angiographically apparent, the disease may progress rapidly.[3,4]

The problem of coronary arterial disease in the allograft occurred in the first medium-term survivor of human heart transplantation,[5] and soon became recognized as a major factor limiting the long term outcome of heart transplantation.[6] The pathological characteristics of the disease differ from spontaneous coronary atheroma and similar vascular lesions have been found in renal allografts.[7,8] This Chapter provides an overview of this clinical problem as a framework for subsequent discussions of the mechanisms underlying the disease.

Terminology

In this Chapter, the term *spontaneous coronary atherosclerosis* is used to describe the coronary disease which is prevalent in the general adult population of the economically developed nations. Coronary disease within the cardiac allograft has been approached from different perspectives. The cardiologist and cardiac surgeon often view it as a special form of coronary arterial disease, whereas the immunologist may regard it as an example of chronic vascular rejection. A variety of terms have been used to describe this disease reflecting particular aspects of the condition. *Transplant-related coronary disease* and *transplant atherosclerosis* emphasize the link with spontaneous coronary atherosclerosis. *Graft vascular disease and transplant vasculopathy* draw attention to the distribution of the disease which also affects the section of the aorta within the allograft, the coronary microvasculature and causes venous intimal sclerosis.[9,10] *Accelerated coronary disease* underscores the aggressive nature of the disease in many

Transplant-Associated Coronary Artery Vasculopathy, edited by Marlene L. Rose.
©2001 Eurekah.com.

patients. *Chronic rejection* implies that the cause is mainly due to an alloimmune response. It also draws attention to the similarity of the process to the chronic pathology seen in other solid organ transplants where progressive functional deterioration may occur in association with gradual vascular obliteration and other morphological changes that ultimately lead to fibrosis.[8] *Chronic graft* dysfunction does not presume that a specific mechanism (such as alloimmunity) has a central role and also does not emphasize that damage is usually centered within the coronary arteries.

Characteristics of Transplant-Related Coronary Disease

Coronary arterial disease within the allograft frequently has features which are not typical of conventional coronary atherosclerosis. Such transplant-related lesions often coexist with lesions which are more typical of spontaneous atherosclerosis (Table 1.1).

Both conventional atherosclerosis and the transplant-related disease are believed to result from the response of the vessel to injurious stimuli[11,12] (also see Chapter 2). The vascular narrowing may be viewed as the manifestation of an inappropriate reparative response within the vessel wall. Transplant disease differs from spontaneous atherosclerosis by virtue of the nature, intensity and distribution of the initiating insult.[11,13]

Endothelial cells play a key role in regulating coronary vasomotor tone and in orchestrating the response to injury.[12,14] A variety of insults may damage the endothelium and activate "response to injury" programs within the cells of the vessel wall.[15] In the allograft, ischemia and reperfusion injury from organ preservation and alloimmune-mediated damage will interact with the factors which predispose to coronary disease in the general population (Fig. 1.1). The resulting disease is more diffuse and has a tendency to affect more distal coronary branches as well as the proximal epicardial vessels. The lesions present in the arteries of transplanted hearts represent a spectrum of features from typical transplant disease to more conventional atherosclerosis.[16,17]

The pathology of transplant disease demonstrates intimal lesions which are relatively uniform containing both cellular and fibrous tissue. Consequently luminal narrowing is similar in adjacent segments, and conventional coronary angiography is relatively insensitive in detecting the disease.[18] The disease is not restricted to the intimal layer: adventitial fibrosis and fibrous infiltration of the adjacent subepicardial tissues may inhibit positive remodeling of the epicardial arteries and may even constrict them ("negative remodeling"). Inflammatory cellular infiltrates are often present in the subepicardial tissues even when myocardial interstitial inflammatory infiltrates were absent.[18] Arteries from hearts examined at autopsy, or at explant because of retransplantation, more than 2 months after the original operation, frequently contain thrombus.[19]

In life, the spectrum of coronary lesions can be demonstrated angiographically. Gao and colleagues described a variety of lesion types (Table 1.2). Type A lesions have the features of conventional coronary atherosclerotic disease whereas Type B and C lesions represent the more typical transplant-related disease. An alternative angiographic method of classification, which emphasizes suitability of disease within a coronary territory for revascularization procedures, is presented in Table 1.3 (Figs. 1.2-1.5). In diffuse disease the physiological significance of a lesion cannot be assessed from the apparent "percentage stenosis" associated with it.

Further insight into the heterogeneity of the coronary lesions within the transplanted heart has been obtained from intravascular ultrasound (IVUS) techniques which allow direct imaging of the vessel wall.[20] Although donor-related (conventional) and acquired (transplant-related) lesions have differing characteristics, there is considerable overlap.[21]

Clinical Features

The clinical presentation of coronary disease in the heart transplant recipient differs from that seen in the general population.

Table 1.1. Stereotypic features of spontaneous coronary atherosclerosis and of transplant-related coronary disease. Pathological examination of the coronary arteries of allografts usually reveals a variety of lesions with features spanning the spectrum between these two extremes

Spontaneous Atherosclerosis	Transplant-Related Atherosclerosis
Epicardial vessels	Large and medium sized arteries
	Microvascular component
Eccentric lesions	Concentric lesions
Slow progression	Rapid progression
Lesions lipid rich	Lesions lipid poor
Calcification early	Calcification relatively late

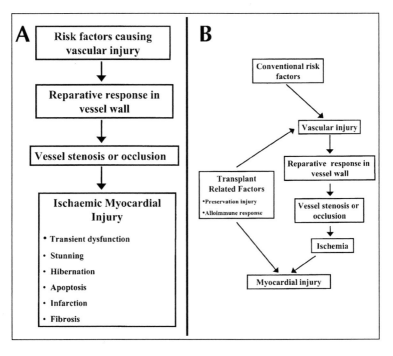

Fig. 1.1. Comparison of the pathways leading to vascular and myocardial injury in spontaneous atherosclerosis and in transplantation. A) In spontaneous disease the primary insult is to the vessel wall and myocardial damage occurs due to ischemia. B) In the allograft additional transplant-related factors contribute to the vascular injury but may also cause direct injury to other structures in the heart including the myocardium, specialized conducting tissues and pericardium.

Table 1.2. The Gao/Stanford Classification of angiographic lesion morphology.[23,42]

Lesion Type	Characteristics
A	Focal, often eccentric, involve proximal epicardial vessels
B1	Abrupt transition from normal to diffusely thinned vessel, often at branch point
B2	Gradual tapering of the distal vessel and its branches due to pathological thinning
C	Narrowed distal vessels with abrupt terminations ("Cut-off" appearance)

Table 1.3. Classification of angiographic disease within a coronary territory, related to potential suitability for revascularization

Disease Type	Characteristics	Comment
1	Proximal, discrete, relatively short lesions	Characteristic of conventional coronary atherosclerosis. Corresponds to Gao Type A lesions. Usually suitable for PTCA/Stent placement or CABG.
2	Diffuse disease and involving distal/branch vessels.	Typical features of transplant-related disease (Corresponds to Gao Types B1,B2 and C). Generally unsuitable for PTCA procedures or CABG.
3	Combination of diffuse Type 2 disease with one of more focal (Type 1) lesions in the proximal part of the vessel.	Proximal lesions may be technically suitable for PTCA/Stent placement but the procedure usually only produces temporary palliation because of progression of the distal disease
4	Total vessel occlusion	Nature of underlying disease indeterminate unless previous angiogram available

PTCA: Percutaneous coronary angioplasty; CABG: Coronary artery bypass graft

Surgical denervation of the heart results in a number of physiological abnormalities and may abolish the *angina pectoris* that often accompanies episodes of myocardial ischemia in other settings.[22,23] With time, however, the majority of patients develop some degree of sympathetic and afferent reinnervation[24] and occasional, well documented, cases of ischemia causing chest pain have been reported.[25-27] The physician should be wary of chest discomfort after heart transplantation although the great majority of patients

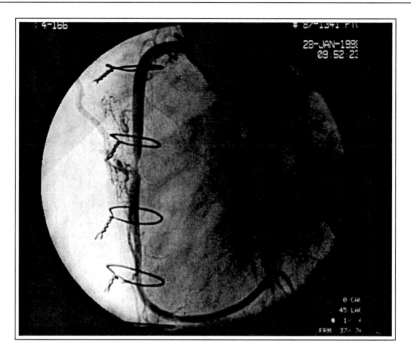

Fig. 1.2. Right coronary artery of a transplanted heart with a focal stenosis 10 years after surgery (type 1 Disease).

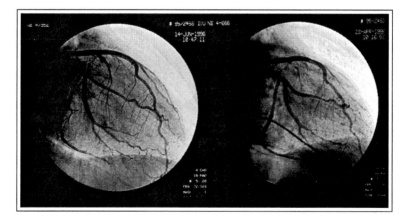

Fig. 1.3. Angiograms of a left coronary artery early after transplantation (left panel) and again at 3 years (right panel). Diffuse branch vessel disease, typical of transplant-related (type 2) disease, has developed affecting the septal and diagonal branches of the anterior descending artery and also the marginal branch of the circumflex.

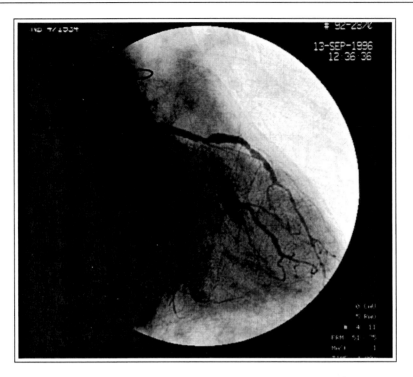

Fig. 1.4. Left coronary artery of transplanted heart at 5 years. The anterior descending shows both discrete proximal narrowings and diffuse disease manifest by pathological irregularity and tapering of the distal vessel (type 3 disease).

with transplant-related coronary disease do not experience pain.[28]

Myocardial infarction presents with pain in only a minority of cases. New onset of fatigue or breathless or palpation is more typical.[29] The conventional ECG changes of a Q-wave infarction are often absent. This may be partly due to pre-existing ECG abnormalities in some patients but also reflects the patchy, multifocal, nontransmural nature of infarcts that occur in the setting of diffuse transplant coronary disease.[29] Patients usually present too late for thrombolysis or revascularization by direct angioplasty to be effective. Myocardial infarction in this setting carries a high risk of heart failure, cardiogenic shock and death.[29,30]

Progressive *myocardial damage and fibrosis* can lead to deteriorating cardiac performance which may present nonspecifically

as effort intolerance, exertional dyspnea and fatigue. Ventricular systolic dysfunction is as important a predictor of cardiac events as the presence of angiographically detectable coronary artery disease.[31] Overt *congestive heart failure* may occur due to systolic or diastolic ventricular dysfunction.[30,32]

Damage to, and fibrosis within, both the myocardium and the specialized conducting tissues can lead to *arrhythmia, bradycardia* and *sudden death.*[33,34] Sudden death may also be a complication of recent coronary thrombosis and myocardial infarction.[30]

Coronary spasm has been documented in number of case reports.[35-38] This phenomenon probably reflects endothelial dysfunction and may be the harbinger of overt coronary disease.[37] Once again, cardiac denervation may prevent chest pain from occurring. Two cases where coronary spasm caused syncope have been reported.[39]

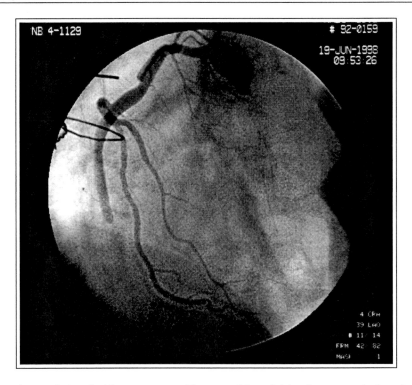

Fig. 1.5. Abrupt occlusion of a right coronary artery. The nature of the underlying disease cannot be determined from this angiogram (type 4 disease).

Clinically apparent spontaneous coronary atherosclerosis is virtually unknown in childhood although early lesions can be found microscopically[11] and clinical disease may occur in certain pathological conditions.[40] Unfortunately, *pediatric recipients* of heart transplants can develop coronary disease which may follow the same accelerated clinical course which occurs in some adult patients.[41]

Incidence and Prognosis

Despite its limitations in detecting transplant-related disease at an early stage, angiography remains the tool most frequently used for clinical surveillance after heart transplantation.[23,42] All series report a progressive increase in the prevalence of angiographically defined disease over time. The reported incidence varies between centers reflecting differences in definition and in patient management protocols. For example, prevalence between 5.8%[43] and 18%[44] at 1 year and between 36%[45] and 67%[46] at 5 years has been reported. A survey of 2607 patients in the Multicentre Cardiac Transplant Research Database who have undergone angiography showed an overall prevalence of 42% at 5 years after transplantation. The disease was graded as mild in 27%, moderate in 8% and severe in 7%.[47] Studies using quantitative angiography and intravascular ultrasound indicate that many patients have disease which is undetected by conventional angiography.[48,49]

Angiographically defined disease predicts subsequent cardiac events and is associated with an increased mortality.[2,31,50-53] In one study, actuarial survival 5 years after the detection of moderate or severe angiographic disease was only 17%.[52] Disease that develops early after transplantation appears to be

Table 1.4. Risk factors contributing to the development of transplant-related coronary disease

Donor	Recipient	Immunological
• Increasing age • Male sex • Hypertension • ACE genotype • Process of brain death • Organ ischemic/ reperfusion injury	• Ischemic heart disease • Overweight • Hyperlipidemia • Increasing age • Male sex • Posttransplant hypertension • Smoking after transplant • Diabetes • Increased plasma insulin • Coagulation abnormalities	• Cellular rejection • Antibody response (donor-specific and antiendothelial) • HLA matching (?) • Low cyclosporin dose • Increased steroid exposure • CMV infection
References: 47,54-61	43,47,57,62-73	43,51,55,56,62,65,68,74-91

more aggressive and to be linked to a worse prognosis.[51] Uretsky found for those with angiographic coronary disease the relative risk of any cardiac event was 3.4 and of death 4.6 compared to patients without such disease.[30]

Factors Contributing to Coronary Disease

Attempts to reduce the incidence and impact of this disease are based on our current understanding of its pathogenesis. A brief outline of our current knowledge about contributory factors is given here (Table 1.4). The mechanisms underlying transplant coronary disease are addressed in detail in subsequent Chapters.

Spontaneous atherosclerosis is a multifactorial disease. Many of the factors which are known to contribute to such atherosclerosis have also been shown to promote transplant-related disease when they are present in either the donor or the recipient. Other factors play a prominent role in the transplant setting including injury during preservation of the transplanted organ at the time of surgery and the alloimmune response of the recipient to the graft.

The process of brain death is known to cause damage to the cardiac allograft prior to it being retrieved for transplantation.[59-61] In kidney transplantation, organs retrieved from live unrelated donors appear to fare better than cadaveric organs suggesting that brain death in the organ donor may prejudice the future function of the graft.[8] In addition, ischemia/reperfusion injury related to organ preservation may damage the vascular endothelium and predispose to chronic rejection.[8] These insults, which are not dependent on the presence of alloantigens within the graft, may explain why kidney isografts gradually can develop chronic graft dysfunction with features identical to chronic allograft rejection.[92] Nonallospecific graft injury may render the graft more "visible" to the immune system and so predispose to subsequent rejection and thus to further damage.

A number of studies have found an association between acute cellular rejection and the subsequent development of transplant coronary disease,[55,56,62,68,75,76,93] although others have not.[57,64,66,94] Individual studies have emphasized various features of acute rejection which may predispose to chronic damage such as early rejection,[43,56,93] recurrent rejection,[44] and a potential link with episodes of acute coronary vasculitis.[95-99] Although there is good evidence to regard acute cellular rejection as a predisposing factor for transplant coronary disease, it is not clear which aspect of the process (severity, duration, recurrence) is most important. In the clinical setting, assessment of acute cellular rejection is made by endomyocardial biopsy and the endomyocardial samples obtained may not be fully representative of the processes occurring near the epicardial surface of the heart or within the coronary arteries.[18]

Although some studies have demonstrated a relationship between donor-recipient HLA matching and the risk of transplant coronary disease,[55,100] many studies have been unable to confirm such a link.[57,66,74,101] Humoral or antibody-mediated rejection has also been linked to the development of transplanted-related coronary disease. Both antibodies to HLA antigens[80,82] and those directed at vascular endothelial cells have been implicated.[77-79,81,83]

Infection may also play a role in the pathogenesis of vascular injury. After heart transplantation, cytomegalovirus (CMV) infection (which is the commonest opportunistic pathogen in this population) appears to be a contributory factor. The majority of studies have found an association between CMV infection and transplant-related coronary disease [51,55,62,84-91] although some have not.[63,66]

The relationship between the immunosuppressive agents used to prevent allograft rejection[102] and transplant-related coronary disease is complex. To the extent that acute rejection appears to predispose to chronic rejection, improved maintenance immunosuppression could reduce transplant coronary disease. However, both cyclosporin and corticosteroids produce adverse effects on lipid metabolism and cyclosporin is injurious to endothelial function.[103-108] Low cyclosporin dose has been related to the development of transplant coronary disease presumably due to less effective immunosuppression.[56] In one report changing from immunosuppression with corticosteroids and azathioprine to corticosteroids and cyclosporin did not reduce the incidence of transplant-related coronary disease despite the improved immunosuppression with

cyclosporin.[109] At our hospital, changing from a corticosteroid and azathioprine protocol to one of cyclosporin and azathioprine, did reduce the incidence of coronary problems.[65] Whether this was due to a beneficial effect of azathioprine or an injurious effect of steroids is not clear. Since double therapy immunosuppression protocols do not produce adequate control of acute rejection and most centers currently use a triple therapy protocol (cyclosporin, corticosteroids and either azathioprine or mycophenolate mofetil), this issue cannot be resolved. Newer immunosuppressive agents, which have antiproliferative properties, such as mycophenolate mofetil and sirolimus (rapamycin) may have a beneficial effect on transplant coronary disease by: more effective immunosuppression; direct actions on the vasculature; or by allowing reduced use of corticosteroids and cyclosporin. The one-year results of the recent trial of mycophenolate in heart transplantation were not conclusive with regards to coronary disease and the three year results are awaited.[110]

A number of risk factors for spontaneous coronary atherosclerosis have been established from epidemiological studies and some can be modified by specific therapeutic interventions. Many of these factors have been found to have a role in the pathogenesis of transplant-related coronary disease when present in either the organ donor[47,55-58] or in the recipient after transplantation.[43,57,62-66,69] Hypertension is common after heart transplantation[111] and is associated with transplant coronary disease.[68] Hyperlipidemia is also common and linked to the disease.[57,65,67,112]

Methods of Detecting Transplant-Related Coronary Disease for Clinical Diagnosis and Research

By the time a heart transplant recipient has developed symptoms related to transplant coronary disease the disease is usually very advanced and therapeutic options are limited. A number of methods can be used to detect the presence of transplant coronary disease

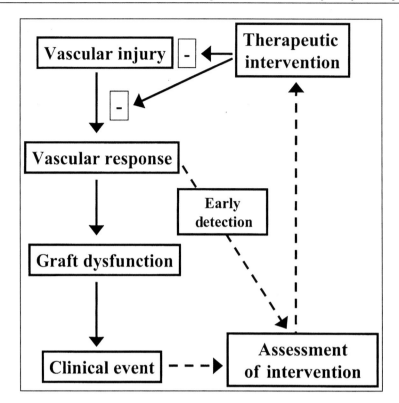

Fig. 1.6. Evaluation of therapeutic interventions for transplant coronary disease. Although the coronary disease observed in the cardiac allograft often runs an accelerated course compared to spontaneous atherosclerosis, the time interval between the initiating events and major clinical events (such as heart failure, myocardial infarction and death) may be many years. This slows the evaluation of therapeutic maneuvers aimed at preventing the disease. Early detection of the disease, using surrogate markers such as coronary changes detected with quantitative angiography or intravascular ultrasound can speed up clinical research. The success of this strategy depends on how well the surrogate end-points used predict future clinical events which would be of significance to the patient.

before symptoms develop. These may be used as diagnostic investigations to guide the clinical management of an individual patient and can also be applied to groups of patients in clinical research protocols aimed at either defining the mechanisms underlying the disease or evaluating therapeutic interventions. Clinical research into chronic conditions such as transplant coronary disease is slowed by the time interval between the initiating factor (or the application of a therapeutic intervention) and the occurrence of clinical events. The use of sensitive detection methods to find early disease can speed up the research process. The strategy of using such methods to provide surrogate end-points for a study depends on

the relationship between such end-points and future clinical events. (Fig. 1.6).

Both noninvasive and invasive methods have been used for diagnosis. In general, noninvasive methods rely on detecting the consequences of transplant coronary disease, such as ischemia and are unlikely to detect early vascular changes (Fig. 1.1). This is because noninvasive imaging does not have the resolution to examine the coronary arterial tree in detail. Ultrafast CT has been used to diagnose coronary disease by detecting coronary calcification but it does not provide detailed information about the vessel lumen or wall.[113,114] Therefore, this review focuses on the invasive approach to diagnosis using coronary angiography and the allied techniques of intravascular ultrasound and intracoronary Doppler measurement.

Qualitative Coronary Angiography

Coronary angiograms are obtained by injecting radio-opaque contrast to produce X-ray images of the vessel lumen. Various projections are used to generate a series of two dimensional radiographic images. In standard clinical practice, the images obtained are interpreted subjectively by the cardiologist. Areas of narrowing are evaluated by comparison with adjacent segments of the vessel. Angiography has an inherent tendency to underestimate atherosclerotic disease because only the lumen of the vessel is seen. The technique is particularly insensitive in transplant-related coronary disease because the vessels are often diffusely narrowed and there is no reference segment in which the normal diameter of the vessel can be assessed. Quite often luminal irregularities are present which give a clue to the presence of disease. In an attempt to compensate for the insensitivity of conventional angiography in the transplant setting, most cardiologists report the presence of even minor irregularities. Despite this approach, a considerable amount of disease is missed. In addition, over-reporting may occur if a branch which is small and tortuous is misinterpreted as having atherosclerosis. Many reports have documented the insensitivity of conventional coronary angiography to diffuse transplant-related coronary disease.[115-119]

The situation can be improved considerably by reviewing a patient's angiograms as a series and making side by side comparisons of vessel caliber and morphology. If an angiogram is performed early after transplantation, any disease present is likely to be focal in nature representing spontaneous atherosclerosis already present in the donor heart. Such disease can be assessed relatively successfully by conventional angiography. Subsequent transplant-related disease, which may produce diffuse thinning of the vessels, can be appreciated more easily by comparison with the original angiogram. Technical factors, including the angiographic projections and magnification used, can interfere with such comparisons and it is advisable to use the same angiographic protocol at each study. This becomes essential if the films obtained are to be used for quantitative angiography.

Quantitative Coronary Angiography

Computer-assisted quantitative image analysis can be applied to a series of coronary angiograms to measure changes in lumen diameter. Edge-detection algorithms remove the subjective element from the measurement (Fig. 1.7).[120-122] High quality angiographic images are required and the radiographic techniques used must be standardized to provide comparable images for measurement from each study. Intracoronary nitrates must be administered to eliminate any variation due to changing vascular tone. Quantitative coronary angiography (QCA) does not provide an absolute measurement of lumen diameter. This is because each image is calibrated against the diameter of the angiography catheter used for the study. The catheter does not lie in the same plane as the artery and a magnification error will occur. Nevertheless, provided the technique is standardized the magnification ratio will remain constant between examinations allowing valid comparisons to be made. Quantitative angiography is more sensitive than conventional angiography for detecting vascular narrowing after transplantation.[123-125] The method has been used to evaluate therapy aimed at preventing transplant-related coronary disease.

Despite the improvements made by computer-assisted quantification, QCA has limitations in assessing coronary disease and is less sensitive than intravascular ultrasound (IVUS). Since the vessel wall is not imaged in QCA, early changes may be missed. As atherosclerotic plaque accumulates the vessel may dilate to preserve the luminal area.[126] This remodeling phenomenon can mask disease and a dilated form of arteriopathy has been described after transplantation.[127]

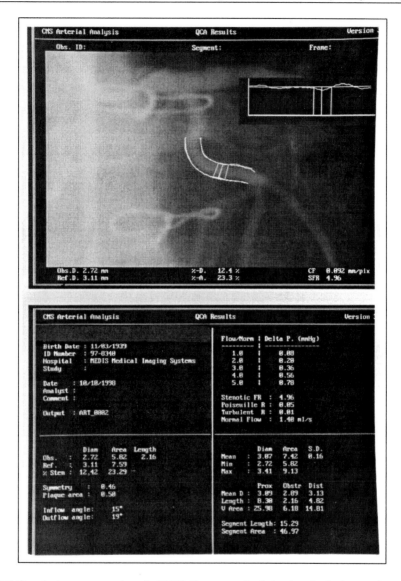

Fig. 1.7. Quantitative coronary angiography (QCA). Computer assisted edge-detection has been used to measure the luminal diameter of a segment of the circumflex coronary artery.

QCA has some advantages over IVUS. It is far less invasive, and requires only minor modifications to the conventional angiographic technique, making it more acceptable to patients and ideal for incorporation into a routine follow-up protocol. The patients do not need systemic anticoagulation, which is required for intravascular ultrasound, and so there is no concern about performing a cardiac biopsy during the same procedure. Most importantly, the distal branch vessels which are so often involved in transplant disease can be easily examined angiographically but are inaccessible to the intravascular ultrasound probes that are currently available. The overall image of the vessel provided by an-

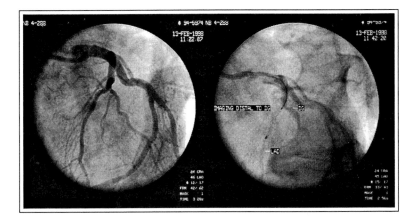

Fig. 1.8. Intravascular ultrasound (IVUS) imaging. The anatomy of the left coronary artery has been demonstrated angiographically (left panel). A 3Fr IVUS catheter has been introduced over an angioplasty guide wire into the anterior descending branch (LAD) and the guide wire has then been withdrawn. The tip of the catheter lies in the mid anterior descending and the ultrasound transducer (which can be moved within the catheter) has been positioned just beyond the diagonal branch (DG).

giography is essential to guide the passage of an intravascular ultrasound probe and to plan therapeutic revascularization procedures.

Intravascular Ultrasound

IVUS is the only technique which can provide cross-sectional images of the coronary arterial wall during life. These images have been compared to those produced by histology,[20] although the resolution is relatively low and ability to distinguish variations in tissue quality rather limited. Ultrasound imaging is performed as part of a coronary angiographic procedure. After anticoagulation with heparin and administration of intracoronary nitrates to reduce the risk of vessel spasm, an angioplasty guide wire is introduced into coronary artery via a guide catheter and then the ultrasound catheter is introduced over the guide wire. Radiographic imaging is used to position the wire, the ultrasound catheter and the transducer (Fig. 1.8). The position of the transducer can also be assessed from the ultrasound images which reveal the location of vascular branch points which can then be identified angiographically. The ultrasound catheters currently in use are between 2.9 and 4.3 Fr in diameter and allow imaging from

within the epicardial vessels as well as large proximal branches. Accessibility may be limited by variations in coronary anatomy, excessive tortuosity or disease.

Imaging at 30 MHz, an axial resolution of 150 mm can be achieved. In normal young people, the thickness of the intima is below the level of resolution and the artery wall has a uniformed single layered appearance. In older adults, and in disease, thickening of the intima allows it be imaged as a separate structure from the media and adventia giving the artery wall a characteristic three-layered appearance. The initima area can be planimetered to measure the plaque burden as a proportion of the vessel area or area inside the external elastic lamina (Fig. 1.9).[128]

The ultrasound beam thickness is approximately 200 mm, so that each image represents a thin cross-section through the vessel wall.[128] Since the degree of intimal thickening is usually not uniform within a segment, serial measurement to evaluate disease progression must either be made in precisely identified positions or averaged over a vessel segment.[129,130] Landmarks such as vessel branch points and coronary veins can be used to identify the positions used in serial

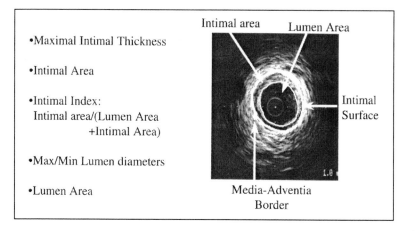

Fig. 1.9. Cross sectional IVUS image of a coronary artery within an allograft which shows marked intimal thickening. The intimal surface bounds the lumen area while the media-adventia border defines the vessel area. A number of parameters of intimal proliferation can be derived by planimetry.

studies[128] but the process can be lengthy if multiple measurements are required. Since atherosclerosis is often more prominent near vessel bifurcations this approach may bias the measurements obtained. Methods to assess the average plaque burden (intimal area) within a vessel segment are therefore required. One approach is to sample, randomly, sufficient images from a continuous pull-back scan to provide a valid estimate of the plaque within a vessel segment.[130]

Intravascular ultrasound has provided a new perspective on the atherosclerotic process in both the transplant and nontransplant settings. The mechanisms of lumen narrowing can now be partitioned into changes in intima/plaque area and remodeling changes in the vessel wall.[131] Unique insights have been obtained into the mechanism of action of therapeutic interventions.[128] In transplant patients, IVUS is more sensitive than angiography.[21,64,132,133]

IVUS has been criticized for being too sensitive to early coronary changes, but it can predict the subsequent development of angiographic disease in transplant patients.[134] Nevertheless, IVUS has some limitations. It is highly invasive, requires anticoagulation and prolongs the cardiac catheterization proce-

dure. It cannot be applied to the smaller branch vessels of the coronary tree. The cost of the single-use ultrasound catheters is relatively high and adds appreciably to the cost of cardiac catheterization.

Intracoronary Doppler Flow Measurement

Doppler ultrasound can be used to measure intracoronary flow velocity. The original technique involved the placement of an intracoronary Doppler catheter of about 3 Fr (1 mm) diameter. However, catheters of this size could significantly disturb the flow within small or diseased vessels. Technological improvement has lead to the introduction of Doppler transducers mounted on an angioplasty guide wire of 0.014-0.018" diameter.[135,136] Spectral analysis provides beat-by-beat measurement of coronary flow velocity. Pharmacological interventions to increase coronary flow, such as the intracoronary injection of adenosine, can then be used to measure maximal coronary flow and calculate coronary flow reserve (Fig. 1.10).[137,138] Significant focal or diffuse coronary disease will reduce maximal flow and reduce the flow reserve which provides a measure of the functional effect of the coronary disease. A

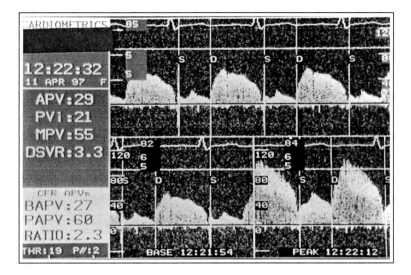

Fig. 1.10. Doppler flow measurements within the left coronary artery. A Doppler ultrasound transducer mounted on an 0.014" angioplasty guide wire (FloWire, Cardiometrics Inc.) has been introduced into the artery. The spectral display shows beat-by-beat coronary flow (which occurs predominately in diastole in the left coronary system). Following injection of adenosine the flow velocity increases from baseline (lower left hand panel within the display) to a peak (lower right hand panel). The ratio of peak to baseline flow velocity defines the coronary flow reserve (CFR).

number of other factors may influence the measurement, however, including those that alter resting myocardial oxygen demand and thus coronary flow (e.g., heart rate), loading conditions (e.g., blood pressure) and other pathological states (e.g., left ventricular hypertrophy).[136,139] Technical factors also require attention, care must be taken to place the Doppler sample volume in the center of the vessel, with a minimal angle of incidence to the flow, to ensure the maximal velocity is recorded. A stable position is required so that the velocity recorded returns to the same baseline between each pharmacological challenge. IVUS and Doppler studies can be combined to provide detailed information about physiology and pharmacology of both large and small coronary vessels.[140] Coronary flow reserve has been found to be reduced in patients with angiographic transplant coronary disease and deteriorates with increasing time after transplantation.[141,142] Measurement of coronary flow reserve reflects changes in

the microvasculature as well as in the epicardial coronary arteries. At present, it is unknown whether this technique provides additional information which is of clinical importance in the transplant setting.

Endothelial Function in the Epicardial Coronary Arteries

The endothelium plays a key role in the regulation of vascular tone.[14] Endothelial dysfunction is an early event in atherosclerosis and functional abnormalities within the endothelium may be found before structural changes are detected within the vessel wall. Glyceryl trinitrate acts directly on vascular smooth muscle to produce vasodilatation whereas some other agents, such as acetylcholine, work indirectly through their action on the endothelium.[14,143] The loss or attenuation of endothelium-dependent vasodilatation can be used to detect endothelial cell dysfunction. Both QCA and IVUS can be used to assess changes in vessel diameter in response to pharmacological interventions.[144] Abnormal responses to acetylcholine have been reported both early and late after heart transplantation.[145,146]

Therapeutic Approaches to Transplant-Related Coronary Disease

As with spontaneous atherosclerosis, therapy may be directed at: preventing the initial vascular injury; interrupting secondary changes within the vessel wall; or attempting to deal with the consequences of the disease by revascularization. In view of the poor prognosis associated with clinically apparent disease, the adage that 'prevention is better than treatment' is very appropriate.

There is a strong evidence linking lipid abnormalities after transplantation to transplant coronary disease.[57,65,67,112] Initially, pharmacological interactions between the HMG Co A reductase inhibitors and cyclosporin caused concern about the vigorous treatment of hypercholesterolemia in transplant patients because of the danger of rhabdomyolysis.[147] However, two prospective clinical trials have confirmed that lipid lowering therapy (pravastatin or simvastatin) can improve survival after heart transplantation and reduce transplant-related coronary disease.[148,149]

Hypertension is strongly linked to spontaneous coronary atherosclerosis, is common in the postheart transplant population, and is often difficult to control in transplant patients. Although no studies have specifically addressed the issue of whether blood pressure control is related to the risk of transplant-related coronary disease, it seems likely that it is. Current guidelines recommend that blood pressure should be vigorously treated in groups with poor cardiovascular risk profiles and in those with renal impairment.[150] Since post transplant hypertension is predominantly a problem of vasoconstriction, vasodilator agents such as calcium antagonists are recommended.[150] One study has examined the use of the calcium antagonist diltiazem for the primary prevent of transplant coronary disease.[151,152] Diltiazem prevented loss of lumen diameter, as assessed by quantitative angiography, but overall survival was similar to that seen in the control group.

Although, there is no direct evidence that treating other 'conventional' coronary risk factors is beneficial after heart transplantation, it is rational to pay attention to the issues of smoking and weight control as well as ensuring careful management of concomitant diabetes.

The link between acute rejection and transplant coronary disease, suggests that more effective immunosuppression protocols may reduce coronary problems. Unfortunately many of the agents available have poor cardiovascular risk profiles and may have adverse coronary effects. Corticosteroids, cyclosporin and tacrolimus can cause weight gain, hypertension, dyslipidemia and diabetes mellitus.[102] Mycophenolate mofetil has a benign cardiovascular profile and also has antiproliferative effects on vascular smooth muscle cells. The preliminary, one year, results of the study of mycophenolate as a primary immunosuppressive agent after heart transplantation did not show definite benefits from the standpoint of coronary disease but the 3 year results are needed before a conclusion can be made.[110] Other new immunosuppressive agents such as sirolimus (rapamycin) and RAD also have antiproliferative properties but, unfortunately, also have adverse effects on lipid metabolism. It is interesting to note that some of the beneficial effects of lipid lowering therapy after transplantation may come from immunomodulatory actions.[153]

Coronary thrombosis is a common cause of infarction and death in patients with advanced transplant coronary disease.[19,154] Aspirin is widely used as a prophylactic agent as it is in spontaneous coronary disease. There is some in vitro evidence that platelets from patients with transplant coronary disease may be resistant to the inhibitory effects of aspirin.[155] There are no clinical trials to assess whether anticoagulation with warfarin or newer antiplatelet agents such as clopidogrel can offer further protection.

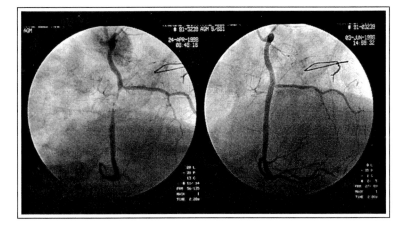

Fig. 1.11. Left: Type 1 disease in the right coronary artery of a patient 6 years after heart transplantation. Right: Lesion successfully treated by angioplasty followed by the implantation of a stent.

Revascularization and Retransplantation

As with conventional coronary artery disease, a number of approaches are available to try to relieve ischemia in patients with transplant coronary disease. Unfortunately the diffuse and distal nature of the disease present in many patients make them unsuitable for such procedures (Type 2 Disease). Percutaneous coronary angioplasty (PTCA) has been used successfully in patients with type 1 disease. The procedural success rates have been high with low procedural risks.[156-162] Restenosis has been a problem and, as with other groups of patients, intracoronary stenting has helped to reduce this (Fig. 1.11).[163-167] Several other interventional procedures (Rotational atherectomy, directional atherectomy and laser therapy) have been reported in small numbers of patients.[157,168,169] The indications for PTCA/coronary stenting have usually been the presence of suitable angiographic lesions with or without objective evidence of ischemia. Since angina is usually absent in this patient population, symptomatic benefits are difficult to assess. By the time interventional procedures are performed, most patients already have advanced disease. Since no randomized controlled trials have been conducted, it is un-

certain whether PTCA/stenting alters the prognosis of transplant coronary disease. In patients with type 3 disease the procedure is purely palliative and progression of distal and branch vessel disease usually limits the benefit obtained.

Patients with multi-vessel type 1 disease and adequate distal vessels may be suitable for coronary bypass surgery (CABG)(Fig. 1.12). Careful case selection is important due to the tendency for conventional angiography to underestimate the extent of disease. Relatively small numbers of procedures have been reported, reflecting both the preferential use of PTCA and coronary stenting and also that many patients do not have suitable distal vessels for bypass grafting.[157,165,170-173]

The problem of distal and branch vessel disease precluding conventional bypass surgery has lead investigators to explore the role of transmyocardial laser revascularization (TMR).[174-176] Theoretically TMR may overcome a paucity of distal vessels suitable for conventional surgery. The method is, however, still experimental and one recent study has cast doubt on its value in the nontransplant setting.[177] Further research is required to establish whether TMR has any role in transplant-related coronary disease.

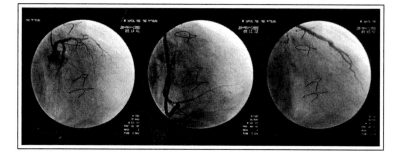

Fig. 1.12. Angiograms of the left coronary artery of a heart transplant patient, performed 1 month after coronary bypass surgery for type 1 disease. The left coronary artery was dominant. Vein grafts had been inserted into the anterior and posterior descending vessels.
Left: injection into native left coronary artery showing that the anterior descending and circumflex vessels have now become occluded. Middle: Vein graft feeding the posterior descending and distal circumflex territories. Right: Vein graft feeding the anterior descending artery.
The patient has done well during the subsequent year with good left ventricular function.

The poor prognosis of patients with advanced transplant coronary disease, coupled with the unsuitability of many cases for PTCA or CABG, leads to the issue of retransplantation. Although this can be successful in selected patients, there are limitations to this approach. Many patients develop complicating medical conditions after their initial transplant, such as cyclosporin-related renal impairment, which are potential risk factors or contraindications to further transplant surgery. Survival rates are lower after second transplants than after first operations.[52,178,179] The shortage of donor organs for transplantation, coupled with the inferior results of second transplants creates an ethical dilemma. From the utilitarian perspective, organs should be allocated to those most likely to benefit. Some have argued that it is better to use organs to give more patients the opportunity of a first transplant rather than allocate two organs to the same individual. Others have held the view that patients needing second transplants should be considered on the same basis as those being evaluated for first transplants giving due consideration to their individual risk-factors.[180] In practice, because of the shortage of donor organs and medical contraindications/risk factors, very few retransplants are performed. Between 1982 and 1995, 22,355 primary heart transplants were registered with the ISHLT Registry compared with only 653 retransplants.[180]

Conclusions

Coronary arterial disease in the cardiac allograft is one of the most important long-term complications of heart transplantation. The disease has many features in common with spontaneous coronary atherosclerosis, but there are also important differences both in pathogenesis and in the nature and distribution of the coronary lesions. In addition to 'conventional' coronary risk factors, preservation injury at the time of surgery and the host alloimmune response are important causative factors. Surgical denervation, and limited subsequent reinnervation, of the transplanted heart make chest pain a relatively uncommon clinical presentation in these patients. Significant technical advances have been made in our ability to diagnose and monitor the disease using computer assisted measurement of coronary angiograms, intravascular ultrasound and intravascular Doppler measurement. Our growing understanding of the pathogenesis of the disease has produced opportunities for pharmacological prevention. The prognosis of patients with advanced transplant-related coronary disease remains poor. In many patients, the diffuse and distal nature of the coronary lesions prevents revascularization. PTCA and stenting of discrete, proximal, coronary lesions is technically feasible but it is uncertain whether this improves prognosis. Retransplantation may benefit small number of carefully selected patients but cannot make a substantial impact on the overall problem. In the future, new immunosuppressive agents, improved methods of organ preservation and other adjunctive therapies may help to reduce the incidence of this disease. Clinical trials, using QCA and IVUS techniques, will be needed to evaluate each new therapeutic approach.

References

1. Hosenpud JD, Bennett LE, Keck BM et al. The Registry of the International Society for Heart and Lung Transplantation: Fifteenth Official Report—1998. J Heart Lung Transplant 1998; 17:656-68.
2. McGiffin DC, Kirklin JK, Naftel DC et al. Competing outcomes after heart transplantation: A comparison of eras and outcomes. J Heart Lung Transplant 1997; 16:190-8.
3. Olivari MT, Kubo SH, Braunlin EA et al. Five-year experience with triple-drug immunosuppressive therapy in cardiac transplantation. Circulation 1990; 82:Iv276-80.
4. Schroeder JS, Gao SZ, Hunt SA et al. Accelerated graft coronary artery disease: Diagnosis and prevention. J Heart Lung Transplant 1992; 11:S258-65.
5. Thomson JG. Production of severe atheroma in a transplanted human heart. Lancet 1969; II:1088-92.
6. Uys CJ, Rose AG. Pathologic findings in long-term cardiac transplants. Arch Pathol Lab Med 1984; 108:112-6.
7. Billingham ME. Cardiac Transplant Atherosclerosis Transplant Proc 1987; 19(4 Sull 5):19-25.
8. Azuma H, Tilney N. Immune and non-immune mechanisms of chronic rejection in allografts. J Heart Lung Transplant 1995; 14 (6 part 2):S136-42.
9. Mehra MR, Ventura HO, Stapleton DD et al. Allograft aortopathy: An in vivo study of donor aorta involvement in cardiac allograft vasculopathy. Am Heart J 1997; 133: 698-702.
10. Oni AA, Ray J, Hosenpud JD. Coronary venous intimal thickening in explanted cardiac allografts. Evidence demonstrating that transplant coronary artery disease is a manifestation of a diffuse allograft vasculopathy. Transplantation 1992; 53:1247-51.
11. Fuster V, Badimon L, Badimon JJ et al. The pathogenesis of coronary artery disease and the acute coronary syndromes (part 1). N Engl J Med 1992; 326:242-50.
12. Fuster V, Badimon L, Badimon JJ et al. The pathogenesis of coronary artery disease and the acute coronary syndromes (part 2). N Engl J Med 1992; 326:310-8.
13. Ip JH, Fuster V, Badimon L et al. Syndromes of accelerated atherosclerosis: Role of vascular injury and smooth muscle proliferation. J Am Coll Cardiol 1990; 15:1667-87.
14. Vane JR, Anggard EE, Botting RM. Regulatory functions of the vascular endothelium N Engl J Med 1990; 323:27-36.
15. Ross R. The pathogenesis of atherosclerosis—an update. N Engl J Med 1986; 314: 488-500.
16. Johnson DE, Gao SZ, Schroeder JS et al. The spectrum of coronary artery pathologic findings in human cardiac allografts. J Heart Transplant 1989; 8:349-59.
17. Liu G, Butany J. Morphology of graft arteriosclerosis in cardiac transplant recipients. Hum Pathol 1992; 23:768-73.
18. Arbustini E, Roberts WC. Morphological observations in the epicardial coronary arteries and their surroundings late after cardiac transplantation (allograft vascular disease). Am J Cardiol 1996; 78:814-20.
19. Arbustini E, Dal Bello B, Morbini P et al. Frequency and characteristics of coronary thrombosis in the epicardial coronary arteries after cardiac transplantation. Am J Cardiol 1996; 78:795-800.
20. Waller BF, Pinkerton CA, Slack JD. Intravascular ultrasound: a histological study of vessels during life. The new 'gold standard' for vascular imaging. Circulation 1992; 85:2305-10.
21. Kapadia SR, Nissen SE, Ziada KM et al. Development of transplantation vasculopathy and progression of donor-transmitted atherosclerosis: Comparison by serial intravascular ultrasound imaging. Circulation 1998; 98:2672-8.
22. Banner NR, Yacoub MH. Physiology of the orthotopic cardiac transplant recipient. Semin Thorac Cardiovasc Surg 1990; 2:259-70.
23. Gao SZ, Hunt SA, Schroeder JS. Accelerated transplant coronary artery disease. Semin Thorac Cardiovasc Surg 1990; 2:241-9.
24. Wilson RF. Reinnervation re-examination. J Heart Lung Transplant 1998; 17:137-9.
25. Stark RP, McGinn AL, Wilson RF. Chest pain in cardiac-transplant recipients. Evidence of sensory reinnervation after cardiac

transplantation. N Engl J Med 1991; 324: 1791-4.

26. Ramsdale DB, Bellamy CM. Angina and threatened acute myocardial infarction after cardiac transplantation. Am Heart J 1990; 119:1195-7.

27. Schroeder JS, Hunt SA. Chest pain in heart-transplant recipients. N Engl J Med 1991; 324:1805-7.

28. Grinstead WC, Smart FW, Pratt CM et al. Detection of asymptomatic myocardial ischemia in a heart transplant patient before sudden death. J Heart Lung Transplant 1991; 10:1026-8.

29. Gao SZ, Schroeder JS, Hunt SA et al. Acute myocardial infarction in cardiac transplant recipients. Am J Cardiol 1989; 64:1093-7.

30. Uretsky BF, Kormos RL, Zerbe TR et al. Cardiac events after heart transplantation: Incidence and predictive value of coronary arteriography. J Heart Lung Transplant 1992; 11:S45-51.

31. Barbir M, Lazem F, Banner N et al. The prognostic significance of noninvasive cardiac tests in heart transplant recipients. Eur Heart J 1997; 18:692-6.

32. Mudge GH, Jr. Strategies to manage the heart failure patient after transplantation. Clin Cardiol 1992; 15 Suppl 1:I37-41.

33. Bharati S, Billingham M, Lev M. The conduction system in transplanted hearts. Chest 1992; 102:1182-8.

34. Grinstead WC, Smart FW, Pratt CM et al. Sudden death caused by bradycardia and asystole in a heart transplant patient with coronary arteriopathy. J Heart Lung Transplant 1991; 10:931-6.

35. Dalal JN, Brinker JA, Resar JR. Coronary artery spasm in the transplanted human heart: A case report. Angiology 1996; 47: 291-4.

36. Goldenberg IF, Levine TB. Coronary artery spasm in a denervated orthotopic transplanted human heart. Cathet Cardiovasc Diagn 1986; 12:44-7.

37. Little T, Macoviak J, Villanueva P et al. Diffuse coronary vasospasm and accelerated atherosclerosis in a transplanted human heart. Am J Cardiol 1989; 64:825-7.

38. Hruban RH, Kasper EK, Gaudin PB et al. Severe lymphocytic endothelialitis associated with coronary artery spasm in a heart transplant recipient. J Heart Lung Transplant 1992; 11:42-7.

39. Colizza F, Pelletier GB, Carrier M et al. Syncope: A symptom of coronary artery spasm after cardiac transplantation. Can J Cardiol 1992; 8:299-302.

40. Kallen RJ, Brynes RK, Aroson AJ et al. Premature coronary atherosclerosis in a 5-year old with corticosteroid refractory nephrotic

syndrome. Am J Dis Child 1977; 131: 976-80.

41. Braunlin EA, Hunter DW, Canter CE et al. Coronary artery disease in pediatric cardiac transplant recipients receiving triple-drug immunosuppression. Circulation 1991; 84:Iii303-9.

42. Gao SZ, Alderman EL, Schroeder JS et al. Accelerated coronary vascular disease in the heart transplant patient: Coronary arteriographic findings. J Am Coll Cardiol 1988; 12:334-40.

43. Narrod J, Kormos R, Armitage J et al. Acute rejection and coronary artery disease in long-term survivors of heart transplantation. J Heart Transplant 1989; 8:418-20.

44. Uretsky BF, Murali S, Reddy PS et al. Development of coronary artery disease in cardiac transplant patients receiving immunosuppressive therapy with cyclosporine and prednisone. Circulation 1987; 76:827-34.

45. Olivari MT, Homans DC, Wilson RF et al. Coronary artery disease in cardiac transplant patients receiving triple-drug immunosuppressive therapy. Circulation 1989; 80: III111-5.

46. von Scheidt W, Ziegler U, Kemkes BM et al. Heart transplantation: Hemodynamics over a five-year period. J Heart Lung Transplant 1991; 10:342-50.

47. Costanzo MR, Naftel DC, Pritzker MR et al. Heart transplant coronary artery disease detected by coronary angiography: A multi-institutional study of preoperative donor and recipient risk factors. Cardiac Transplant Research Database. J Heart Lung Transplant 1998; 17:744-53.

48. O'Neill BJ, Pflugfelder PW, Singh NR et al. Frequency of angiographic detection and quantitative assessment of coronary arterial disease one and 3 years after cardiac transplantation. Am J Cardiol 1989; 63: 1221-1226.

49. Pflugfelder PW, Boughner DR, Rudas L et al. Enhanced detection of cardiac allograft arterial disease with intracoronary ultrasonographic imaging. Am Heart J 1993; 125:1583-91.

50. Park SJ, Pifarre R, Sullivan H et al. Natural history of allograft coronary arteriopathy: A retrospective study of 54 patients over a 8 1/2-year period. Cardiovasc Surg 1996; 4:37-41.

51. Gao SZ, Hunt SA, Schroeder JS et al. Early development of accelerated graft coronary artery disease: Risk factors and course. J Am Coll Cardiol 1996; 28:673-9.

52. Keogh AM, Valantine HA, Hunt SA et al. Impact of proximal or midvessel discrete coronary artery stenoses on survival after

heart transplantation. J Heart Lung Transplant 1992; 11:892-901.

53. Carrier M, Pelletier G, Leclerc Y et al. Accelerated coronary atherosclerosis after cardiac transplantation: major threat to long-term survival. Can J Surg 1991; 34:133-6.

54. Weis M, von Scheidt W. Cardiac allograft vasculopathy: A review. Circulation 1997; 96:2069-77.

55. Brunner La Rocca HP, Schneider J, Kunzli A et al. Cardiac allograft rejection late after transplantation is a risk factor for graft coronary artery disease. Transplantation 1998; 65:538-43.

56. Mehra MR, Ventura HO, Chambers RB et al. The prognostic impact of immunosuppression and cellular rejection on cardiac allograft vasculopathy: Time for a reappraisal. J Heart Lung Transplant 1997; 16:743-51.

57. Gao SZ, Schroeder JS, Alderman EL et al. Clinical and laboratory correlates of accelerated coronary artery disease in the cardiac transplant patient. Circulation 1987; 76: V56-61.

58. Cunningham DA, Crisp SJ, Barbir M et al. Donor ACE gene polymorphism: A genetic risk factor for accelerated coronary sclerosis following cardiac transplantation. Eur Heart J 1998; 19:319-25.

59. Novitsky D, Wicomb WN, Cooper DKC et al. Electrocardiographic, haemodynamic and endocrine changes occurring during experimental brain death in the Chacma baboon. J Heart Transplant 1984; 4:63-9.

60. Shanlin RJ, Sole MJ, Rahimifar M et al. Increased intracranial pressure elicits hypertension, increased sympathetic activity, electrocardiographic abnormalities and myocardial damage in rats. J Am Coll Cardiol 1988; 12:727-36.

61. Greenhoot JH, Reichenbach DD. Cardiac injury and subarachnoid haemorrhage. A clinical, pathological and physiological correlation. J Neurosurg 1969; 30:521-31.

62. Stovin PG, Sharples L, Hutter JA et al. Some prognostic factors for the development of transplant-related coronary artery disease in human cardiac allografts. J Heart Lung Transplant 1991; 10:38-44.

63. Sharples LD, Caine N, Mullins P et al. Risk factor analysis for the major hazards following heart transplantation-rejection, infection, and coronary occlusive disease. Transplantation 1991; 52:244-52.

64. Rickenbacher PR, Kemna MS, Pinto FJ et al. Coronary artery intimal thickening in the transplanted heart. An in vivo intracoronary ultrasound study of immunologic and metabolic risk factors. Transplantation 1996; 61:46-53.

65. Barbir M, Banner N, Thompson GR et al. Relationship of immunosuppression and serum lipids to the development of coronary arterial disease in the transplanted heart. Int J Cardiol 1991; 32:51-6.

66. Dresdale AR, Kraft PL, Paone G et al. Reduced incidence and severity of accelerated graft atherosclerosis in cardiac transplant recipients treated with prophylactic antilymphocyte globulin. J Cardiovasc Surg Torino 1992; 33:746-53.

67. Barbir M, Kushwaha S, Hunt B et al. Lipoprotein(a) and accelerated coronary artery disease in cardiac transplant recipients. Lancet 1992; 340:1500-2.

68. Radovancevic B, Poindexter S, Birovljev S et al. Risk factors for development of accelerated coronary artery disease in cardiac transplant recipients. Eur J Cardiothorac Surg 1990; 4:309-12.

69. Ladowski JS, Kormos RL, Uretsky BF et al. Heart transplantation in diabetic recipients. Transplantation 1990; 49:303-5.

70. de Lorgeril M, Dureau G, Boissonnat P et al. Platelet function and composition in heart transplant recipients compared with nontransplanted coronary patients. Arterioscler Thromb 1992; 12:222-30.

71. de Lorgeril M, Loire R, Guidollet J, Boissonnat P et al. Accelerated coronary artery disease after heart transplantation: The role of enhanced platelet aggregation and thrombosis. J Intern Med 1993; 233: 343-50.

72. Hunt BJ, Segal H, Yacoub M. Hemostatic changes in heart transplant recipients and their relationship to accelerated coronary sclerosis. Transplantation 1993; 55:309-15.

73. Hunt BJ, Segal H, Yacoub M. Endothelial cell hemostatic function after heart transplantation. Transplant Proc 1991; 23:1182-3.

74. Hornick P, Smith J, Pomerance A et al. Influence of acute rejection episodes, HLA matching, and donor/recipient phenotype on the development of 'early' transplant-associated coronary artery disease. Circulation 1997; 96:II-148-53.

75. Schutz A, Kemkes BM, Kugler C et al. The influence of rejection episodes on the development of coronary artery disease after heart transplantation. Eur J Cardiothorac Surg 1990; 4:300-7.

76. Itescu S, Tung TC, Burke EM et al. An immunological algorithm to predict risk of high-grade rejection in cardiac transplant recipients. Lancet 1998; 352:263-70.

77. Ationu A, Collins A. Molecular cloning and expression of 56-58 kD antigen associated with transplant coronary artery disease. Biochem Biophys Res Commun 1997; 236: 716-8.

78. Ferry BL, Welsh KI, Dunn MJ et al. Anticell surface endothelial antibodies in sera from cardiac and kidney transplant recipients: Association with chronic rejection. Transpl Immunol 1997; 5:17-24.

79. Jurcevic S, Dunn MJ, Crisp S et al. A new enzyme-linked immunosorbent assay to measure antiendothelial antibodies after cardiac transplantation demonstrates greater inhibition of antibody formation by tacrolimus compared with cyclosporine. Transplantation 1998; 65:1197-202.

80. Rose EA, Pepino P, Barr ML et al. Relation of HLA antibodies and graft atherosclerosis in human cardiac allograft recipients. J Heart Lung Transplant 1992; 11:S120-3.

81. Rose ML. Role of antibody and indirect antigen presentation in transplant-associated coronary artery vasculopathy. J Heart Lung Transplant 1996; 15:342-9.

82. Rose EA, Smith CR, Petrossian GA et al. Humoral immune responses after cardiac transplantation: Correlation with fatal rejection and graft atherosclerosis. Surgery 1989; 106:203-7.

83. Dunn MJ, Crisp SJ, Rose ML et al. Antiendothelial antibodies and coronary artery disease after cardiac transplantation. Lancet 1992; 339:1566-70.

84. Arkonac B, Mauck KA, Chou S et al. Low multiplicity cytomegalovirus infection of human aortic smooth muscle cells increases levels of major histocompatibility complex class I antigens and induces a proinflammatory cytokine milieu in the absence of cytopathology. J Heart Lung Transplant 1997; 16:1035-45.

85. Koskinen PK, Nieminen MS, Krogerus LA et al. Cytomegalovirus infection and accelerated cardiac allograft vasculopathy in human cardiac allografts. J Heart Lung Transplant 1993; 12:724-9.

86. Min KW, Wickemeyer WJ, Chandran P et al. Fatal cytomegalovirus infection and coronary arterial thromboses after heart transplantation: A case report. J Heart Transplant 1987; 6:100-5.

87. Grattan MT, Moreno Cabral CE, Starnes VA et al. Cytomegalovirus infection is associated with cardiac allograft rejection and atherosclerosis. JAMA 1989; 261:3561-6.

88. Scott JP, Large SR, Schofield P et al. Association of coronary artery disease in cardiac transplant recipients with cytomegalovirus infection. Am J Cardiol 1990; 66:1025-6.

89. Loebe M, Schuler S, Zais O et al. Role of cytomegalovirus infection in the development of coronary artery disease in the transplanted heart. J Heart Transplant 1990; 9:707-11.

90. Everett JP, Hershberger RE, Norman DJ et al. Prolonged cytomegalovirus infection with

91. viremia is associated with development of cardiac allograft vasculopathy. J Heart Lung Transplant 1992; 11:S133-7.

91. Kendall TJ, Wilson JE, Radio SJ et al. Cytomegalovirus and other herpesviruses: Do they have a role in the development of accelerated coronary arterial disease in human heart allografts? J Heart Lung Transplant 1992; 11:S14-20.

92. Tullius SG, Heemann UW, Hancock WW et al. Long-term kidney isografts develop functional and morphological changes which mimic those of chronic allograft rejection. Ann Surg 1994; 220:423-35.

93. Hornick P, Smith J, Pomerance A et al. Influence of donor/recipient phenotype and degree of HLA mismatch on the development of transplant-associated coronary artery disease in heart transplant patients. Transplant Proc 1997; 29:1420-1.

94. Stovin PG, Sharples LD, Schofield PM et al. Lack of association between endomyocardial evidence of rejection in the first six months and the later development of transplant-related coronary artery disease. J Heart Lung Transplant 1993; 12:110-6.

95. Frazier OH, McAllister HA, Jammal CT et al. Occlusive coronary arteritis: A cause of early death in a cardiac transplant patient. Ann Thorac Surg 1987; 43:554-6.

96. Yowell RL, Hammond EH, Bristow MR et al. Acute vascular rejection involving the major coronary arteries of a cardiac allograft. J Heart Transplant 1988; 7:191-7.

97. Davies H, Tikriti S. Coronary arterial pathology in the transplanted human heart. Int J Cardiol 1989; 25:99-117.

98. Gravanis MB. Allograft heart accelerated atherosclerosis: Evidence for cell-mediated immunity in pathogenesis. Mod Pathol 1989; 2:495-505.

99. Normann SJ, Salomon DR, Leelachaikul P et al. Acute vascular rejection of the coronary arteries in human heart transplantation: Pathology and correlations with immunosuppression and cytomegalovirus infection. J Heart Lung Transplant 1991; 10:674-87.

100. Radovancevic B, Birovljev S, Vega JD et al. Inverse relationship between human leukocyte antigen match and development of coronary artery disease. Transplant Proc 1991; 23:1144-5.

101. Zerbe T, Uretsky B, Kormos R et al. Graft atherosclerosis: Effects of cellular rejection and human lymphocyte antigen. J Heart Lung Transplant 1992; 11:S104-10.

102. Banner N. Immunosuppressive agents. Surgery 1998; 16:30-3.

103. Becker DM, Chamberlain B, Swank R et al. Relationship between corticosteroid exposure

and plasma lipid levels in heart transplant recipients. Am J Med 1988; 85:632-8.

104. Taylor DO, Thompson JA, Hastillo A et al. Hyperlipidemia after clinical heart transplantation. J Heart Transplant 1989; 8:209-13.

105. Ratkovec RM, Wray RB, Renlund DG et al. Influence of corticosteroid-free maintenance immunosuppression on allograft coronary artery disease after cardiac transplantation. J Thorac Cardiovasc Surg 1990; 100: 6-12.

106. Haug C, Duell T, Lenich A et al. Elevated plasma endothelin concentrations in cyclosporine-treated patients after bone marrow transplantation. Bone Marrow Transplant 1995; 16:191-4.

107. Huang LQ, Whitworth JA, Chesterman CN. Effects of cyclosporin A and dexamethasone on haemostatic and vasoactive functions of vascular endothelial cells. Blood Coagul Fibrinolysis 1995; 6:438-45.

108. Bloom IT, Bentley FR, Spain DA et al. An experimental study of altered nitric oxide metabolism as a mechanism of cyclosporin-induced renal vasoconstriction. Br J Surg 1995; 82:195-8.

109. Gao SZ, Schroeder JS, Alderman EL et al. Prevalence of accelerated coronary artery disease in heart transplant survivors. Comparison of cyclosporine and azathioprine regimens. Circulation 1989; 80:III100-5.

110. Kobashigawa J, Miller L, Renlund D et al. A randomized active-controlled trial of mycophenolate mofetil in heart transplant recipients. Transplantation 1998; 66:507-15.

111. Ozdogan E, Banner NR, Fitzgerald M et al. Factors influencing the development of hypertension after heart transplantation. J Heart Transplant 1990; 9:548-553.

112. Rickenbacher PR, Pinto FJ, Lewis NP et al. Prognostic importance of intimal thickness as measured by intracoronary ultrasound after cardiac transplantation. Circulation 1995; 92:3445-52.

113. Barbir M, Lazem F, Bowker T et al. Determinants of transplant-related coronary calcium detected by ultrafast computed tomography scanning. Am J Cardiol 1997; 79: 1606-9.

114. Lazem F, Barbir M, Banner N et al. Coronary calcification detected by ultrafast computed tomography is a predictor of cardiac events in heart transplant recipients. Transplant Proc 1997; 29:572-5.

115. Johnson DE, Alderman EL, Schroeder JS et al. Transplant coronary artery disease: Histopathologic correlations with angiographic morphology. J Am Coll Cardiol 1991; 17:449-57.

116. Alderman EL, Wexler L. Angiographic implications of cardiac transplantation. Am J Cardiol 1989; 64:16e-21e.

117. Johnson TH, McDonald K, Nakhleh R et al. Allograft vasculopathy and death in a cardiac transplant patient with angiographically normal coronary arteries. Cathet Cardiovasc Diagn 1991; 24:37-40.

118. Dressler FA, Miller LW. Necropsy versus angiography: How accurate is angiography? J Heart Lung Transplant 1992; 11:S56-9.

119. Young JB, Smart FM, Lowry RL et al. Coronary angiography after heart transplantation: Should perioperative study be the "gold standard"? J Heart Lung Transplant 1992; 11:S65-8.

120. Reiber JHC, Serruys PW, Kooijman CJ et al. Assessment of short-, medium-, and long-term variations in arterial dimensions from computer assisted quantitation of coronary cineangiograms. Circulation 1985; 71:280-8.

121. de Feyter PJ, Serruys PW, Davies MJ et al. Quantitative coronary angiography to measure progression and regression of coronary atherosclerosis: Value, limitations and implications for clinical trials. Circulation 1991; 84:412-23.

122. Van der Linden MMJM, Balk AHMM, de Feyter PJ. Short- and long-term quantitative angiographic follow up after cardiac transplantation. In: Serruys PW, Foley DP, de Feyter PJ, eds. Quantitative Coronary Angiography in Clinical Practice. Dordrecht: Kluwer, 1994:667-79.

123. O'Neill BJ, Pflugfelder PW, Singh NR et al. Frequency of angiographic detection and quantitative assessment of coronary arterial disease one and three years after cardiac transplantation. Am J Cardiol 1989; 63: 1221-6.

124. Gao SZ, Alderman EL, Schroeder JS et al. Progressive coronary luminal narrowing after cardiac transplantation. Circulation 1990; 82:Iv269-75.

125. Mills RM, Jr., Hill JA, Theron HD et al. Serial quantitative coronary angiography in the assessment of coronary disease in the transplanted heart. J Heart Lung Transplant 1992; 11:S52-5.

126. Glagov S, Weisenberg E, Zarins CK et al. Compensatory enlargement of human atherosclerotic coronary arteries. N Engl J Med 1987; 316:1371-5.

127. von Scheidt W, Erdmann E. Dilated angiopathy: a specific subtype of allograft coronary artery disease. J Heart Lung Transplant 1991; 10:698-703.

128. Metz JA, Yock PG, Fitzgerald PJ. Intravascular ultrasound: Basic interpretation. Cardiology Clinics 1997 (Part 1); 15:1-15.

129. Pinto FJ, Chenzbraun A, Botas J et al. Feasibility of serial intracoronary ultrasound imaging for assessment of progression of intimal proliferation in cardiac transplant recipients. Circulation 1994; 90:2348-55.

130. Johnson JA, Kobashigawa JA. Quantitative analysis of transplant coronary artery disease with use of intracoronary ultrasound. J Heart Lung Transplant 1995; 14:S198-202.
131. Pethig K, Heublein B, Wahlers T et al. Mechanism of luminal narrowing in cardiac allograft vasculopathy: Inadequate vascular remodeling rather than intimal hyperplasia is the major predictor of coronary artery stenosis. Working Group on Cardiac Allograft Vasculopathy. Am Heart J 1998; 135:628-33.
132. St. Goar FG, Pinto FJ, Alderman EL et al. Intracoronary ultrasound in cardiac transplant recipients. In vivo evidence of "angiographically silent" intimal thickening. Circulation 1992; 85:979-87.
133. Valantine H, Pinto FJ, St. Goar FG et al. Intracoronary ultrasound imaging in heart transplant recipients: The Stanford experience. J Heart Lung Transplant 1992; 11:S60-4.
134. Liang DH, Gao SZ, Botas J et al. Prediction of angiographic disease by intracoronary ultrasonographic findings in heart transplant recipients. J Heart Lung Transplant 1996; 15:980-7.
135. Bach RG, Kern MJ. Practical Coronary Physiology. Clinical application of the Doppler flow velocity guide wire. Cardiology Clinics 1997; 15 (Part 1):77-99.
136. Joye JD, Schulmann DS. Clinical application of coronary flow reserve using an intracoronary Doppler guide wire. Cardiology Clinics 1997; 15 (part 1):101-29.
137. Wilson RJ, Wyche K, Christensen BV et al. Effects of adenosine on human coronary arterial. Circulation 1990; 82:1595-1606.
138. Kern MJ, Bach RG, Mechem CJ et al. Variations in normal coronary vasodilatory reserve stratified by artery, gender, heart transplantation and coronary artery disease. J Am Coll Cardiol 1996; 28:1154-60.
139. McGinn AL, White CW, Wilson RF. Interstudy variability of coronary flow reserve. Influence of heart rate, arterial pressure, and ventricular preload. Circulation 1990; 81:1319-30.
140. Sudhir K, MacGregor JS, Barbant SD et al. Assessment of coronary conductance and resistance vessel reactivity in response to nitroglycerin, ergonovine and adenosine: In vivo studies with simultaneous intravascular two dimensional and Doppler ultrasound. J Am Coll Cardiol 1993; 21:1261-8.
141. Mullins PA, Chauhan A, Sharples L et al. Impairment of coronary flow reserve in orthotopic cardiac transplant recipients with minor coronary occlusive disease. Br Heart J 1992; 68:266-71.
142. Mazur W, Bitar JN, Young JB et al. Progressive deterioration of coronary flow reserve after heart transplantation. Am Heart J 1998; 136:504-9.
143. Furchgott RF, Zawadzski JV. The obligatory role of endothelial cells in the relaxation of arterial smooth muscle cells by acetylcholine. Nature 1980; 288:373-6.
144. Vassalli G, Gallino A. Endothelial dysfunction and accelerated coronary artery disease in cardiac transplant recipients. Microcirculation Working Group, European Society of Cardiology. Eur Heart J 1997; 18:1712-7.
145. Hartmann A, Mazzilli N, Weis M et al. Time course of endothelial function in epicardial conduit coronary arteries and in the microcirculation in the long-term follow-up after cardiac transplantation. Int J Cardiol 1996; 53:127-36.
146. Fish RD, Nabel EG, Selwyn AP et al. Responses of coronary arteries of cardiac transplant patients to acetylcholine. J Clin Invest 1988; 81:21-31.
147. Southworth MR, Mauro VF. The use of HMG-CoA reductase inhibitors to prevent accelerated graft atherosclerosis in heart transplant patients. Ann Pharmacother 1997; 31:489-91.
148. Kobashigawa JA, Katznelson S, Laks H et al. Effect of pravastatin on outcomes after cardiac transplantation. N Engl J Med 1995; 333:621-7.
149. Wenke K, Meiser B, Thiery J et al. Simvastatin reduces graft vessel disease and mortality after heart transplantation: A four-year randomized trial. Circulation 1997; 96:1398-402.
150. JNC. The Sixth Report of the Joint National Committee on prevention, detection, evaluation and treatment of high blood pressure. Arch Intern Med 1997; 157:2413-46.
151. Schroeder JS, Gao SZ, Alderman EL et al. A preliminary study of diltiazem in the prevention of coronary artery disease in heart-transplant recipients. N Engl J Med 1993; 328:164-70.
152. Schroeder JS, Gao SZ. Calcium blockers and atherosclerosis: Lessons from the Stanford Transplant Coronary Artery Disease/Diltiazem Trial. Can J Cardiol 1995; 11:710-5.
153. Katznelson S, Wang XM, Chia D et al. The inhibitory effects of pravastatin on natural killer cell activity in vivo and on cytotoxic T lymphocyte activity in vitro. J Heart Lung Transplant 1998; 17:335-40.
154. Mullins PA, Cary NR, Sharples L et al. Coronary occlusive disease and late graft failure after cardiac transplantation. Br Heart J 1992; 68:260-5.

155. de Lorgeril M, Dureau G, Boissonnat P et al. Increased platelet aggregation after heart transplantation: Influence of aspirin. J Heart Lung Transplant 1991; 10:600-3.

156. Fiane AE, Klow NE, Simonsen S et al. Percutaneous transluminal angioplasty and retransplantation due to transplant coronary artery disease. Scand Cardiovasc J 1997; 31:223-7.

157. Patel VS, Radovancevic B, Springer W et al. Revascularization procedures in patients with transplant coronary artery disease. Eur J Cardiothorac Surg 1997; 11:895-901.

158. Mullins PA, Shapiro LM, Aravot DA et al. Experience of percutaneous transluminal coronary angioplasty in orthotopic cardiac transplant recipients. Eur Heart J 1991; 12:1205-7.

159. Prewitt KC, Wortham DC, Banks AK. Coronary angioplasty following cardiac transplantation: A case report and review of the literature. Cathet Cardiovasc Diagn 1991; 22:25-7.

160. Sandhu JS, Uretsky BF, Reddy PS et al. Potential limitations of percutaneous transluminal coronary angioplasty in heart transplant recipients. Am J Cardiol 1992; 69: 1234-7.

161. Halle AAd, Wilson RF, Massin EK et al. Coronary angioplasty in cardiac transplant patients. Results of a multicenter study. Circulation 1992; 86:458-62.

162. Swan JW, Norell M, Yacoub M et al. Coronary angioplasty in cardiac transplant recipients. Eur Heart J 1993; 14:65-70.

163. Cusick D, Davidson C, Frohlich T et al. Coronary artery stenting postcardiac transplant: A report of two cases. Cathet Cardiovasc Diagn 1997; 40:92-6.

164. Jain SP, Ramee SR, White CJ et al. Coronary stenting in cardiac allograft vasculopathy. J Am Coll Cardiol 1998; 32: 1636-40.

165. Parry A, Roberts M, Parameshwar J et al. The management of postcardiac transplantation coronary artery disease. Eur J Cardiothorac Surg 1996; 10:528-32.

166. Wong PM, Piamsomboon C, Mathur A et al. Efficacy of coronary stenting in the management of cardiac allograft vasculopathy. Am J Cardiol 1998; 82:239-41.

167. Vetrovec GW, Cowley MJ, Newton CM et al. Applications of percutaneous transluminal coronary angioplasty in cardiac transplantation. Preliminary results in five patients. Circulation 1988; 78:III 83-6.

168. Strikwerda S, Umans V, van der Linden MM et al. Percutaneous directional atherectomy for discrete coronary lesions in cardiac transplant patients. Am Heart J 1992; 123: 1686-90.

169. Topaz O, Bailey NT, Mohanty PK. Application of solid-state pulsed-wave, mid-infrared laser for percutaneous revascularization in heart transplant recipients. J Heart Lung Transplant 1998; 17:505-10.

170. Barstad RM, Fosse E, Geiran OR et al. Minimally invasive direct coronary artery bypass grafting without cardiopulmonary bypass in combination with intraoperative percutaneous transluminal coronary angioplasty for palliative coronary revascularization in a heart transplant recipient. J Heart Lung Transplant 1998; 17:629-34.

171. Miller LW, Donohue TJ, Wolford TA. The surgical management of allograft coronary disease: A paradigm shift. Semin Thorac Cardiovasc Surg 1996; 8:133-8.

172. Copeland JG, Butman SM, Sethi G. Successful coronary artery bypass grafting for high-risk left main coronary artery atherosclerosis after cardiac transplantation. Ann Thorac Surg 1990; 49:106-10.

173. Frazier OH, Vega JD, Duncan JM et al. Coronary artery bypass two years after orthotopic heart transplantation: A case report. J Heart Lung Transplant 1991; 10:1036-40.

174. Frazier OH, Kadipasaoglu KA, Radovancevic B et al. Transmyocardial laser revascularization in allograft coronary artery disease. Ann Thorac Surg 1998; 65:1138-41.

175. Malik FS, Mehra MR, Ventura HO et al. Management of cardiac allograft vasculopathy by transmyocardial laser revascularization. Am J Cardiol 1997; 80:224-5.

176. McFadden PM, Robbins RJ, Ochsner JL et al. Transmyocardial revascularization for cardiac transplantation allograft vasculopathy. J Thorac Cardiovasc Surg 1998; 115:1385-8.

177. Schofield PM, Sharples LD, Caine N et al. Transmyocardial laser revascularisation in patients with refractory angina: A randomized controlled trial. Lancet 1999; 353: 519-24.

178. Schnetzler B, Pavie A, Dorent R et al. Heart retransplantation: A 23-year single-center clinical experience. Ann Thorac Surg 1998; 65:978-83.

179. Gao SZ, Schroeder JS, Hunt S et al. Retransplantation for severe accelerated coronary artery disease in heart transplant recipients. Am J Cardiol 1988; 62:876-81.

180. Novick RJ. Heart and lung retransplantation: Should it be done? J Heart Lung Transplant 1998; 17:635-42.

CHAPTER 2

Atherosclerosis: Response to Injury

Nick Gall and Robin Poston

Introduction[1]

The consequences of atherosclerosis are the commonest cause of death in the Western World. It causes ischemic heart disease, cerebrovascular accidents and peripheral vascular disease and is therefore also a serious, significant and costly cause of morbidity. For example, 13 million people in the United States suffer from ischemic heart disease, of whom 500,000 will die each year.[2] Although we now know of over 246 independent risk factors for its genesis and can modify at least a few of these, we can achieve only modest reductions in its incidence and consequences. Thus a detailed study of its pathology is vital if we are to conquer, the largest current challenge to Western medicine.

The word *'atherosclerosis'*, first coined by Thoma in 1886, is derived from two Greek words—*'athero'*—meaning gruel and *'sclerosis'* meaning hardening. This is descriptive of the advanced form of atherosclerosis—a fibrous thickening of the arterial wall made up of connective tissue and smooth muscle cells, surrounding a soft cholesterol-filled center with associated macrophages. In essence, it is a process in which the arterial wall becomes damaged and subsequently attempts to heal itself.

In this Chapter we will discuss what are felt to be some of the most important aspects of the pathology of atherosclerosis. In the first section we will give an overview of its pathology, both in its stable form and in the special case of acute coronary syndromes. After that, we will discuss, in detail, certain molecules which are believed to be central to atherogenesis. This will be followed by a discussion of how these molecules interact with various parts of the plaque. We will then deal with how these small molecules and the individual aspects of the plaque are affected by some of the most important risk factors for atheroma formation. The final section will then attempt to bring all of the information together to form an all-encompassing hypothesis for atherogenesis.

General Pathology[3-12]

The Response to Injury

Atherosclerosis appears to be a chronic inflammatory reaction to intimal damage, particularly of the endothelium. This insult can take many forms, one of the commonest being cholesterol. A series of reactions then draw macrophages, T cells and smooth muscle cells into the area in an attempt to heal the damage. It appears, however, that as in many chronic inflammatory conditions, the reaction is in itself detrimental, further enhancing the disease process. This view is known as *'the response to injury hypothesis'* and was first mooted by Russell Ross in 1973.

Not all arteries are susceptible to the process. Only the aorta, together with large and medium sized muscular arteries, other than

Transplant-Associated Coronary Artery Vasculopathy, edited by Marlene L. Rose.
©2001 Eurekah.com.

the internal thoracic arteries, are affected. The disease is particularly severe near branch points. These are areas of low shear stress and blood flow direction reversal, where leukocytes and cholesterol have time to interact with the endothelium and endothelial adhesion of leukocytes can occur[13] (shear forces are forces exerted at the interface of a moving and a nonmoving phase i.e., between the blood and the subendothelial matrix). There is also evidence that some atheroma may form at sites of pre-existing intimal thickening. These areas are known as intimal cushions and are made up of smooth muscle cells.

Atherogenesis is the name given to the process by which the disease evolves. The progression of lesions has been classified by the Committee on Vascular Lesions of the Council on Arteriosclerosis of the American Heart Association, as reported by Stary et al.[14] Their classification will be used throughout this Chapter.

The earliest signs of atheroma can be seen only under the electron microscope. In animal models, small vesicles appear in the intima after two weeks of a high cholesterol diet. They contain cholesterol and are known as extracellular liposomes. They have been observed in a number of different species, including man. At the same time endothelial adhesion molecule expression, prior to monocyte involvement, is also seen.

The first step visible at light microscope level is monocytes trafficking through the endothelium to form a *type I lesion*. This is a collection of subendothelial macrophages that are ingesting cholesterol. The cholesterol uptake occurs in an uncontrolled manner producing lipid-laden cells called foam cells. These are the characteristic cells of the early atherosclerotic plaque. No distortion of the lumen is seen.

As more macrophages accumulate due to the continuing toxic insult, they spread deeper into the arterial wall. At the same time, smooth muscle cells migrate through the internal elastic lamina into the intima where they are transformed from a contractile to a synthetic phenotype. This change is associated with the accumulation of a large number

of endoplasmic reticula and Golgi bodies. The function of the smooth muscle cells is to produce connective tissue matrix proteins such as collagen, proteoglycans and elastic fibers in an attempt at healing. A few of the smooth muscle cells may become foam cells. T lymphocytes also accumulate in the intima. This form of plaque is known as the *fatty streak* or *type II lesion* (Fig. 2.1).

Detailed post mortem studies of children dying with unrelated conditions show that 45% of infants have fatty streaks in their coronary arteries by the age of eight months. Fatty streaks have even been observed in the arteries of fetuses.

Some fatty streaks will regress and may even disappear. Foam cells have been seen to emigrate from the intima back into the blood stream in primate experiments.

As the lesion progresses more macrophages and smooth muscle cells accumulate. Extracellular lipid droplets also appear, leading to some disruption of the coherence of the smooth muscle cells. These are known as *type III lesions* (Fig. 2.2).

A lipid core may then accumulate, partly composed of cholesterol released from dying macrophages and smooth muscle cells and partly of cholesterol derived directly from plasma proteins that have bound to intimal proteoglycans. A space in which the core can exist is produced by the local action of macrophage-derived metalloproteinases and cell death. This disrupting lipid core can produce clinical sequelae (see later). This lesion is known as the *type IV lesion*.

In an attempt at healing, many of the intimal smooth muscle cells produce connective tissue elements, both to strengthen the weakened wall and to isolate the toxic stimulus of oxidized lipid from the endothelium and vascular lumen. Thus the plaque becomes a localized area, covered by endothelium, beneath which the connective tissue elements form a cap over the lipid core and foam cells. The fibrous cap is not an inert body but is constantly being laid down by smooth muscle cells and broken down by locally produced enzymes such as metalloproteinases. At the shoulders of the plaque macrophages

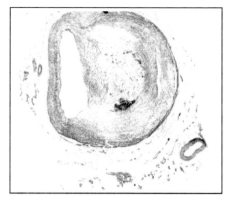

Fig. 2.1. A fatty streak (Stary II) lesion stained with the macrophage-reactive antibody HAM56. A large collection of sub-endothelial macrophages is demonstrated. However the lesion is not associated with any focal thickening of the wall. Potential positive feedback loop in the growth of the atherosclerotic plaque, based on the self-perpetuation of monocyte traffic. Macrophages in the artery wall facilitate the entry of more monocytes.

Fig. 2.3. A Stary Type V-b lesion, stained for smooth muscle cells with antiactin antibody HHF35, showing a large lesion (right side of arterial wall) with a massive lipid core containing some calcification (stained red). The fibrous cap is well developed (brown stained, immediately to right of lumen). Note the thinning of the well stained media outside the plaque (far right).

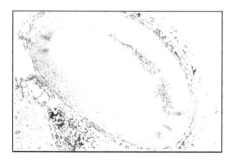

Fig. 2.2. A Stary Type III lesion, stained with the macrophage-reactive antibody HAM56. The arterial intima is focally thickened (right side of arterial wall), and the antibody reveals a focal collection of macrophages in the intima outside a fibrous cap of smooth muscle cells adjacent to the lumen. They remain close to the lumen at the shoulders of the lesion (particularly at the top of the lesion in this artery), where further monocyte traffic into the wall occurs. There is no substantial accumulation of extracellular lipid.

dominate near the arterial lumen; it is here that they enter, if there is continued stimulus. These are known as *type V lesions*. Some may also contain calcium (*type Vb*) (Fig. 2.3). In addition new blood vessels may infrequently be formed inside the lesion in a further attempt at healing, known as neovascularization. This process, when it occurs, is very important in lesion growth and progression, as some intraplaque hemorrhage may originate from these vessels and cause a rapid increase in plaque size, possibly precipitating an acute coronary syndrome.

The majority of acute coronary syndromes however, arise from thrombus formation inside the plaque, initiated by the mechanical cracking of the weakened intima. Lesions containing fissure, hematoma or thrombus are known as *type VI lesions*.

In some cases more advanced lesions will burn themselves out, the macrophages disappear, just leaving the fibrous cap—known as a *type Vc lesions*.

Lipid Metabolism

The most important risk factor for atherosclerosis is cholesterol.[15] It is known that in populations with very low levels of plasma cholesterol, the incidence of ischemic heart disease is also extremely low. As the population's average cholesterol increases, so does the incidence of atheroma.

When fat is absorbed into the blood stream[16] from the gut it is formed into chylomicrons. These are micelles or lipoproteins consisting of triglycerides predominantly, with some cholesteryl-ester at their center, surrounded by a surface polar coat of phospholipids. The purpose of the coat is to enable the water-insoluble lipids to be transported in the blood. The coat also contains apolipoproteins whose roles are to stabilize the surface and to enable recognition by cell surface receptors. These receptors make lipid transport a guided delivery system.

The chylomicrons circulate in the blood. In adipose and other tissues some of the triglyceride is broken down by the enzyme lipoprotein lipase, assisted by interactions with the apolipoproteins Apo A1, A2, A4, and B48 contained in the chylomicron. The free fatty acids formed are absorbed. The remaining cholesterol-rich lipid, chylomicron remnants, is then transported to the liver for processing, being recognized by liver cell receptors for Apo B48 and Apo E.

Endogenously produced triglycerides are transported from the liver as **V**ery **L**ow **D**ensity **L**ipoprotein (VLDL). Smaller quantities of cholesterol, phospholipids, Apo B100, Apo C and Apo E are also found in VLDL. This lipoprotein circulates in the plasma. It interacts with lipoprotein lipase in tissues and is broken down to free fatty acids and **I**ntermediate **D**ensity **L**ipoprotein (IDL). The fatty acids are absorbed and the IDL either returns to the liver, or is transformed into **L**ow **D**ensity **L**ipoprotein (LDL) in the blood. This latter complex has a higher content of cholesterol. As it is present in high concentration in the blood, it carries most of the serum cholesterol to areas of need. It is recognized in tissues using the LDL receptor which binds to Apo B100.

As the LDL circulates, because of its lipid-soluble nature, it can diffuse into arterial walls both through and between endothelial cells, down a concentration gradient. The LDL can then diffuse back into the blood stream either by going directly back into the vessel from whence it came or by traversing the arterial wall and thence into the *vasa vasorum*. These are the blood vessels of the blood vessels and are situated in the adventitia. All cells require cholesterol for plasma membrane and sterol synthesis and with the exception of liver cells that can synthesize it, have to absorb it from LDL, via the LDL receptor. Once enough cholesterol has been absorbed, cellular cholesterol synthesis is switched off and the LDL receptor is downregulated by internalization. Cholesterol released from dying cells or from membrane turnover is returned to the liver in the form of another cholesterol-rich lipoprotein, **H**igh **D**ensity **L**ipoprotein (HDL).

If the LDL remained unaltered and pristine in the artery wall, then atheroma would not form. However, because of the Western diet, being rich in cholesterol and low in antioxidants, the LDL becomes modified. This can occur in a number of ways, the most common being by oxidation. It is modified by endothelial cells and once oxidized can damage them, increasing their permeability. It is perceived as a toxin and activates various chemical-signaling systems. Monocytes are thus attracted to the area, bind to the endothelial cells and track between them. Native LDL can be oxidized in the arterial wall by monocytes and smooth muscle cells, with a similar end result. Once through into the wall the monocyte can then mature into the macrophage.[17]

Some modified LDL will remain sequestered on the subendothelial proteoglycan matrix while most will be absorbed. It is not absorbed by the LDL receptor however, but instead is taken up avidly by the *scavenger receptor* found primarily on monocytes/macrophages. This receptor is not down regulatable and thus the more oxidized LDL present the more that is absorbed. The cholesterol accumulates in the cell cytoplasm in small vesicles, eventually forming a foam cell. In the process of lipid absorption, the macrophage becomes activated, releasing further factors which attract monocytes, T lymphocytes and smooth muscle cells to the area. These factors increase the ability of circulating monocytes and T cells to adhere to the endothelium in response to the inflammatory stimulus.

In this way, if there is enough oxidized LDL to keep the system going, it is likely that a positive feedback loop will be set up causing the initial fatty streak to progress to the more complicated lesions described above. Evidence supporting this positive feedback hypothesis comes from rabbit experiments in which early lesions could be induced by high cholesterol feeding.[17] On returning the animals to a low cholesterol diet the lipid continued to accumulate forming classical atheromatous lesions.

Aneurysms

The atheroma associated with abdominal aortic aneurysms differs from the situation described above in that it occurs in an older age group that tends not to have clinically apparent atheroma elsewhere. It is common even in nonaneurysmal atherosclerosis for there to be some degree of medial thinning adjacent to lesions. When the thinning is severe as it is in aneurysms, there is disorganization of elastin fibers and disruption of the media. This is probably due to excessive metalloproteinase production. Large numbers of other inflammatory cells, secreting different proteolytic enzymes, with much the same effect, are also seen. This difference in balance between matrix destruction on the one hand and cellular proliferation and matrix production on the other explains the different clinical presentations, rupture as opposed to ischemia and infarction.

Acute Coronary Syndromes[19,20]

As has been discussed above, the atherosclerotic plaque causes symptoms by nonocclusive obstruction to blood flow causing distal tissue ischemia under stressful conditions. In the normal circulation the endothelium is resistant to platelet adhesion due to several factors. However when the endothelium is damaged or lost, as in atherogenesis, platelets can adhere and become activated. These activated platelets produce chemical messengers which further activate the cells of the atherosclerotic plaque. The platelets also activate each other causing aggregation and thrombus formation. This thrombus can then become incorporated into the lesion which may then be significant enough to cause complete obstruction to the arterial lumen, causing ischemic complications (unstable angina, acute myocardial infarction and sudden coronary death). Thrombus will also form over a crack or rupture in the intima resulting in hemorrhage into the wall, with similar consequences. This is the usual mechanism of the acute coronary syndrome. Both type IV and type V lesions are susceptible to this process.

The pathogenesis of this aspect of plaque natural history is of great importance. For example, in the USA and Western Europe over 4,000,000 people are hospitalized for acute coronary syndromes annually. It has been estimated that one quarter of fatal coronary thromboses occur due to endothelial denudation, perhaps due to the toxic effects of oxidized LDL, the remainder because of plaque rupture. Some feel that the former process may be more common in women. Plaques that rupture have certain characteristics (Fig. 2.4)—over half of their volume is made up of the lipid core, they contain numerous macrophages and have a thin and disorganized fibrous cap with few stabilizing smooth muscle cells. These soft lipid-rich plaques are more likely to rupture when subjected to bending, changes in coronary artery perfusion pressure and high shear stress, for example around significant stenoses.

Important Molecules in Atherogenesis

This section of the Chapter discusses the biology of certain macromolecules thought to be of key importance in atherogenesis. Much atherosclerosis research in recent years has centered on three groups of such proteins:

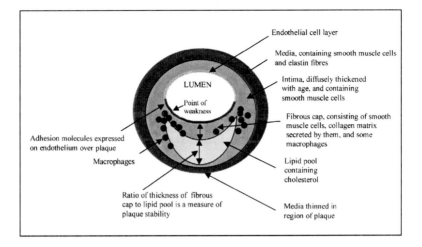

Fig. 2.4. Diagram showing the major features of an advanced fibro-fatty atherosclerotic plaque. Macrophages in reality are smaller and more numerous than shown. The greater the thickness of the lipid pool in relation to the fibrous cap, the more unstable the plaque will be.

adhesion molecules, cytokines and growth factors.

Adhesion Molecules[21]

General Pathology

The endothelium is an extremely complex organ with many regulatory functions, including defense. This function is effected by the expression of inducible adhesion molecules.

Over the last decade or so the detailed processes by which circulating blood cells adhere to the endothelium and other cells have become apparent. Cells such as monocytes, lymphocytes and polymorphonuclear leukocytes have both constitutively expressed and inducible ligands on their outer surface. On the endothelial surface and on other cells there are complementary structures, receptor molecules, which enable interactions to occur. The complementary ligands, likewise, can be either constitutively expressed or inducible and both sets of ligands are referred to as adhesion molecules. Their interaction may effect inter-cell signaling or cellular migration.

They have been shown to be vital for inflammatory responses, including chronic conditions e.g., rheumatoid arthritis, psoriasis and chronic periaortitis. Atherosclerosis, currently also viewed as a chronic inflammatory process, is therefore a prime target for research. (Table 2.1).

Six classes of adhesion molecules have been identified:[22] selectins, members of the immunoglobulin supergene family, integrins, cadherins, proteoglycans and mucins. Certain selectins, integrins and immunoglobulin supergene family members appear to have a primary role in atherogenesis.

In rabbit models of atheroma, induced by a high cholesterol diet, the earliest event seen at the light microscope level is the adhesion of both monocytes and T cells to the endothelium. Indeed, the earliest detectable vascular change in these rabbits is the focal appearance of P-selectin and VCAM-1 on aortic endothelial cells.[23] Interest in atherosclerotic circles has focused therefore on adhesion molecules in general and on four molecules specifically.

P-Selectin

Selectins are relatively low-affinity adhesion molecules which slow leukocytes down considerably as they pass over the endothelium, by causing them to roll along the endothelial surface. Subsequently, if the

Table 2.1. Comparison of the expression of adhesion molecules in normal and inflamed small vessels with those in normal arteries and atherosclerosis. The expression of adhesion molecules is indicated semi-quantitatively

Adhesion Molecule	ICAM-1	VCAM-1	P-selectin	E-selectin
Normal artery	±	−	−	−
Atherosclerosis	+++	−	+++*	+
Normal small vessel	+++	−	+++	−
Inflammation	+++	+++	+++*	+++

* P-selectin is present on the endothelial surface in atherosclerosis and inflammation; otherwise it is within the cytoplasm of normal small vessel endothelial cells.

signals are correct, the leukocytes can be completely arrested and preparation for ingress made. This arrest is brought about by other adhesion molecules, which mediate the next stage of the interaction (see below).

Selectins are made up of an N-terminal lectin-like domain at their outer extremity, their binding function being located here. An Epidermal Growth Factor (EGF) region then follows linked to several consensus repeats, similar to complement binding proteins, joined to a transmembrane domain and a cytoplasmic terminal.

Selectins differ from other adhesion molecules in that rather than interacting directly with the amino acid chain of other proteins, their ligands are carbohydrate side-chains. In the case of P-selectin, these molecules include Lewis x antigen, sialyl Lewis x antigen, sulphatides and sulphated glycans.

P-selectin, previously known as GMP-140, is 130 kD in size. It is found in the endothelial cell cytoplasmic Weibel-Palade bodies. It is constitutively-expressed on the endothelium of venules and small veins. It is not routinely found on the surface of arterial endothelium but its presence can be induced by thrombin, complement, PMA, histamine and oxygen free-radicals. On stimulation, the endothelial cell granules are released to the surface in coated pits. Monocytes can then bind via their surface carbohydrates. This binding is sufficient to

activate the monocyte and cause production of Macrophage Chemoattractant Protein-1 (MCP-1), Tumor Necrosis Factor-α (TNFα) and the transcription factor NF-κβ.[24] The cells roll along the surface of the endothelium until they are brought to a halt by stronger interactions with other adhesion molecules, ICAM-1 for instance (see later). This is the first stage in the trafficking of inflammatory cells through the endothelium.

P-selectin is also found in platelet α-granules and is involved in their adhesion. Its expression on their surface can be stimulated by agonists such as thrombin and lysophosphatidylcholine, a major phospholipid component of oxidized LDL.[25] The latter is at least partially mediated by intracellular signaling through protein kinase C.

Immunohistochemical experiments,[26] quantitated using color image-analysis, have shown that P-selectin is expressed over macrophage-positive plaques, especially over the active macrophage-rich shoulder regions (Fig. 2.5). Little expression was detectable over normal artery or over quiescent, fibrous plaques, thus suggesting a role for P-selectin in atherogenesis.

To examine endothelial binding in a more physiological way we have used the *Stamper-Woodruff* assay.[27] This uses a tissue section derived from a sample of atherosclerotic artery with a suspension of white cells layered on top. It is then placed on a moving

Fig. 2.5. Strong expression (brown staining) of the adhesion molecule P-selectin in the endothelium over an atherosclerotic plaque shown by immunohistochemistry with the antibody LYP20. P-selectin is not normally expressed in arterial endothelium, unlike its presence in normal small blood vessels, and indicates that the endothelial cells have been activated.

platform and the white blood cells are washed back and forth across the slides to facilitate binding. Any cells bound can then be fixed, stained and identified. Antibodies to adhesion molecules can be used to block binding. These experiments have shown that cultured, activated macrophages will adhere well to the endothelium overlying plaques. This binding can be partly inhibited by antibodies to P-selectin, and double immunofluorescence immunohistochemistry has confirmed the presence of P-selectin on the endothelium.

Some adhesion molecules are released into the plasma, particularly after engagement of their receptors—their levels can be measured indicating their level of activation. Circulating levels of soluble P-selectin have been found to be increased in acute myocardial infarction, thus suggesting that this adhesion molecule may also play a role in acute coronary syndrome pathology as well as in the early events of atherogenesis.[28] In the same study levels of thrombomodulin were also found to be raised. This compound is stored in the alpha granules of platelets as well as P selectin, suggesting that platelets are extensively activated in this situation and are the source of the P-selectin.

E-Selectin

This molecule, previously known as ELAM-1, is also a selectin, 115 kD in size. Its role is to allow neutrophil, T helper cell and monocyte traffic into areas of inflammation. The slightly differing functions of P-selectin and E-selectin become clear in knockout mice.[29] In mice that cannot express P-selectin, leukocytes will not roll at sites of inflammation, but the white cell count remains normal. In contrast, E-selectin-deficient animals are virtually normal. If both selectins are missing, no rolling of leukocytes occurs and there is an excessive leukocytosis associated with severe bacterial infections.

E-selectin is not always present in atherosclerotic plaques and is usually only weakly expressed. However in contradistinction to ICAM-1, its expression is specific for plaques.[30] whereas ICAM-1 can be found, albeit in smaller quantities, in normal arteries. Staining for E-selectin has also been noted in adventitial vessels,[31] suggesting that it may be more important for inflammatory cell ingress via this route.

In one study soluble E-selectin was found to be increased in those with mild hypertension,[32] a condition in which there is known to be endothelial dysfunction. It has also been shown that soluble E-selectin is increased in patients with stable angina as compared to normal controls. Levels may fall during episodes of ischemia.[33] This raises the possibility that E-selectin may be involved in the chronic but not the acute stages of atherogenesis.

ICAM-1

Inter-Cellular Adhesion Molecule-1 (ICAM-1), CD54, is 95 kD in size and a member of the immunoglobulin supergene family. This family share structural similarities with immunoglobulins in their domain structure. ICAM-1 can be expressed on many cells with appropriate stimuli but its main functional role is to assist in the interaction between endothelial cells and monocytes or lymphocytes. It is constitutively-expressed on the endothelium of small blood vessels but arterial expression is more variable, reduced or even nonexistent.

It binds to the molecule LFA-1, an integrin of the β_2 family, found on monocytes and lymphocytes. It also binds to MAC-1, another β_2 integrin found on monocytes. Integrins are membrane glycoproteins made up of two subunits, alpha and beta. Both active and inactive forms exist, their activation involving conformational change. This increases their affinity for their ligands. β_2 integrins are typically expressed on leukocytes. Studies have suggested that fibrinogen can act as an intermediary binder between leukocytes and endothelial cells.[34] Unexpectedly the fibrinogen binding receptor was also found to be ICAM-1.

ICAM-1's expression can be upregulated in vitro by stimulating cultured endothelial cells with interleukin-1β (IL1β), TNFα,[35] lysophosphatidylcholine[36] and interferon-γ (IFNγ). Other studies have shown that treatment of cultured endothelial cells with scavenger receptor ligands[37] will induce the expression of ICAM-1, VCAM-1 (see later) and E-selectin. This expression could be inhibited by protein kinase C inhibitors.

When monocytes are exposed to LDL[38] that has been copper oxidized, they release several factors stimulating the endothelial

expression of ICAM-1, VCAM-1 and E-selectin. Shear stress can also induce endothelial ICAM-1 expression, as does smoking. ICAM-1 expression may also be enhanced by hypoxia.[39] In recent experiments it was shown that in culture hypoxia alone would not induce ICAM-1 expression. However in the presence of lipopolysaccharide (LPS), hypoxia would further increase its expression. This was found to be associated with a significant rise in the transcription factor NF-κβ.

Studies[40-42] in our laboratory and elsewhere have shown that although low levels of ICAM-1 can be found on the luminal surface of some normal arterial endothelium, its expression is significantly upregulated over active atherosclerotic plaques, that is those with recently entered intimal macrophages. There is no expression over those plaques that are apparently quiescent, type Vc lesions. ICAM-1 can also be found expressed on the surface of some plaque macrophages and smooth muscle cells suggesting that these cells may be able to adhere to and interact with other cells in the intima using this molecule.

Using the Stamper-Woodruff assay, monocytes were shown to actively adhere to both the endothelium and the intima. This adhesion was inhibitable by anti-ICAM-1 antibodies. Adventitial vessels and intimal neovascularization[43] also stain positively for ICAM-1 in a manner typical of small vessels.

If one examines the blood of patients during myocardial ischemia,[44] the levels of soluble ICAM-1 are increased, suggesting that it may also play a role in acute events, as well as in long-term pathogenesis. The levels of circulating soluble ICAM-1 have also recently been shown to have predictive value for the occurrence of myocardial infarction.[45]

VCAM-1[46]

This molecule, Vascular Cell Adhesion Molecule-1, 110 kD in size is, like ICAM-1, a member of the immunoglobulin supergene family. Its expression is stimulated by IL1, TNFα,[47] LPS, IL4, IFNγ and lysophosphatidylcholine. Low shear stress[48] has also been shown to induce its expression. The transcription factor NF-κβ has been implicated in its regulation. Its receptor is very late antigen-4 (VLA-4) which is expressed on mononuclear leukocytes. VLA-4 is a member of the integrin family, a β1-integrin, expressed on monocytes and B and T lymphocytes.

Recent studies[49] have suggested that the antioxidants pyrrolidine dithiocarbamate and o-phenantrolene can inhibit VCAM-1's expression in vitro, when induced by IL1 or TNFα. Similar results were found for E-selectin suggesting that oxidative mechanisms are involved in both these forms of upregulation.

Rabbits fed a hyperlipidemic diet show increased endothelial expression of VCAM-1.[50] This was found to occur within the first four weeks of the diet, preceding monocyte ingress. Interestingly, it was also stimulated in diabetic normolipidemic rabbits. Other rabbit studies[51] have shown that some smooth muscle cells can also express VCAM-1 and that this can be stimulated in both human and rabbit cell cultures by IFNγ and IL4, both products of activated T cells.

Studies in human specimens have not confirmed, completely, that the observations in the rabbit are applicable to man. VCAM-1 has been detected in human lesions in some studies, but not in others. No inhibition of adhesion occurred using a VCAM-1 antibody in the Stamper-Woodruff assay. However, VCAM-1 has been found in large quantities in the vasa vasorum of affected arteries. It is also present in areas of neovascularization and inflammatory infiltrate in the basal areas of plaques. A significant part of the VCAM-1 expression is also found in plaque macrophages and smooth muscle cells.[52] Thus the possibility arises that it may be involved in the formation of complicated plaques and in acute coronary syndromes. Indeed soluble VCAM-1 has been found in increased quantities in the serum of patients with unstable angina, as compared to those with stable angina,[53] and the levels correlate with the extent of atherosclerosis in comparative studies. It is not clear at present from which cell type this soluble VCAM-1 is derived.

Interestingly, in some species it is expressed in cardiac muscle.

Others

Studies performed using a bank of monoclonal antibodies have discovered a further adhesion molecule thought to be involved in atherosclerosis.[54,55] The antibody known as mAb-IG9 identified an adhesion molecule induced by minimally-modified LDL, found not to be one of the other known adhesion molecules. Its expression is also induced by stimulation with IL1 and TNFα. It appears to specifically assist the binding of monocytes to endothelium.

Antibodies against CD14, when used in the Stamper-Woodruff assay, also inhibited binding. CD14 is a 55 kD mucin-like molecule on monocytes that acts as an endotoxin receptor and possibly as an adhesion molecule. It also has an intercellular signaling role and can activate β2 integrins on monocytes. Very recently, CD14 has been shown to be a receptor by which monocytes bind to apoptotic cells.[56]

In addition, recent work with the Stamper-Woodruff assay has shown that monocytes bind to atherosclerotic plaques via the αvβ3 integrin. A role for this molecule in monocyte-endothelial adhesion has been established previously by several laboratories, with the endothelial ligand being found to be CD31 (PECAM, the platelet endothelial cell adhesion molecule).[57] This interaction is inhibitable by peptides containing the RGD (arg-gly-asp) sequence, which is involved in integrin binding. Interestingly the Stamper-Woodruff showed inhibition of binding to both endothelium and intima matrix with this peptide. Like CD14, αvβ3 integrin is a receptor also implicated in monocyte/macrophage interaction with apoptotic cells.[58] One could speculate that apoptotic activity in the plaque might extend to the endothelium and be significant in enhancing monocyte adhesion.

The above studies therefore suggest that adhesion molecules are intimately involved in the pathogenesis of atheroma. The possibility therefore arises that due to endothelial cell activation, possibly due to locally present minimally-modified LDL, P-selectin initially and then ICAM-1 are expressed on the luminal endothelial surface. Circulating monocytes bind on, roll and are then brought to a halt. At this stage they can then migrate into the intima. During the ingestion of oxidized LDL present in the arterial wall, macrophages, being derived from migrating monocytes, become activated and may then produce cytokines which can further stimulate adhesion molecule expression on the luminal surface of the artery, monocyte chemoattractant factor release etc.

Furthermore, these activated macrophages are probably highly effective at further oxidizing the local extracellular LDL. This may thus set up a positive-feedback cycle. Those same cytokines may also cause adhesion molecule expression on other cells contained in the plaque, smooth muscle cells for instance, thus trapping them and enabling further interactions.

This hypothesis is given credence by recent studies[59] using gene deleted C57BL/6 mice. In these animals when fed a high fat diet, deletion of the genes for ICAM-1, β_2-integrins or P-selectin produced a 50-75% reduction in atheroma. Similarly, in animals with LDL receptor gene deletions and therefore naturally atheroma-prone, there was a 50% reduction in atheroma when the genes for P or E-selectin were also removed.

Growth Factors[60]

General Pathology

Growth factors are small proteins, often released from cells after activation, that promote cellular growth in either a paracrine or an autocrine fashion. Extensive research over recent years has suggested that certain members of this broad group of compounds may be instrumental in atherogenesis. Platelet Derived Growth Factor (PDGF), Insulin-like Growth Factor-1 (IGF-1), basic Fibroblast Growth Factor (bFGF), Heparin Binding Epidermal Growth Factor (HB-EGF) and Transforming Growth Factor beta (TGFβ) are all found in atheroma. However they are also present to some extent in the normal arterial wall.

Growth factors are carefully regulated by their very short half lives, often less than five minutes. This is effected by their binding to various proteins. One family of molecules subserving this binding and regulatory function is the glycosaminoglycans, such as heparin. Heparin binds PDGF, HB-EGF and acidic FGF (aFGF) but can also release TGFβ from its carrier protein. A further binding protein present in atheroma is SPARC—Secreted Protein, Acidic and Rich in Cysteine—also known as osteonectin. This binds PDGF-BB or PDGF-AB and inactivates it. In other cases growth factors bind to soluble forms of their receptors or alternatively to large plasma proteins such as alpha$_2$-macroglobulin. This latter form of regulation occurs with PDGF, TGFβ, EGF and bFGF.

Not only do these molecules stimulate growth, some of them also cause chemotaxis: TGFβ for monocytes and smooth muscle cells, PDGF for smooth muscle cells and IGF-1 for smooth muscle and endothelial cells.

Although there are many growth factors present in plaques they may not all be active. Some need postrelease enzymatic cleavage, in some cases the receptors may not be present and in others multiple simultaneous stimuli are required for a particular effect.

There has been long-standing interest in the role of PDGF in atherosclerosis, through the work of Ross and collaborators but rather less is known of other growth factors. Several immunohistochemical studies have reported PDGF-β in atheroma macrophages and have found that PDGF-α is expressed widely.[61] However some uncertainty must remain over these results, as in situ hybridization studies have not shown PDGF expression in macrophages. Smooth muscle cells in the plaque are responsive to PDGF-β, as they have high levels of the β-receptor.[62]

Despite the uncertainty over PDGF, the atheroma macrophage appears to be the cell producing the largest range of growth factors, as we have also found expression of IGF-1 and EGF. Interestingly the macrophages express receptors for IGF-1, FGF and probably several more growth factors and so maintain their viability by autocrine stimulation.

It is therefore probable that the macrophage is the principal cell type responsible for the increased production of growth factors in the atherosclerotic plaque. Its role in this site appears equivalent to that seen in wound healing, where macrophages have been shown to synthesize a wide range of growth factors.

PDGF

This molecule is secreted by macrophages, platelets, smooth muscle cells, fibroblasts and endothelial cells. It causes proliferation of smooth muscle cells, T cells and fibroblasts, acts as a chemotactic agent for macrophages, smooth muscle cells and fibroblasts and will stimulate collagen synthesis. It is made up of two chains, A and B, the genes for which are located on different chromosomes. It can exist in three isoforms, AA, AB and BB, with slightly differing functions.

PDGF production[63,64] in endothelial cells and smooth muscle cells can be stimulated by oxidatively-modified LDL, specifically the lysophosphatidylcholine component but not by native LDL or HDL. Interleukin-1, TNFα and TGFβ cause gene expression of PDGF-AA in smooth muscle cells also. The same molecules stimulate endothelial cells in culture to produce PDGF-BB. In this way smooth muscle cells may be stimulated to proliferate in either an autocrine or a paracrine fashion.

It is known that during the process of atherogenesis medial smooth muscle cells change phenotypically from a contractile state to a synthetic one and migrate into the intima. These intimal cells have been shown to express de novo, a protooncogene, *c-fms*,[65] which is more commonly found in monocytes. The protein produced by this gene encodes for the Macrophage-Colony Stimulating Factor (M-CSF) receptor, a cell surface receptor that has intrinsic tyrosine kinase activity which can thereby activate multiple intracellular pathways. A combination of PDGF-BB and either EGF or FGF will stimu-

late the production of c-fms, whereas it is reduced by IFNγ or M-CSF. This suggests that the stimulating growth factors may be involved in smooth muscle cell phenotypic transformation. It is also a possibility that platelets binding to damaged endothelium release PDGF, thus causing smooth muscle cell migration and transformation.

PDGF-BB stimulates endothelial-dependent vasodilatation via NO[66] and causes endothelium-independent constriction. Thus it may contribute to atherogenesis by smooth muscle cell chemotaxis and proliferation and cause vasoconstriction in the presence of dysfunctional endothelium, as occurs in acute coronary syndromes.

IGF-1

IGF-1 is present in significant quantities in many cell types in the arterial wall including arterial smooth muscle cells. It causes endothelial cells and fibroblasts to divide and acts in concert with FGF or PDGF to promote smooth muscle cell multiplication.

FGF[67,68]

Fibroblast growth factors (FGFs) have only limited expression in the normal arterial wall. They exist as a family of related proteins, of which the best known have been studied in atherosclerosis. These are acidic FGF (aFGF or FGF-1) and basic FGF (bFGF or FGF-2). They are each 16 kD molecules and share 57% homology. They both bind to a family of high affinity receptors, of which FGFR1 (flg gene) and FGFR2 (bek gene) are the best studied and the most expressed.

Fibroblast growth factors are found in many cells of the body. In normal arteries bFGF is found in medial smooth muscle cells and adventitial vessels. Acidic FGF and the FGF receptor on the other hand are only found in adventitial vessels.

Fibroblast growth factors are not secreted directly but are found in the cytoplasm of cells and also bound to matrix proteins and the basement membrane, having strong affinity for heparin-like glycosaminoglycans. They are released from cells as a consequence of cell injury or from their binding to extracellular matrix proteins by certain macrophage enzymes such as heparanase and by acidic conditions. Fibroblast growth factor therefore acts as a stored growth factor.

FGF both attracts and causes division of fibroblasts, endothelial cells and smooth muscle cells. In addition it causes the deposition of extracellular matrix proteins. In normal circumstances it therefore acts as a repair-inducing growth factor. In endothelial cell culture bFGF maintains the thrombo-resistant surface. It can also act as an angiogenic factor.

In atheromatous lesions some studies have suggested that there is increased expression in both intimal smooth muscle cells and macrophages. Both forms of FGF have been associated with atherosclerosis. Acidic FGF has been shown to be increased in plaques, and bFGF and the FGFR-1 are particularly present in areas of neovascularization. However the overall levels of bFGF are decreased. Basic FGF has been shown to be involved in the migration and proliferation of early intimal thickening-associated smooth muscle cells in rat postangioplasty restenosis models. It is not however required for their chronic proliferation[69] and may therefore only be involved in early lesion formation. Its release from damaged tissues may play a role therefore in restenosis after angioplasty.

EGF[70]

EGF is derived from macrophages and platelets and causes fibroblasts and endothelial cells to divide. There are a number of growth factors belonging to the EGF group. One such molecule is HB-EGF. It is 86 amino acids long, being derived from a transmembrane precursor. Lysophosphatidylcholine can induce its production. HB-EGF attracts and causes proliferation in smooth muscle cells by binding to the EGF receptor. Both HB-EGF and its receptor are found in plaque smooth muscle cells and macrophages, but not in normal arterial wall, suggesting that it may play a role in the smooth muscle cell proliferation seen. EGF itself is also found in the macrophages of the plaques.

TGFβ

This growth factor is a homodimeric protein, 25 kD in size and is known to be produced by T cells, smooth muscle cells, aggregating platelets, endothelial cells and macrophages. It causes extracellular matrix production, especially fibronectin, collagens I,III,IV and V, elastic fibers, glycosaminoglycans and proteoglycans. It also inhibits smooth muscle cell proliferation and has variable effects on PDGF release by smooth muscle cells depending on its concentration.

TGFβ has been identified in macrophages in atherosclerotic lesions,[71] and the finding of collagen gene expression in nearby smooth muscle cells suggests that it mediates fibrous cap formation. As it is also a potent chemotactic factor for monocytes, its appearance in the lesions could theoretically enhance plaque development by attracting further cells. Curiously, the levels of circulating active TGFβ are greatly decreased in patients with atherosclerosis.[72] The significance of this observation is not yet completely clear, but it is noteworthy that TGFβ has been shown to bind to endothelial cells and inhibit their expression of adhesion molecules.[73] By this means it may be able to inhibit monocyte traffic into the plaque. Furthermore, excess TGFβ in the blood could generate a negative concentration gradient with the tissues and thus prevent migration towards any arterial source.

A further important fact which has emerged very recently is that TGFβ is sequestered into an inactive pool by lipoproteins.[74] It will need to be elucidated whether this means that the inverse association of TGFβ levels with atherosclerosis is only an epiphenomenon of lipoprotein levels or whether it implies that lipoproteins have a further atherogenic mechanism by inhibition of TGFβ.

In summary, although the detailed processes which underlie the role of growth factors in atherogenesis are still unclear, it seems that toxic insults to the endothelium and intima allow various factors to be released as part of inflammatory and repair mechanisms. In some cases this works successfully ending with an inactive fibrous type Vc plaque, in others it causes worsening luminal obstruction and distal ischemia. Much more will have to be discovered before we can understand the regulation of these processes.

Cytokines

Introduction

These molecules were originally known as lymphokines, as those first discovered were produced by white blood cells during immunological reactions. Later it became clear that they are produced by many cells for the purpose of inter-cell signaling, hence cytokines. They are small proteins with specific receptors that exhibit pleiotropy and redundancy; they often have similar effects to each other and each one has several different functions.

A vast range of cytokines has been discovered and characterized. However, it is clear from experimental evidence that a small group are very important in the pathogenesis of atheroma.

Interleukin-1

Interleukin-1 was one of the first cytokines to be discovered. It is 17 kD in size and exists in two forms, alpha and beta. Most cell types seem capable of entering an activation state with appropriate stimuli, with the subsequent production of IL-1. It can be produced by stimulated macrophages, endothelial cells, smooth muscle cells, epithelial cells and keratinocytes. Its role in atherogenesis has been extensively investigated because of its vast range of functions:

- The induction on endothelial cells of adhesion molecules: ICAM-1, VCAM-1 and E-selectin. The evidence for the production of IL-1 in atherosclerotic lesions is somewhat conflicting but it is possible that monocytes close to the endothelium may produce it and thereby induce endothelial adhesion molecules. TNFα may also be important in this respect (see below). IL1β will also induce ICAM-1 and VCAM-1 on smooth muscle cells, which may be important in their intercellular com-

munication.[75] Intimal smooth muscle cells in human atherosclerosis express ICAM-1 well, but it is not clear which stimulus is responsible.

- Release of acute phase proteins, fibrinogen for example, which increases the likelihood of thrombus formation in acute coronary syndromes.
- Proliferation of antigen-stimulated T cells and activated B cells. Thus, IL-1 assists in immune responses to various antigens that may play a role in atherogenesis (see below).
- The production and release of other cytokines, for instance IFNγ, colony stimulating factors and interleukin-6 (IL-6).
- The proliferation of smooth muscle cells, endothelial cells and fibroblasts. IL-1 stimulates smooth muscle cells to produce PDGF-AA which can then act in an autocrine or paracrine manner. It will also increase their production of FGF and its receptor.
- IL-1, in vitro, has been shown to reduce endothelial cell production of tissue-type plasminogen activator (t-PA) and to increase the production of its inhibitor, plasminogen activator inhibitor-1 (PAI-1).[76] In this way IL1 may increase the ability of fibrin to deposit on the endothelial surface. Tissue factor procoagulant activity is also increased by IL-1, with the subsequent deposition of clotting factors IX and XI and hence thrombin and fibrin. The thrombin may then stimulate endothelial cell production of platelet activating factor and P-selectin. Prostacyclin production is also reduced by IL-1 with the same end result.
- It can, when bound to the platelet surface, stimulate the endothelial production of interleukin-8 (IL-8),[77] an effect shared with TNF. IL-8 can then cause neutrophil and T lymphocyte chemotaxis.

Tumor Necrosis Factor-alpha[78-81]

This is an inflammatory cytokine, of 17 kD molecular weight, with a very similar range of action to IL-1. It belongs to a family of similar molecules and is made up of three chains, all identical. Its two receptors, 55 kD and 75 kD in size respectively, are dimeric transmembrane proteins which contain a cysteine-rich area extracellularly. There is a cytoplasmic sequence of 60 amino acids, known as the death domain, in the 55 kD receptor which communicates a signal required for apoptosis.

TNF is secreted by monocytes/macrophages, T cells, granulocytes and smooth muscle cells. It can:

- cause tumor cell death,
- induce angiogenesis,
- cause thrombosis by stimulating the expression of thrombomodulin,
- stimulate collagenase production,
- act as a pyrogen,
- inhibit lipoprotein lipase,
- cause endothelial cell activation with the concurrent expression of adhesion molecules and major histocompatibility (MHC) class I molecules and the production of IL1 and PGE_2,
- stimulate smooth muscle cells to produce ICAM-1 which might be capable of interacting with macrophages,
- induce smooth muscle cells to produce IL1, TNFα and certain prostaglandins.

Smooth muscle cells and macrophages will produce TNF-α by incubation with oxidized LDL. Studies have therefore looked for its presence in arterial walls. It was found by Barath et al[78] to be present in normal intima and fibrous plaques but to be significantly increased in intimal thickening. Its presence has been further confirmed in atheroma by other groups.[82] Further characterization revealed its presence in the cytoplasm of macrophages and both in the cytoplasm and on the surface of endothelial cells and smooth muscle cells. Medial smooth muscle cells do not express it however.

It is known that in atheromatous arteries the production of nitric oxide (NO) is reduced (see later). In vitro work has suggested that this may be, at least in part, due to TNFα.[83] The mechanism seems to be due to the increased degradation of one of the nitric oxide synthase enzymes, eNOS, usually a stable enzyme.

Monocytes placed in contact with endothelial cells in vitro induce adhesion molecules on them.[84] This effect requires TNF-α and possibly IL-1. It therefore seems likely that once monocyte traffic starts at a point on an artery, it will tend to self-perpetuate, generating a lesion.

IL-1β_2 and TNFα levels can be measured in the blood. One study[85] showed that in patients with stable angina, IL-1β levels are lower than in unstable angina. The reverse was true for TNFα. This suggests that different cytokines may be important at different times in the progression of the plaque.

Interferon gamma

This cytokine is produced by T cells and Natural Killer (NK) cells that have been activated by an antigen-presenting cell. It causes the expression of MHC class II on the stimulated cell. This enables the cell to become an antigen-presenting cell and hence be involved in immune responses. IFNγ also inhibits collagen synthesis.

In vitro studies have shown that IFNγ will reduce the expression of the scavenger receptor on monocyte-derived macrophages, thus reducing foam cell formation.[86] The effect was mediated by a reduction in the number of receptors on the surface of the macrophage. Interferon γ has also been shown to promote NO production by smooth muscle cells, inhibit smooth muscle cell proliferation and inhibit lipoprotein lipase.

Immunohistochemical studies of atherosclerotic plaques have shown that a proportion of the cells present, from even the earliest stages, are T cells. Surface markers reveal that the majority of these are T helper cells and that some are activated, not just passively trapped by nonspecific adhesion molecule interaction. Other studies have shown that IFNγ is present in the plaque, especially around activated macrophages. As no NK cells are found, it can be concluded that the origin of the IFNγ is the T cell and may therefore result from an immune response. Oxidized LDL is an antigen that may be involved (see below).

Macrophage Chemoattractant Protein-1[87]

It is known that macrophages enter the intima by binding to endothelial cells. The stimulus for attracting these cells from the blood appears, at least partly to be due to MCP-1 acting in concert with endothelial adhesion molecules. Macrophage chemo-attractant protein-1 is a 76 amino-acid protein that can be produced by monocytes, endothelial cells and smooth muscle cells in vitro. Its release from endothelial cells can be triggered by IL1β and TNFα and to a lesser extent by IFNγ.[88] Most importantly, it can also be induced by minimally-modified LDL in both endothelial cells and smooth muscle cells.[89]

Studies have shown that its expression is increased in atherosclerotic plaques as compared to normal arteries.[90] This occurs mostly in intimal macrophages associated with modified LDL and macrophages around organizing thrombi and around the lipid-rich core. In early plaques it appears that it is mainly produced by endothelial cells and subintimal macrophages.[91] Later on most of the expression appears in macrophages and some in smooth muscle cells, whereas expression is lost from the endothelium. Its generation by oxidized LDL can be inhibited by HDL or antioxidants. Although the chemokine (chemotactic cytokines) family, of which MCP-1 is a member, is now large, there is little information available on the involvement of other members in atherosclerosis.

Mechanisms in Atherogenesis

It is clear that certain 'small molecules' appear to be very important in atherogenesis. In the next section of this Chapter individual aspects of the plaque are discussed, with reference to these molecules.

Lipids[92,93]

Lipid Metabolism

As has been discussed above, one of the principle initiating causes of atheroma appears to be cholesterol. Research has therefore been undertaken to discover the details of its atherogenic effect.

The main form in which cholesterol is transported in the blood is LDL. In the normal situation, LDL either passes through or between the endothelial cells, down a concentration gradient into the intima and beyond. No receptor is needed for this process. Studies have examined lipoprotein entry into the arterial wall. After the injection of labeled lipoproteins, the arterial wall concentration increases to a peak between 1 and 8 hours. It then falls as the lipoproteins equilibrate with those left in the blood. In the steady state, the molecular weight of a plasma protein determines its distribution between the intima and plasma.[94] The higher the molecular weight, the greater the relative intimal concentration. For this reason, LDL, with a molecular weight of several millions, is selectively concentrated in the intima compared to most plasma proteins. In addition, some LDL is trapped and bound in the intima by a three-dimensional matrix of extracellular connective tissue proteins.

Using de-endothelialized and re-endothelialized rat aorta, Falcone et al[95] showed that in the latter, lipoprotein was preferentially concentrated in the arterial wall and did not further equilibrate with the plasma, thus increasing its likelihood of oxidation. The significance of this interesting observation is uncertain however.

Once in the tissues, the LDL is then absorbed into cells by binding to the LDL receptor which uses Apo B100 or Apo E as its ligand. The receptor is an 839 amino acid multidomain protein. Its amino-terminus has seven 40-residue cysteine-rich repeats, with a group of acidic residues near the carboxy-terminus. The first repeat binds calcium. Repeats two to seven bind Apo B100 and Apo E with their very basic residues.

Once LDL is absorbed, the cholesterol is internalized in lysosomes. Here it is hydrolyzed and is used for steroid synthesis and the formation of plasma membrane units. LDL may also be absorbed using the dextran sulphate receptor as it can become complexed to this molecule. Once sufficient cholesterol has been absorbed the LDL receptors are downregulated. Thus with LDL in its native form, the process is controllable.

It has become apparent that modification of extracellular LDL can profoundly affect its interaction with cells in the arterial wall. The prime means of modification, in vivo, is oxidation, although in diabetes glycosylation is also important.

LDL can be affected in many ways by oxidation. The major pathways can be summarized as follows. Some LDL will be damaged by the release of oxygen free-radicals. In other cases, the lipid will become damaged by cellular lipoxygenases. These enzymes can either act on the LDL itself or on cellular lipids which are then incorporated into the LDL. Initially with only minor changes to the structure, the minimally-modified LDL formed is still recognized by the LDL receptor. However, it is noted to be abnormal and its binding stimulates adhesion molecule and cytokine production.

Further LDL damage can occur: lecithin is transformed into lysolecithin by the intrinsic phospholipase A_2 activity of LDL and oxidation of fatty acids produces short-chain aldehydes such as malondialdehyde and 4-hydroxynonenal. These latter molecules then can bind to the amino groups of the lysine residues of Apo B, causing structural changes. Malondialdehyde can also be produced by platelet degradation of arachidonic acid and by macrophages phagocytosing lipids. Once Apo B is affected sufficiently, with more than 16% of the lysine residues being modified, a conformational change occurs in the molecule, possibly increasing the number of aminoacyl groups displayed on the outer surface. In this state, recognition by the LDL receptor can no longer occur and scavenger receptor recognition takes place in its stead.

The oxygen supply of the arterial wall may also be important in the generation of oxidized LDL.[96] Oxygen destined for the intima and inner media is derived, by direct diffusion, from the arterial lumen. Oxygen required by the adventitia and outer media comes from the vasa vasorum. There is therefore a zone in the outer intima and inner media that is relatively hypoxic. As the thickness of the arterial wall increases, due to cellular recruitment and matrix glycosaminoglycan deposition, this zone increases in size, as does the oxygen requirement. This thickening occurs not only because of atheroma but also because of diffuse intimal thickening secondary to hypertension and aging. This causes steep gradients in the partial pressure of oxygen. It is hypothesized that an increased propensity to cause oxygen free-radicals, superoxide, hydrogen peroxide and hydroxyl radicals, thus occurs. The hypoxia can be further exacerbated by smoking. In rabbit models of atheroma, hypoxia has been shown to exacerbate and hyperoxia to reduce plaque formation.

The scavenger receptor was first described in 1979 by Goldstein. It has been characterized in humans, from the phorbolester treated monocytic cell line THP-1.[97] It consists of three subunits of 70 kD size, linked by three disulfide bonds, each subunit containing six domains: I—cytoplasmic, II—membrane spanning, III—spacer, IV—alpha-helical coiled coil, V—collagen-like triple helix and VI—type-specific carboxyterminal. Two different cDNAs were isolated, derived from RNA transcripts encoding two alternatively spliced forms of the scavenger receptor—I and II. Both types I and II have been shown to be present in atheroma.[97] They differ in their sixth domain: type I receptors have a set of six cysteine residues separated by various gaps. This structure is conserved across a wide range of species. Type II receptors have more cysteine residues.

These scavenger receptors are found mainly on macrophages, monocytes, Küpffer cells and endothelium, especially in the liver. They have also been found on smooth muscle cells in atheromatous plaques. The scavenger receptor does not become downregulated by cholesterol absorption and continues to absorb modified LDL as long as it is present, and the cell continues to function, eventually leading to foam cell formation. Another molecule, CD36, has also been implicated as a scavenger receptor and is classified as a class B receptor.[99] Current research suggests that there are more to find.

Lipid Oxidation
i. General Pathology

The next question to answer is where does the oxidation occur? It is unlikely that significant oxidation occurs in the plasma because of antioxidants such as vitamin C, urate, bilirubin and HDL. Others in the lipoprotein molecule, such as vitamin E and betacarotene are also important. Furthermore, oxidized LDL would be quickly removed by scavenger receptors in the liver. It is more likely therefore that the oxidation occurs in the arterial wall itself. LDL can be oxidized in culture with macrophages, smooth muscle cells and endothelial cells. Eighty-five per cent of the LDL that gets into the intima will leave. Some will undoubtedly become minimally-modified at this time. It is possible that on subsequent trips to the intima these molecules may be more susceptible to further damage, thus culminating in maximally-modified LDL.

There is a significant amount of evidence now to show that oxidized LDL is present in atheromatous plaques and may be etiological. Yia-Herttuala's group[100] has extracted LDL from both rabbit and human atheroma using saline extraction and density gradient ultracentrifugation in the presence of antioxidants and protease inhibitors and examined its physical properties.

The following pieces of evidence have been found:

- Lesional LDL's physical and chemical characteristics are more like in vitro oxidized LDL than native LDL. For example, lesional LDL is of higher density, shows greater electrophoretic mobility and contains more cholesterol, sphingomyelin and lysophospatidylcholine.
- Apo B exists in lesional LDL as complete molecules and fragments. These molecules were shown to react with antibodies raised against oxidized LDL and not native LDL extracted from normal arteries.
- Macrophages degrade oxidized LDL. LDL extracted from lesions was degraded by macrophages. This process could be competitively inhibited by oxidized LDL but not by native LDL.
- Lesional LDL is chemotactic for macrophages in the same way as in vitro oxidized LDL. Native LDL shows no propensity to do this.
- Lesional LDL can be recognized using antibodies directed against malondialdehyde-conjugated lysine residues, among others. Immunohistochemical experiments using the same antibodies showed oxidized LDL to be present within foam cells and not in the walls of normal arteries. This has been confirmed by other groups. The possibility arises that as the antibodies used were in fact autoantibodies against the patients' own oxidized LDL, that some of the absorption of oxidized LDL in vivo may be accomplished by the macrophage Fc receptor.[101,102]
- Immunohistochemical studies using antibodies directed against oxidized LDL, not cross reactive with native LDL or other modified forms of LDL, have shown oxidized LDL to be present in rabbit atheroma.[103]
- Patient studies have revealed that there is a correlation between the severity of a patient's disease and the ease with which their LDL is oxidizable in vitro with copper ions.[104] This correlates with the findings that lipid peroxide concentrations are increased in patients with vascular disease and that levels of vitamin E are reduced in such patients.
- Indirect evidence comes from human experiments in which the intimal thickness was measured in carotid arteries

using Doppler ultrasound scanning.[105] An increase in thickness was noted over a period of two years and was found to be proportional to serum copper, important in oxidation reactions, and LDL levels. It was also found to be inversely proportional to plasma selenium, a cofactor in antioxidant enzymes.

- Some groups have shown that in smooth muscle cell culture, LDL taken from healthy subjects will not cause foam cell transformation,[106] whereas that taken from patients with atheroma caused transformation.

- It is a possibility that other LDL derivatives may play a role. Acetylated LDL will also be taken up by the scavenger receptor with similar results.[107] However acetyl-LDL has not been isolated in human atheroma. Glycosylated LDL also binds to the scavenger receptor and may be important in diabetic atherosclerosis.

- Oxidized LDL can elicit immune responses. autoantibodies have been shown to be present in patients' serum. It has also been shown that 10% of the CD4 cells in lesions proliferate in response to oxidized LDL.

Thus there is compelling evidence that oxidized LDL is found and maybe etiological in atheroma, in humans and in animal models. It is likely to exert its effects on a number of elements of the plaque, as follows.

ii. Endothelial Effects of Oxidized LDL

The endothelium is usually antiatherogenic due to the production of vasodilatory molecules such as NO and prostacyclin, together with thrombus-inhibiting molecules, such as NO, prostacyclin, thrombomodulin and t-PA. Endothelial function is known to be disturbed in atheroma, leading to a reduced vasodilatory ability. This state, which has been termed endothelial dysfunction, can be assessed clinically and as discussed later, may be important in acute coronary syndromes.

Studies have therefore examined the effect of oxidized LDL on the production of NO and of the vasoconstrictor endothelin.[108] These have revealed that the levels of NO fall and endothelin rise in the presence of oxidized LDL. It has also been shown that with cholesterol reduction, the endothelial dysfunction can be reversed.

Highly oxidized LDL has also been shown to inhibit endothelial cell migration thus reducing their ability to repair damage. Eventually this form of LDL is cytotoxic to the endothelium.

iii. Cytokine Production

In vitro studies have shown that minimally-modified LDL can activate monocytes and endothelial cells and induce them to secrete various inflammatory mediators found in atheroma, in particular M-CSF and MCP-1. This does not occur with native LDL or fully oxidized LDL. Significant levels of mRNA encoding both proteins are found in human fatty streaks, the increase being mediated via cyclic AMP (cAMP).[109] It is therefore possible that the MCP-1 then attracts monocytes to the area where they are stimulated to proliferate by the M-CSF, although macrophage proliferation in plaques seems to be a minor phenomenon.

Oxidized LDL also reduces the expression of TNFα[110] and can cause macrophages to release IL-8,[111] independent of IL-1β, which may then act as a T cell chemoattractant.

It therefore seems likely that oxidized LDL has potential importance as a stimulator of cytokine production, that in turn promotes the development of the plaque.

iv. Adhesion Molecules[112]

In many animal models the first event in atherogenesis is the adhesion of monocytes to endothelial cells. Thus many groups have examined the effects of cholesterol on monocyte adhesion. Endemann et al,[113] using human monocytes, showed that by treating bovine endothelial cells with βVLDL from cholesterol fed rabbits or with VLDL from cebus monkeys, increased adhesion between the monocytes and the endothelial cells occurred. βVLDL is felt to be the atherogenic lipoprotein in rabbits, being equivalent to LDL in humans.

Frostegård et al[114] have used in vitro oxidized LDL to examine the situation in humans. They oxidized LDL using copper ions or by incubating it with monocytes or endothelial cells. They found that after incubating the oxidized LDL with peripheral blood monocytes and cultured endothelial cells, the adhesion of the monocytes to the endothelial cells was increased. It was also shown that while this effect was as good as using IL1β, it was not in itself due to IL1β secretion. By using cycloheximide, an inhibitor of protein synthesis, they showed that the increased adhesion was due to new protein formation.

Further confirmation comes from experiments performed in rat mesentery,[115] in which endothelial dysfunction and leukocyte adherence to endothelial cells was stimulated using oxidized LDL. This adherence was shown to be inhibitable by monoclonal antibodies to ICAM-1, P-selectin and L-selectin.

Gebührer and colleagues[116] showed that LDL oxidized by culture with endothelial cells or by treatment with copper ions induced the expression of P-selectin on cultured human umbilical vein endothelial cells. The degree of stimulation was related to the degree of LDL oxidation. Native LDL had no effect at similar concentrations, although some expression was induced at very high levels. The expression of P-selectin correlated with monocyte binding. These findings have been confirmed in studies performed on rat aortae.[117] In addition P-selectin expression was also increased using L-NAME, an inhibitor of eNOS, one of the enzymes used to synthesize NO. This enzyme is known to be reduced in endothelial cells after treatment with oxidized LDL.

An important connection can therefore be made that oxidized LDL is likely to stimulate endothelial adhesion molecule expression that can then promote the migration of monocytes into the plaque.

v. Eicosanoids

These molecules, derived from arachidonic acid after release from phospholipids, appear to be intimately linked to the formation of atheroma. LDL, when incubated with endothelial cells for a prolonged period, has been shown to stimulate the release of some arachidonic acid epoxides.[118] The latter can increase vascular permeability, are chemotactic for macrophages and promote adhesion between macrophages and endothelial cells.

Thorin et al[119] have shown that chronic exposure to oxidized LDL inhibits the release of prostacyclin from endothelium thus affecting its resistance to platelet adhesion. Prostacyclin also inhibits platelet aggregation, cell proliferation and monocyte adhesion. It has been shown that oxidized LDL, as opposed to native LDL, acetyl-LDL and βVLDL, increases the production of proinflammatory eicosanoids such as PGE_2 and LTC_4 and decreases the production of 6-keto-PGF_{1alpha} (a metabolite of prostacyclin). The active stimulus was found to be the lipoprotein-lipid peroxide complex.[120] In other studies[121] showing leukocyte adhesion induced by oxidized LDL in vivo (in hamsters), use of a leukotriene biosynthesis inhibitor, inhibited adhesion, suggesting that the adhesion event involves leukotrienes as intermediaries in the signaling pathway.

It is fair to conclude that there are two sides to the action of the eicosanoid mediators. Prostacyclin is a distinctly beneficial substance, is produced by normal endothelial cells and inhibits thrombosis and atherogenic pathways. By contrast, the other eicosanoids, being proinflammatory, are potential mediators of atherogenesis and the production of some can be induced by oxidized LDL.

vi. Macrophages

There is good evidence that macrophages are the central cell type in atherosclerosis. These cells are very rare in the normal arterial wall but concentrate in large numbers in most atherosclerotic lesions. Macrophages have potent activities that are likely to be atherogenic. They oxidize LDL, produce inflammatory cytokines, induce adhesion molecules on endothelial cells and synthesize a wide range of growth factors that can induce smooth muscle proliferation. These topics are dealt with in other sections of this Chapter. By contrast, smooth muscle cells are found both in the aged arterial intima and in the lesions. In a recent study in our laboratory we found that the expression of the smooth muscle cell marker, alpha actin, is significantly reduced per unit area when compared to control diffusely thickened intima.

A few very fibrous type Vc lesions contain very few or no macrophages and interestingly they do not express endothelial adhesion molecules. These are likely to represent healed lesions that have lapsed into inactivity. The proliferation of smooth muscle cells and the deposition of collagen by them seems likely to have produced a barrier that has walled off the macrophages from the arterial lumen or vasa vasorum.

Hypercholesterolemia activates monocytes, as described above. In matched controls, Bath et al[122] showed that monocytes derived from patients with raised cholesterol were larger and showed greater propensity to react to chemotactic agents and to adhere to endothelial cells than those from controls. Other groups have confirmed this.[123]

It has been shown that lysophosphatidylcholine is one of the most monocyte chemoattractant compounds present in oxidized LDL. Once they arrive, their departure is inhibited by another component of the lipid fraction, so far uncharacterized. The macrophages can then cause further LDL oxidation, which may cause a self-perpetuating reaction to be set up, which may be important in inducing focal plaque formation. Modified LDL will also cause macrophage proliferation, thus enhancing the process.

Thus oxidized LDL has a very wide range of effects on cells involved in the atherogenic process. Research[124] has found that many of these effects are mediated by an increase in cAMP and the subsequent induction of the group of transcription factors known as NF-κβ. The latter can also be activated by IL1 and TNFα. Normally these transcription factors are found in the cytoplasm as dimers bound to an inhibitory molecule-Iκβ. Once activated, the inhibitor dissociates and the transcription factors are free to enter the nucleus to initiate gene transcription. Binding sites for NF-κβ exist in the promoters for the genes encoding IL6, IL8, MCP-1, M-CSF and various adhesion molecules.

Lipoprotein Pathology[125-128]

Lipoproteins have been classified according to their electrophoretic mobility; VLDL, LDL, HDL etc. Recent work, however, has shown this to be an over-simplification. LDL can exist in three broad forms-I, II and III. These range from light LDL (I) to small, dense LDL (III). Light LDL is usually in the majority. High triglyceride levels, say above 2.1 mmol/L, cause more small, dense LDL to form. It has been shown that this latter form is more atherogenic; it is cleared less well and oxidized and removed by macrophages more easily. Small dense LDL formation is brought about by an exchange of cholesterol from the cholesterol-rich molecules HDL and LDL, to the cholesterol-poor molecules VLDL and chylomicrons. This is effected by cholesterol ester transfer protein, CETP, which interestingly is increased by smoking. The cholesterol-enriched lipoproteins, VLDL and chylomicrons, may then be taken up by macrophages, enhancing foam cell formation.

HDL is also made up of a number of different subgroups. HDL contains two apolipoproteins, A-I and A-II and can exist in at least five different forms. It is generally accepted that HDL is atheroma protective. For every 1mg/dL increase in plasma HDL concentration there is a 2-3% reduction in the risk of coronary heart disease; it seems that HDL subfractions that contain only apo-AI are the most active. HDL$_2$, being one of these subfractions, has been found to be more protective than HDL$_3$. This is confirmed by the finding that HDL that has had Apo AI substituted by serum amyloid A enhances LDL modification rather than inhibiting it. Serum amyloid A is an acute phase protein that becomes substituted in HDL in certain conditions, postmyocardial infarction and postcardiothoracic surgery for instance. Much of the protective effect of HDL is related to reverse cholesterol transport which involves removing cholesterol from macrophages and esterifying it. Its resorption is thereby prevented and it can therefore be transported directly to the liver for disposal. HDL has, however, other antiatherogenic effects: promoting endothelial cell proliferation, endothelial cell prostacyclin production and by a favorable regulation of postprandial lipemia. Studies have also shown, in vitro, that HDL may exert part of its antiatherogenic action by inhibiting the expression of VCAM-1, ICAM-1 and E-selectin induced by TNF and IL1.[129]

Some studies, however, have not confirmed HDL's protective effect, for example in some Russian population studies and genetic studies involving those with either HDL deficiency or excess. HDL is found in combination with two enzymes:[130,131] platelet activating factor acetylhydrolase (also found to some extent in LDL) and paraoxanase. These enzymes are capable of degrading oxidized lipids. Only a small proportion of HDLs contain these enzymes. Their presence is determined in part by genetic influences and in part by the environment e.g., diet and the acute phase response. Their existence in varying quantities is a possible explanation for some of the variability of HDL protection in studies.

It may be that differences in the relative quantities of each HDL subfraction also exert differing protective influences. For example, small, dense HDL is more rapidly cleared by the liver thus reducing its ability to engage in reverse cholesterol transport.

Other lipids also exert influences on plaques. In addition to the proatherogenic effects of triglycerides mentioned above, these lipids also exert procoagulant effects by stimulating the synthesis of PAI-1 (see below). Not all fatty acids are atherogenic, however. A group known as the omega-3-fatty acids belong to the nonatherogenic group. One member is docosahexenoic acid. It has been shown that its presence in cellular lipids reduces the expression of VCAM-1, ICAM-1, E-selectin, IL6 and IL8 in endothelial cells when those cells are stimulated in vitro with cytokines such as IL1, TNFα or IL4.[132] There are other forms of LDL which contain different apolipoproteins; for instance lipoprotein a is LDL containing the apolipoproteins B

and a. This molecule is independently associated with premature atheroma.

In conclusion, there is considerable evidence that oxidized LDL can be found in atheroma. Experiments in vitro have shown that it can exert effects that are consistent with a role in atherogenesis. The final key would be to show that by either reducing cholesterol levels or reducing the likelihood of oxidation, the incidence of plaque formation falls. Cholesterol reduction has indeed been shown to reduce significantly the acute effects of atheroma, in numerous studies; 4S[133] being the seminal trial. It is not clear whether this favorable clinical effect is mediated by reduced atheroma or plaque stabilization. The following sections discuss this aspect in greater detail.

Antioxidants
i. Vitamins

The role of antioxidant vitamins has been examined extensively. Vitamins C, E and β-carotene all act as antioxidants when either free in the plasma or linked with LDL. Vitamin E has been shown to reduce the oxidation of oxidized LDL and also to improve the impaired endothelium-dependent relaxing response.[134] Other in vitro studies[135] using human derived cells have determined the effect of antioxidants on the interaction of LDL, endothelial cells and monocytes without the vitamin in the culture medium. Increased monocyte-endothelial cell adhesion was noted, as was an increase in soluble ICAM-1 present in the medium. With the addition of vitamin E, both the adhesion of monocytes and the level of soluble ICAM-1 fell. These results therefore indicate that antioxidants reduce the important leukocyte-endothelial cell interaction.

Adhesion of leukocytes to endothelium and the clumping of leukocytes and platelets in the circulation has been shown by Lehr at al to occur as a consequence of cigarette smoke exposure.[136] Adhesion molecules are found to be increased on both endothelium and leukocytes. Studies in animals have shown that, both in vivo and in vitro, pretreatment with vitamin C will prevent this. Although this extremely important topic has only been studied in a few laboratories, it is clear that smoking has a profound action on both leukocytes and endothelial cells. More work is required, particularly in human studies but it seems highly probable that these mechanisms are the route by which smoking induces some of its deleterious vascular effects.

Epidemiological studies[137] have linked antioxidant vitamin ingestion with a reduced level of risk for clinically relevant atheroma. This has been confirmed in animal experiments looking at atheroma burden. However randomized therapeutic trials so far have not confirmed these findings. Confounding factors such as vitamin dose or the use of vitamins in those with established atheroma may cause this disparity.

ii. Enzymes

There are naturally occurring antioxidant enzymes present in the arterial wall, especially concentrated in the endothelium: superoxide dismutase, glutathione peroxidase and catalase. Studies in rabbits[138] have shown that the expression of these enzymes is increased in hypertensive and hyperlipidemic animals, which may therefore help to reduce the effects of the atheromatous stimulus.

iii. Drugs[139]

Probucol has both a cholesterol lowering and an antioxidant effect. It reduces macrophage accumulation in cholesterol-fed rabbits whose arteries have been damaged by balloon injury, as a model for percutaneous coronary angioplasty (PTCA).[140] Recent studies have shown that it will also reduce restenosis rates after human PTCA.[141] All of the findings with probucol have been shown to be independent of its cholesterol lowering effects. Others have confirmed these findings in atheroma, as opposed to post-PTCA restenosis, using other antioxidants.

As mentioned above, drugs that inhibit the enzyme HMG coA reductase, an enzyme important in cholesterol synthesis, have been shown to have a profound effect on the consequences of atheroma. The 'statins', as these drugs are known, have been shown to significantly reduce death, myocardial infarction and the need for revascularization in both those with raised and those with more normal cholesterol levels, both in primary as well as in secondary prevention.

On perhaps a lighter note, there is increasing evidence that alcohol in general and red wine in particular have an antioxidant and thus an antiatherogenic effect. It is felt that this explains Professor Renaud's 'French Paradox'. This is the finding that, despite average levels of smoking, hypertension and saturated fat intake, the French and other Southern Mediterranean peoples have a lower cardiovascular mortality than would be expected. It is likely that most of this effect is due to antioxidants in their diet, one of the most important being ethanol. There is some evidence that red wine may confer additional benefits because of the presence of certain chemicals: flavonoids, quercetin and reservatrol for instance.[142]

The therapeutic antiadhesion antibody *ReoPro* is discussed in the platelet section of this Chapter.

Endothelial Dysfunction[143-146]

Vascular Tone

The endothelium has been considered in the past as solely an inert barrier. However, it has become clear more recently, that it has a vast range of functions and is involved in the intimate control of the body's homeostasis. It is already clear that it is a most important link in the atherogenic process.

Maintenance of vascular tone is a vitally important function. This occurs in the normal situation as a balance between dilating and constricting factors, there being a resting dilator tone. Nitric oxide, previously known as Endothelium Derived Relaxing Factor or EDRF, is probably the most important dilating factor. It is derived from the amino acid L-arginine by nitric oxide synthase. This enzyme exists in three isoforms. There are two constitutively expressed forms, one in neurons and one in the endothelium (eNOS). In addition there is an inducible form found in macrophages. These enzymes are very large and share homology with cytochrome P_{450}. The endothelial form is found in vascular endothelium, platelets and on endo- and myocardium. Nitric oxide has a very short half life and acts therefore in a paracrine manner. It increases cyclic GMP (cGMP) in vascular smooth muscle cells, affecting their intracellular calcium levels, thus causing relaxation and hence vasodilatation, as long as the NO is present. It also inhibits smooth muscle cell proliferation, even to the extent of inducing apoptosis. Furthermore, platelet and monocyte adhesion is inhibited, as well as monocyte chemotaxis. Nitric oxide can also be formed by macrophages and vascular smooth muscle cells when stimulated by IFNγ, IL-1, TNFα and lipopolysaccharide.[147]

Both Acetylcholine, via muscarinic receptors and increased blood flow, will produce dilatation in the normal vessel in an endothelium-dependent manner i.e., by the local release of NO. Other than acetylcholine, serotonin, thrombin, substance P, β-adrenergic agonists, shear stress, bradykinin and aggregating platelets also cause its release. The vessel can also be stimulated to dilate by using nitric oxide donors such as glyceryl trinitrate. When the endothelium becomes damaged, production of NO decreases causing the vasomotor balance to swing toward constriction. In addition, many of the chemicals that, with an intact endothelium, cause vasodilatation e.g., acetylcholine and thrombin, directly stimulate smooth muscle cells to contract in the absence of nitric oxide.

The compound L-NAME inhibits NO production. When given to cholesterol-fed rabbits, increased atheroma formation is seen. Some studies, however, have suggested that in atheroma there is in fact increased production of NO. It may be that there is increased breakdown of NO by superoxide anions, which are also produced as a by-product by NO synthase, in the presence of LDL. Thus the evidence from animal studies suggests that if the relative level of NO falls, atherogenesis appears to be accelerated.

Studies have shown that in patients with overt coronary disease, the normal dilatation with acetylcholine is lost and paradoxical constriction occurs. The standard method of measurement is forearm blood flow and the atypical constrictor response has been attributed to a state of endothelial dysfunction.

Similar findings of endothelial dysfunction have been found in young people and children with risk factors for atheroma, for example hypercholesterolaemia,[148] smoking or homocystinuria, and also in adults with hypercholesterolemia. Endothelial dysfunction may also occur in those who have diabetes, hypertension or who are old.[149] In vitro studies[150] have shown that the best predictor for endothelial dysfunction in isolated internal thoracic arterial grafts is serum cholesterol. It has also been shown that in patients treated for their hypercholesterolemia the endothelial dysfunction reverses.[151,152]

It has long been recognized that some smokers show a greater predisposition form to atheroma than others. Some of this may be explained by differences in eNOS, as homozygotes with the rare eNOS 4a allele show a significantly increased risk of severe atheroma.[153]

Thus risk factors for ischemic heart disease cause endothelial dysfunction as evidenced by reduced NO function and this abnormality is associated with atherogenesis. Nitric oxide deficiency may also be important in acute coronary syndromes as, by reducing vessel diameter, shear forces increase which may activate platelet adherence and cause atherosclerotic plaque rupture.

Prostacyclin can also produce vasodilatation, differing from NO in that its effect is transient even in its continued presence. The endothelium also produces other vasodilating chemicals such as endothelium-dependent hyperpolarizing factor. These molecules are also reduced in atheroma.

Endothelium also produces vasoconstrictors such as endothelin-1 (ET-1) and thromboxane A_2. Endothelin-1, which is derived from preproET-1 and big ET-1, is released by the stimuli of thrombin, TGFβ, angiotensin-II, hypoxia, shear stress, oxidized LDL or endotoxin. On its own, IFNγ will not cause ET-1 release.[154] However if given at the same time as TNFα, ET-1 release increases. This contrasts with the finding that if IFNγ is given before the TNFα, ET-1 production is reduced.

Endothelin acts locally by increasing intracellular calcium concentration in smooth muscle cells via an increase in the quantity of c-fos and c-myc RNA. Endothelin-1 is also produced by smooth muscle cells and macrophages, in particular foamy macrophages,[155] in areas of necrosis. Levels of tissue ET-1 have been shown to be increased in patients with unstable coronary syndromes as compared to those with stable angina.[156] Its level in the blood correlates with the prevalence of atheroma, although in levels too low to produce constriction. Thus it appears atheroma in general, and acute coronary syndromes in particular, are associated with endothelial dysfunction and a generalized imbalance of vasomotor agents.

Coagulation

One of the other main functions of the normal endothelium is thromboresistance. All parts of the clotting pathway are affected by endothelial products. Platelet aggregation is inhibited by the production of NO[157] and prostacyclin. Thrombosis is inhibited by the production of antithrombin III, which inactivates thrombin and factor X among others. Both Protein C and Protein S are produced by the endothelium. These act to block factors Va and VIIIa, other members of the clotting cascade. Heparin-like glycosaminoglycans, such as heparin sulphate, are produced, which can assist in the inactivation of thrombin and factor X, among others. Finally the fibrinolytic pathway can be activated by local endothelial production of t-PA and urokinase. If these effects are lost, platelet aggregates and thrombus can form. These can then become incorporated into the plaque, increasing its effect on blood flow, possibly critically in acute coronary syndromes. Platelet products may also promote the growth of other aspects of the plaque; for instance, smooth muscle cell growth is potentially enhanced by PDGF and the synthesis of connective tissue matrix by TGFβ. Likewise the endothelium can also act as a source of connective tissue elements and growth factors.

Permeability

The final function of endothelium is to act as a barrier to uncontrolled macromolecular traffic into the vessel wall. Oxidized LDL increases endothelial permeability to macromolecules such as cholesterol. Permeability is also increased by thrombin, oxygen free radicals, IL-1 and TNFα and decreased by cAMP and cGMP elicited by NO. It is in the macrophage-rich areas in the shoulders of the plaques that significant permeability is located. However the endothelium is at best an imperfect barrier, as most plasma proteins normally leak into the intima (see lipoprotein section).

Thus it can be envisaged that by the action of the risk factors for ischemic heart disease, oxidized LDL, hypertension, smoking, homocysteine, Advanced Glycosylation Endproducts (AGEs—see later) and local hemodynamic factors, endothelial cells become dysfunctional and activated. Endothelial cell activation occurs in two phases.[158] Type I is fast, transcription independent and involves the expression of von Willebrand factor (vWf) and P-selectin on the luminal surface. Endothelial cell retraction, thereby exposing the procoagulant subendothelium also occurs. Later on, in type II activation, protein synthesis occurs resulting in the production of other adhesion molecules and cytokines via gene transcription stimulated by NF-κβ. This transforms endothelial cells from their normal nonadhesive, vasodilatory, semipermeable, thromboresistant state to an activated adhesive, vasoconstrictor, permeable, thrombogenic condition, thus promoting atherogenesis initially and later acute coronary syndromes. It seems highly probable that activation states are the structural and molecular correlates of endothelial dysfunction. In view of the specificity of monocyte/macrophage traffic into the artery wall to the atherosclerotic lesions, it seems likely that the induction of monocyte adhesiveness is a highly important atherogenic consequence of endothelial dysfunction.

Coagulation[159-162]

Platelets

As has been touched on earlier, the coagulation cascade plays a significant role in the pathogenesis of atheroma and acute coronary syndromes. Thrombi can cause expansion of the atherosclerotic plaque to produce either significant stenosis or obstruction. In normal circumstances, coagulation is inhibited by an intact endothelium producing platelet inhibitors such as prostacyclin and NO, working via the secondary messengers cAMP and cGMP respectively. Prostacyclin production is diminished in the endothelium over atheromatous plaques.[163] As discussed above the production of NO is also reduced. Some studies have suggested that the prothrombotic arachidonic acid metabolites, PGE_2 and thromboxane, may be increased in the same areas. All of these factors may therefore shift the balance towards thrombogenesis.

Platelets may therefore bind to dysfunctional endothelium. If plaque rupture occurs, the highly thrombogenic subendothelium is exposed and platelets can then bind to matrix components via their integrin adhesion molecules. Von Willebrand factor is a molecule which circulates in the plasma as a large multimer, the subunits being 220 kD in size, containing 2050 amino acids. The multimers can extend up to 2 μm in length and have an average molecular weight of 10×10^6 kD. It is synthesized in considerable quantities by endothelium cells and stored in intracytoplasmic vesicles, the Weibel-Palade bodies. The endothelium-derived molecule is secreted both on the luminal and the subendothelial surfaces. When subendothelially located it is bound to components of the matrix such as collagen VI. It is also produced by megakaryocytes, being then found in the alpha granules of platelets. Von Willebrand factor has a tendency to bind to matrix surfaces and platelets can then bind to it, the vWf therefore becoming the sticky filling in an adhesion sandwich! Platelet adhesion will now be considered in detail.

Once contact has occurred between the platelet and the surface to which it is sticking, activation and spreading occur, eventually leading to platelet aggregation. The initial contact is determined by shear forces. Once the shear forces increase above a certain level, 50 dynes/cm², the factors involved in initial platelet contact become activated. Shear forces are increased by the disruption of laminar blood flow as occurs in arteries narrowed by atheroma or vasospasm. Initial platelet adhesion to the vWf in the tissues is mediated by the GP1b receptor on the platelet, once it has assumed an active conformation, in response to excess shear. GP1b is a nonintegrin glycoprotein made up of two chains, alpha (143 kD) and beta (22 kD), noncovalently linked together. This molecule is associated with other glycoproteins, glycoproteins IX and V. The binding site for vWf is found on GP1b-alpha.

After contact, the platelets become activated—involving oxygen consumption, pseudopod formation, shape change and granule release. Stimulators of platelet activation include thrombin, ADP, thromboxane, noradrenaline, platelet activating factor and collagen. These act via secondary messengers such as inositol-triphosphate and diacylglycerol. Activated platelets release ADP, 5-hydroxy tryptamine and thromboxane A_2, thus causing other platelets to be recruited and activated. The GP1b-IX-V receptors are then internalized.

At the same time the platelet membrane GPIIb/IIIa receptor, also known as the integrin $\alpha_{IIb}\beta_3$, undergoes conformational change becoming active. The alpha subunit of GPIIb/IIIa is 136 kD in size and is made up of a light and a heavy chain. The light chain exists mostly extracellularly with short transmembrane and cytoplasmic sections. The heavy chain is only found extracellularly. The beta subunit is 92 kD in size and is made up of 762 amino acids. It, like the light chain of the alpha subunit, is made up of a large extracellular part and short transmembrane and cytoplasmic domains. As with other integrins the two chains are bound together noncovalently, an interaction dependent on the presence of calcium. This receptor is the most highly expressed integrin on the platelet, there being approximately 50,000 copies per cell. This activated receptor then binds to vWf, fibrinogen, vitronectin or fibronectin to cause platelet aggregation.

Fibrinogen has several binding sites for GPIIb/IIIa and therefore acts as a bridge between two molecules of GPIIb/IIIa on separate platelets. The interaction with fibrinogen is particularly important for platelet aggregation and with the other ligands also results in adhesion to the extracellular matrix. It may also form a bridge between platelet and endothelium, using the endothelial $\alpha v\beta_3$ integrin molecule.[164] These factors go some way to explaining why fibrinogen is an independent risk factor for ischemic heart disease, as plasma levels are an important determinant of platelet aggregability. Two short amino acid sequences are important for GPIIb/IIIa binding:

- Arg-Gly-Asp or 'RGD', found in all four ligands
- Lys-Gln-Ala-Gly-Asp-Val, the most important site for fibrinogen binding

Von Willebrand factor can also bind to GPIIb/IIIa, its binding being more important for platelet aggregation in areas of high shear stress as compared to fibrinogen-GPIIb/IIIa in areas of low shear stress.

There has been great recent interest in the humanized monoclonal antibody ReoPro, which has useful therapeutic effect in preventing restenosis after angioplasty and in patients with unstable angina. It is presumed to act by blocking the platelet GPIIb/IIIa receptor but as it also has some activity against $\alpha v\beta_3$ and β_2 integrins, blockage of these molecules may also play a part. Other low molecular weight GPIIb/IIIa inhibitors, such as Integrilin, are also effective but less so than ReoPro.[165]

Other receptors are also involved in adhesion; the integrin GPIa/IIa binds to collagen, as does the nonintegrin glycoprotein IV. It appears that atheromatous plaques contain more collagen I and III than normal arterial wall, and it is suggested that this makes them more thrombogenic.[166] By these mechanisms, a platelet thrombus can form.

Recently, activated platelets have been found able to activate umbilical vein endothelial cells in vitro so that they express adhesion molecules and chemokines. The platelets translocate stored CD40 ligand (CD154) to their surface on activation and the binding of it to CD40 on the endothelial cells results in the endothelial activation reaction.[167] It will be important to confirm that this reaction occurs in vivo, particularly in arteries.

Clotting Factors

The coagulation cascade is also important. Tissue factor[168,169] is an integral membrane glycoprotein of between 261 and 263 amino acids length, found in association with anionic phospholipids in the plasma membranes of various cells including fibroblasts and the pericytes of blood vessels. It is usually only found in the adventitia of arteries. Once exposed to blood, tissue factor binds to clotting factor VII. Once bound, the latter can then be activated to form factor VIIa. This complex can then activate clotting factors IX and X and thus set off the extrinsic pathway of the clotting cascade. Thrombin is eventually generated. This molecule binds well to the subendothelial extracellular matrix,[170] and once there remains active and resistant to degradation by its inhibitor antithrombin III. It can then form fibrin from fibrinogen thus producing a thrombus. Not only is thrombin part of the clotting cascade but it can also activate platelets, attract monocytes and cause lymphocyte and smooth muscle cell proliferation. It is a generalized activator of endothelial cells and causes them to synthesize prostacyclin, platelet activating factor, PAI-1, vWf, fibronectin and PDGF and to express P-selectin; all of which have additional roles in atherogenesis. It should also be emphasized that thrombus formation, by producing a certain amount of obstruction, increases the local shear stress, thus activating further platelets.

In atheroma, tissue factor can be found extensively in the intima, especially in the lipid core. It is produced by macrophages, smooth muscle cells and endothelial cells[171] after stimulation with mildly oxidized LDL.[172] In endothelial cell or monocyte culture, IL-1, thrombin and TNFα will also induce tissue factor, via NF-κβ.

Smooth muscle cells can be induced to produce tissue factor by PDGF, EGF, angiotensin II, MCP-1 and thrombin. Macrophages are the most likely source of the majority of tissue factor. Smooth muscle cells produce tissue factor later, once they have migrated from the media.

The thrombin receptor has been shown to be expressed in increased quantities in atheroma.[173] In normal arteries it is usually only found on endothelial cells. However in plaques it is very widespread and is produced by both macrophages and smooth muscle cells. In normal circumstances, the body can disassemble formed thrombus by using the fibrinolytic system. This relies on plasmin, an enzyme produced from plasminogen by t-PA, that degrades fibrin. Tissue plasminogen activator has a regulator itself, plasminogen activator inhibitor-1. The levels of PAI-1 are increased in diabetics. Oxidized LDL also affects the coagulation system, t-PA and thrombomodulin being reduced, the levels of PAI-1 being increased.[174]

All of these effects thus tip the balance towards thrombogenesis and are probably another result of endothelial activation. Furthermore, fibrin on the endothelial surface may predispose to monocyte binding via their integrin receptors.

Not all episodes of endothelial denudation or plaque rupture cause calamitous thrombus formation. Some are more prone to this than others, for a number of reasons. It has been shown that polymorphisms in the GPIIb/IIIa receptor may affect the predisposition to myocardial infarction,[175,176] possibly by increasing the ease with which platelets bind to endothelium. The PLA$_2$ allele of the GPIIIa gene appears to make myocardial infarction more likely when compared to the PLA$_1$ allele, especially if the patient is below the age of 60. This polymorphism interacts with other more conventional risk factors. Tissue factor shows variable expression among atheromatous plaques, being found significantly more in those causing unstable coronary syndromes.[177]

Immune Response

Introduction

Various lines of evidence suggest that the immune response may play a significant role in atherogenesis. While few B cells, granulocytes and natural killer cells are found, T cells, both helper (CD4) and cytotoxic/suppressor (CD8), are present in significant quantities in plaques, both in the intima and in the adventitia around vasa vasora. Some have estimated that as much as 20% of the fibrous cap may be made up of T cells. They can accumulate in other forms of chronic inflammation, such as those found in rheumatoid arthritis, without being antigen-specific and without taking part in an active immunological process, present there by dint of passive trapping. Some animal studies have lent credence to a similar nonspecific view of the inflammation in atheroma; lesions can be produced after cholesterol feeding in mice that are unable to mount immune responses.[178] Atherogenesis may even be enhanced in these animals. Others have shown that only a small proportion of the T cells present in the plaque stain for interleukin-2 (IL-2) receptors; IL-2 is a cytokine required for T cell activation.

However it has been shown that cyclosporin A,[179] an inhibitor of T cell activation, can reduce the size of plaques in experimental animals. Thus it has been postulated that at least some T cells may be attracted there to mount an immune response to antigen or antigens and may therefore be important in promoting the inflammatory aspects of the plaque. A number of antigens have been postulated and are discussed below.

The levels of lymphocytic infiltrates are very high in some atherosclerotic aneurysms. It may be that a cell-mediated response can weaken the wall, as IFNγ, the major T lymphocyte cytokine product, inhibits matrix synthesis.

Immune Complex Disease[180]

It has been known for many years that immune reactions forming immune complexes of antigen and antibody in the circulation can exacerbate atherosclerosis. In both rabbits and monkeys, immune complexes can be induced by the intravenous injection of foreign proteins, bovine serum albumin for instance. The injected proteins produce antibodies, which when combined with the antigens in the circulation, form immune complexes. When cholesterol is also fed to the animals a marked exacerbation of atherosclerosis over the controls on diet only is seen.

Immune complexes are known to deposit in blood vessel walls and will induce inflammation if in sufficient quantities. Although arteries may be less susceptible than small vessels, they can be affected and it is possible to induce arteritis of the aorta and large vessels by immune complexes alone. This inflammatory damage almost certainly results in activation of the endothelial cells and leukocyte emigration into the arterial wall, triggering the atherosclerotic process. There is little evidence to suggest, however, that this is a major mechanism in man but it is possible that an excess of immune complexes may have a role in a minority of patients.

Oxidized LDL[181]

As discussed above, oxidized LDL is likely to be a significant player in atherogenesis. Some researchers consider that it may exert some of its effect by immunological means. In vitro studies have shown that T cells can oxidize native LDL.[182] This oxidized LDL can then cause T cells to increase their expression of HLA-DR and IL-2 receptors, increase their DNA synthesis and proliferate. This effect requires the presence of macrophages. Apo E containing LDL can also activate T cells. The lysophosphatidylcholine part of oxidized LDL is known to be chemoattractant for T cells.[183]

The humoral immune system may also react to lysophosphatidylcholine. Furthermore, circulating immune complexes of lipoperoxides have been found in elderly patients and IgG specific for relevant epitopes has been shown to be present in atheroma. Low levels of autoantibodies to oxidized LDL are common in the UK population and their incidence increases with age. Antibodies to

oxidized LDL proved to be an independent predictor of the progression of carotid atherosclerosis in a Finnish population.[184]

Infections[185-188]

Acute infections can increase the likelihood of acute coronary events by an acute phase response which causes an increase in fibrinogen (see above). There has been considerable interest, recently, in the role that certain chronic infections may play in atherogenesis, as it is known that some infections can cause endothelial dysfunction.

One such infection is caused by *Chlamydia pneumoniae*, an obligate intracellular bacteria, which usually causes an atypical pneumonia or other respiratory infections, often subclinically. Elevated antibody titers to this organism have been shown to be associated with stroke, carotid arterial disease and ischemic heart disease. The organism has been isolated from coronary lesions and shown not to be present to the same extent in the normal arterial wall. It has been found that the organism can reproduce in smooth muscle cells, endothelial cells and macrophages in cell culture. It is hypothesized that the organism is carried into the atherosclerotic lesion by macrophages which have become infected in the lung.

Several recently published clinical trials have added support to the evidence of a possible pathogenic role for *Chlamydia pneumoniae*. Treatment of patients suffering acute coronary syndromes with an antichlamydial antibiotic, Roxithromycin,[189] reduced the risk of further cardiovascular events. Similar results have been found with another antichlamydial antibiotic Azithromycin.[190] While these trials were all too small to be conclusive individually, they are suggestive of a pathogenic role for Chlamydia. There is also some evidence that antibody titers may be increased at the time of acute coronary events, suggesting that reactivation of the infection may be instrumental in the pathogenesis of the acute event.[191]

Other studies have suggested that gastric infection with *Helicobacter pylori* infection may play a similar role, although recent evidence has contradicted this finding. Any chronic infection may cause an acute phase response similar to acute infections but the role of *H. pylori* is not convincing.

Cytomegalovirus (CMV), herpes virus and chronic dental infections have also been implicated. CMV is able to infect smooth muscle cells and induce the production of cytokines, adhesion molecules and clotting factors.

CMV can also affect the recessive oncogene p53 and cause its loss or inactivation. Some studies have suggested that this effect may be instrumental in exacerbating restenosis postangioplasty. Periodontal disease has been shown to be an independent risk factor for myocardial infarction;[192] although there are many confounding factors e.g., smoking, that are common associations with both. The jury, however, on these infections is still out. As with the immune response, infections are likely to be only modulatory in the pathogenesis of atherosclerosis.

Heat Shock Proteins[193]

Heat shock proteins (HSPs) are produced by cells in response to metabolic stress and enhance the cell's ability to withstand it, particularly by preventing protein denaturation. A number of pieces of evidence have suggested that they may play a role in atherogenesis.

Atheroma can be induced in normocholesterolemic rabbits by immunization with HSP-65.[194] This heat shock protein is derived from mycobacteria. Immunization with other components of atheroma and other antigens do not produce a similar response. The strength of the response can be exacerbated by the combination of immunization and hypercholesterolemia. Even in rabbits fed just a high cholesterol diet and not immunized with HSP-65 a, population of T cells is found that have a receptor for this antigen, concentrated in the plaques.[195]

The human version of HSP-65 is called HSP-60 and has a similar structure but a smaller molecular weight, of 60 kD. It is expressed in high quantity in human atheroma by macrophages and to a lesser extent in smooth muscle cells and endothelial cells.[196,197] Its expression appears to correlate with the severity of the atheromatous process. In clinical studies, a positive correlation has been shown between antibodies to HSP-65 and the presence of carotid atherosclerosis.[198] Interestingly both *Chlamydia pneumoniae* and *H. pylori* contain HSP-60 like subunits. It can be hypothesized therefore that shear stress induces heat-shock protein expression to which T cells might then react causing inflammation and atherogenesis.

Other groups have looked at another heat-shock protein, HSP-70. Its expression has been shown to be increased by oxidized LDL and hypertension. It has been found to be mostly distributed in the center of plaques around the lipid-rich necrotic core, being especially produced by macrophage-derived foam cells.[199] Smooth muscle cells nearby produce very little.

The animal experiments and the human data make an impressive body of evidence that immunity to heat shock proteins can play a role in atherosclerosis. The evidence suggests that while HSP-70 probably has a protective role against oxidative stress, HSP 60/65 may be etiological in some patients due to an immune or autoimmune effect. However, further work is required and the effect of knocking out the HSP genes in a mouse atherosclerosis model would be worthy of study.

It is fair to conclude that immune responses and infections can have a modulatory role in atherosclerosis, but that they are a secondary process. The major inflammatory mechanism driving atherogenesis is almost certainly a nonimmune process mediated by oxidized lipid.

Acute Coronary Syndromes

Not only is the immune response thought to be important in atherogenesis but there is some evidence to suggest that it may be involved in the pathogenesis of acute coronary syndromes. Titers of immunoglobulins and IL-2 are increased in patients presenting with unstable angina suggesting that an immune response may be important.[200] It is a well-recognized phenomenon that during or after flu epidemics, postbacteremic episodes and postabdominal surgery there is a significant increase in acute cardiac events.[201] This effect may be brought about in a number of ways. Fibrinogen is raised in acute inflammation and may increase the probability of significant thrombus formation over damaged plaques. Alternatively, endothelial function may be altered by the infection, either directly through endothelial infection, or indirectly via inflammatory cytokines. Bacterial endotoxin can also affect the release of NO. Initially NO's production may be increased causing transient vasodilatation, followed later by a more profound loss of production, termed endothelial stunning. This loss of NO thus tips the balance towards procoagulation, precipitating an acute event.

Apoptosis

Cells can die in one of two ways, by necrosis, an essentially passive and disorganized process, and by apoptosis, which is controlled and can be physiological. The latter process occurs in tissues that have an innate ability to proliferate, to maintain stable cell numbers. Apoptotic cells can be identified by transmission electron microscopy and by a process called in situ Terminal deoxynucleotidyl transferase mediated dUTP Nick End-Labeling (TUNEL), which detects DNA fragmentation.

It is known that at the center of atheromatous plaques there is significant foam cell death, allowing the release of cholesterol into the extracellular matrix. This cholesterol, together with some that has accumulated by dint of passive diffusion of lipoproteins from the blood, forms the central lipid core of the plaque. This process is important as the development of this lipid core is a prerequisite for plaque rupture and acute coronary syndromes. It may be that the cell death alters the stability of the plaque and may therefore increase the likelihood of plaque rupture. This is confirmed by the finding that signs of apoptosis are present in atherosclerosis[202-204] and are significantly more prevalent in

unstable as compared to stable plaques.[205] Apoptosis of monocytes can also cause the release of large amounts of growth factors. Other studies have shown that smooth muscle cells can also undergo apoptosis in the fibrous cap, again affecting plaque stability.

The exact mechanism for apoptosis is still unclear but it may be induced by a number of factors including oxidized but not native LDL.[206,207] Interestingly, one group has shown that mildly oxidized LDL acts as a growth promoter but that further oxidation will then cause cell death.[208] It appears to be the 7-ketocholesterol portion of oxidized LDL that is important in apoptosis, at least in smooth muscle cells.[209] Lysophosphatidylcholine does not affect cell death.

IL-1β converting enzyme is known to be a mammalian cell death gene and has been found to be present in a proportion of atheromatous apoptotic cells. Others[210] have shown that, in vitro, if PDGF or IGF-1, but not EGF or bFGF, are removed from growth medium, smooth muscle cell apoptosis is accelerated. The regulation of gene products is most important in apoptosis, as persistence of the bcl-2 proto-oncogene protects against smooth muscle cell apoptosis. Similarly, deregulation of c-myc expression causes excessive smooth muscle cell proliferation and eventual cell death.[211] A further trigger for apoptosis may be falling levels of M-CSF.

It therefore seems likely that apoptosis is an important mechanism in the pathogenesis of the plaque particularly in the death of foamy macrophages and the consequent formation of the extracellular lipid gruel. It is currently a very active field of research.

Calcification[212]

Calcium is an important component of atheromatous plaques that has only been touched upon so far. In chronic inflammation as a whole, calcium normally plays a protective role in that it is laid down surrounding an inflamed area and may therefore help to isolate the focus, as in tuberculosis. Calcification tends to occur in the more advanced plaques and thereby contributes to the protrusion of the cellular mass. It is important as it creates an interface of widely differing mechanical properties inside the plaque, thus increasing the likelihood of rupture.

Two gene products are involved in calcification elsewhere in the body: osteopontin and osteonectin. The former is 44 kD in size and is a glycoprotein that contains the RGD sequence, mentioned above in the context of integrin binding. It binds to hydroxyapatite and is produced by osteoclasts. Immunohistochemical and in situ hybridization studies have shown that as atheroma becomes more severe, lesional macrophages express more osteopontin.[213]

Osteonectin is also a glycoprotein, 38 kD in size and shows a high degree of affinity for hydroxyapatite, calcium and type I collagen. Osteonectin, appears to be produced by smooth muscle cells deep in the intima, its expression decreasing with the severity of atherosclerosis. It is well known that in chronic inflammation calcification occurs in areas of cell necrosis. The situation in the gruel of plaques is similar, as there is cell loss, probably by both necrosis and apoptosis.

In conclusion it is probably fair to say that the direct adverse effects of calcium are usually slight. However it can sometimes be a problem for vascular surgeons, as rocks make sewing difficult!

Risk Factors for Atherosclerosis

Thus far the small molecules important in atherogenesis and their effects on individual elements of the plaque have been discussed. The following section discusses how some of the major risk factors for atheroma, not discussed so far, interact with those small molecules and plaque elements.

Diabetes[214,215]

The Insulin Resistance Syndrome

In noninsulin-dependent diabetes mellitus (NIDDM) cardiovascular disease is the commonest cause of death. Two thirds of these patients will die of cardiovascular disease, a relative risk of between two and five times that of an equivalent nondiabetic population. Diabetes exacerbates the effects of other risk factors important in atherogenesis. This is highlighted by the fact that in a low risk population, such as the native Japanese, although their risk is doubled, it is still less than nondiabetic people in Western countries.

Diabetes can be induced in rabbits by the injection of alloxan.[216] It has been shown that within just a few weeks of becoming diabetic, their endothelial cells show signs of damage, replicate more and become adherent for white blood cells and platelets. Intimal hypertrophy is also seen, most marked in those rabbits with associated hypercholesterolemia. Peripheral vessels in human diabetics show similar changes.

A significant part of the increased risk is imparted by the association of NIDDM with a metabolic syndrome previously known as Syndrome X but now called the 'Insulin Resistance Syndrome'. This comprises insulin resistance, either NIDDM or impaired glucose tolerance, hypertension, raised plasma triglycerides, central adiposity/obesity, microalbuminuria and hyperuricemia. The insulin resistance syndrome leads to high levels of plasma insulin. This can then increase smooth muscle cell proliferation. The production of extracellular matrix proteins, such as type IV collagen, is also increased. However insulin can also cause vasodilatation, possibly partly mediated by NO. Thus insulin can have both atherogenic and antiatherogenic properties. There is some evidence that the vasodilating actions of insulin are impaired in NIDDM. It may be that in this disease state the atherogenic actions outweigh the antiatherogenic ones.

AGEs

Extensive research has revealed other reasons for the atherogenic propensity.[217] Proteins exposed to the high level of glucose found in diabetics become modified by glycation, initially in a reversible fashion but later irreversibly. The glucose reacts nonenzymatically with amine groups to form Schiff bases. The irreversibly affected proteins are known as Advanced Glycosylation

End-products (AGEs). In the process of AGE formation, reactive oxygen species are produced. Many proteins are affected in this way, including connective tissue matrix proteins which become crosslinked and protease resistant. These modified proteins, in their soluble form, act as macrophage chemoattractants and when insoluble, can trap macromolecules such as lipoproteins. Cells such as macrophages can also be ensnared causing their persistent activation. Apolipoproteins are also affected by glycation causing them to become more susceptible to oxidation, thus increasing their atherogenicity.

AGEs are detected by a receptor, RAGE, one of the immunoglobulin supergene family. This receptor has been found on monocytes and causes their activation. It is also increased on the endothelium in atherosclerosis. Its stimulation can cause the production of NF-$\kappa\beta$, the release of oxygen free radicals, increased endothelial permeability and the increased expression of adhesive molecules and cytokines.

Levels of AGE proteins have been found to be increased in the atheroma of diabetics. AGEs tend to accumulate on long-lived extracellular matrix proteins and on basement membranes.[218] This leads to increased vascular permeability, growth factor and cytokine release and the inactivation of NO. Vlassara et al used normoglycemic rabbits and injected them with AGEs with or without a cholesterol-rich diet.[219] They showed that AGE treatment led to increased endothelial expression of VCAM-1 and ICAM-1 and the development of atheroma. These changes were enhanced by a diet rich in cholesterol.

Lipid Effects

Diabetes has many effects on plasma lipids that can promote atheroma formation. Normally, in the postprandial state, when plasma triglyceride levels are high, fatty acid release from adipose tissue is suppressed. This suppression is impaired in diabetics. Thus more fatty acids are transported to the liver resulting in increased VLDL production. This is further exacerbated by the reduced activity of the enzyme lipoprotein lipase. The high levels of plasma triglyceride thus formed cause the level of HDL to drop and the balance to swing towards smaller denser HDLs. Thus there is a more atherogenic lipid profile in the diabetic. Small, dense LDL particles are also found in increased quantities in diabetes.

Other Effects

Diabetes can also affect another important process involved in atheroma formation, the coagulation cascade.[220] Insulin resistance is associated with increased levels of PAI-1, thus tipping the balance away from fibrinolysis. This effect may also promote connective tissue deposition, as metalloproteinases that assist in tissue breakdown need to be activated by plasmin. The increased propensity to thrombus formation is also increased by higher levels of fibrinogen and enhanced platelet aggregability and adhesiveness. This may lead to the formation of multiple small thrombi, thus causing the arterial wall to be repeatedly stimulated by growth factors released from these microthrombi.

Chronic, but not acute exposure, to high levels of glucose, in vitro, has been shown to increase monocyte adhesion to endothelial cells.[221] Although antibodies to β2-integrins, the molecules used to bind to ICAM-1, could reduce this binding, the binding was not associated with the increased expression of ICAM-1, VCAM-1 or E-selectin.

Diabetes also induces changes in plasma levels of other compounds thought to be important in atherogenesis. TGFβ, ET-1 and angiotensin II are all increased. The increase in the latter two, occurring in association with reduced NO release, may explain at least part of the association of diabetes and hypertension in Syndrome X. The role of increased TGFβ is more enigmatic, however, as it might be expected to be antiatherogenic (see growth factor section).

Overall, diabetes has wide ranging effects on all of the intimate processes contributing to atherogenesis.

Hypertension

Hypertension is one of the major risk factors for atheroma. The mechanisms of this connection are still not completely clear. However it is likely that increased hemodynamic shear stress and possibly increased pulse pressure can activate the endothelium and render it dysfunctional. Nitric oxide bioactivity is known to be reduced in hypertension.[222]

In many patients with hypertension, angiotensin II levels are increased, and it is known that treatment of patients and WHHL rabbits with angiotensin converting enzyme (ACE) inhibitors will reduce the progression of atherosclerosis and the incidence of myocardial infarction. Angiotensin II has been shown to increase the ability of macrophages to oxidize LDL, and also a complex of angiotensin II and LDL can be taken up by the scavenger receptor rather than the LDL receptor.[223] Angiotensin II is also known to be a mitogen for smooth muscle cells and an inducer of cytokine production. It will also increase endothelial-leukocyte interactions. Interestingly, tissue ACE is known to be present in increased quantities in atheroma, thus tipping the balance towards angiotensin II and away from NO. As a side effect the breakdown of bradykinin is increased, again reducing NO production.

Thus the effect of hypertension in atheroma is complex and is the subject of much current research.

Estrogens

It has long been recognized that premenopausal women have a lower rate of ischemic heart disease than their age-matched male counterparts. This benefit is lost at the menopause and by the age of 75 there is no difference in the rates of ischemic heart disease between the sexes. However, the female advantage can be preserved by hormone replacement therapy (HRT), the active ingredient being estrogen. A similar protective effect has been noted in male transsexuals taking estrogen. Furthermore, a male subject was recently reported with a defect in his estrogen receptor that led to endothelial dysfunction.[224] Apparently, in males, estrogens are synthesized from precursor androgens within endothelial cells. No endothelium-dependent vasodilatation occurred in this patient after an ischemic challenge, suggesting that estrogen is important for normal endothelial function.

Estrogen appears to exert its antiatherogenic effect in a number of ways:

- Estrogen can reverse endothelial dysfunction in atheroma, a finding that has been confirmed in those taking HRT. This may be related to effects on endothelial cell NO production or by acting as an antioxidant. Estrogens have been shown to reduce both copper and monocyte-induced oxidation of LDL in vitro.[225]

- Serum cholesterol begins to rise in women after the age of 50 and much earlier in those who smoke. If HRT is initiated, the levels stay low and indeed remain lower than their male counterparts. Smoking also reduces HDL levels in the postmenopausal woman.

- Recent studies have revealed that estrogen may exert at least some of its protective influence against atheroma by inhibiting the IL1 stimulated increase in adhesion molecules: ICAM-1, VCAM-1 and E-selectin.[226] This again suggests that estrogen is normally inhibitory to endothelial activation and dysfunction.

- Bearing the above information in mind, it is easy to understand how women are protected against atherosclerosis by their circulating estrogens.

Homocysteine

This amino acid occurs in large quantities in the serum of homocystinuria patients, an inborn error of metabolism. These individuals are known to suffer from premature vascular disease and to die early from thromboembolic events. It is thought that this is due to endothelial dysfunction.

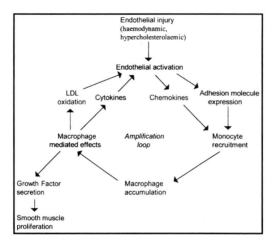

Fig. 2.6. Diagrammatic representation of the positive feedback hypothesis or amplification loop leading from endothelial injury to atherosclerosis.

Acute hyperhomocysteinemia has been shown to induce endothelial dysfunction in vitro.[227] Epidemiological studies have shown that plasma homocysteine levels, in those without homocysteinuria, are also correlated with the presence of vascular disease and act as a strong predictor of mortality in those who have established coronary disease.[228] There is only a very small correlation between homocysteine levels and the extent of atheroma, suggesting that it is the endothelial dysfunction and therefore its effect in acute coronary syndromes that is etiological.[229] It has been calculated that plasma homocysteine levels above 12μmol/L double the risk of myocardial infarction, cerebral infarction or peripheral vascular disease.

The Positive Feedback Hypothesis (Figs. 2.6, 2.7)

The intensive investigation of atherosclerosis over the last decade, as seen by the large amount of experimental and clinical work cited in this Chapter, has produced the dividend of a new understanding of the disease process. The productive starting point has been the *'response to injury'* hypothesis, which has led to the realization that atherosclerosis operates by similar cellular and molecular processes to those involved in inflammation and

repair. It can therefore be considered as a modified form of the inflammatory response operating in a specific tissue, the arterial wall.

A pathway of pathological, cellular and molecular events can now be outlined by which the disease arises and complications occur. Although there are some uncertainties and some parts remain hypothetical, overall a coherent picture emerges (Fig. 2.6):

- An insult to the arterial endothelial cells initiates the process. The most important insult is probably hyperlipoproteinemia but many others exist. In particular, hemodynamic stress, immune and chemical injury may contribute. It also seems likely that the contents of the plaque itself may also add to this endothelial insult, allowing self-perpetuation of the lesion, once started. This self-perpetuation may therefore lead to the plaque morphology of the disease.

- The endothelial cells activate and change their expression of surface molecules. Leukocyte adhesion molecules are expressed and molecules promoting thrombosis appear. Nitric oxide production is diminished. These changes can be observed clinically as endothelial dysfunction, which mirrors the decreased nitric oxide levels.

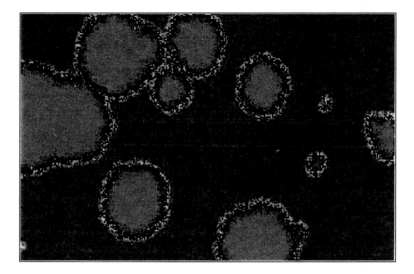

Fig. 2.7. Computer simulation of the positive feedback theory of atherosclerotic plaque formation. The computer screen represents an area of an artery wall. Monocytes (yellow) arrive initially at random at low probability. Increasing number of cells at a point is shown by increasing intensity of color. The presence of a macrophages at a point proportionally increases the probability of further macrophages arriving at that or adjacent points. Similarly the presence of macrophages increases the probability of smooth muscle cells (blue) being recruited into the intima, which are displayed in preference to macrophages even though both can be present. The simulation is run through multiple cycles. Lesions develop at random in classic plaque like forms, with smooth muscle cells at their centers. The majority of large lesions have become highly fibrotic. Note the few smaller lesions without many smooth muscle cells, resembling fatty streaks – they have arisen later during the simulation.

- Monocytes and T lymphocytes adhere to the endothelium and migrate into the arterial wall. The macrophages derived from these monocytes certainly have a major role in subsequent events but the contribution of the T cells is probably smaller and less certain. In the advanced stages of the disease, in a subset of lesions, leukocytes also enter via small vessels in the intima, which appear as a result of neovascularization.
- The macrophages produce a range of further effects: (i) they oxidize LDL, particularly when levels are increased in the wall through hyperlipoproteinemia, (ii) as a consequence of the oxidation they take up the oxidized LDL and become foam cells; they eventually die, liberating their cholesterol into a pool of gruel deep in the plaque, (iii) they produce growth factors which probably have a major role in stimulating smooth muscle cell proliferation, (iv) they produce metalloproteinases, which may weaken the arterial wall and lead to cracking and thrombosis, (v) factors from the macrophages feed back to the endothelium, activating it and causing self-perpetuation of the process, resulting in the formation of a focal plaque.
- Interaction of platelets, the coagulation system and leukocytes with the abnormal arterial surface, particularly when cracked or ulcerated, triggers thrombus formation and the major clinical sequelae of atherosclerosis.

- Sometimes degeneration of the media, as a consequence of exposure to toxic materials from the plaque, will result in weakening and aneurysm formation.

This knowledge (computer simulated in Fig. 2.7) now presents a great challenge to the medical profession and the pharmaceutical industry: how to make good use of it and rid the world of this deadly plague, or should that be plaque!

References

1. Wilhelmson L, Marmot M. Ischemic heart disease: Risk factors and prevention. Diseases of the heart. 2nd Ed. Saunders 1996: 911-31.
2. Neuhaus KL. Coronary thrombosis—defining the goals, improving the outcome. Clin Cardiol 1997; 20(suppl I):I-8-I-13.
3. Ross R. Atherosclerosis: A defense mechanism gone awry. Am J Pathol 1993; 143: 987-1002.
4. Steinberg D, Parthasarathy S, Carew TE et al. Beyond cholesterol: Modifications of low-density lipoprotein that increase its atherogenicity. N Eng J Med 1989; 320:915-24.
5. Ross R. The pathogenesis of atherosclerosis: A perspective for the 1990s. Nature 1993; 362:801-9.
6. Schachter M. Mechanisms of atherosclerosis I: Introduction and current concepts. An Introduction to vascular biology. 1st Ed. Cambridge University Press 1998: 112-118.
7. Poston RN. Mechanisms of atherosclerosis II: Relations between monocytes, macrophages and the endothelium. An introduction to vascular biology. 1st ed. Cambridge University Press 1998: 118-126.
8. Berliner JA, Navab M, Fogelman AM et al. Atherosclerosis: basic mechanisms—oxidation, inflammation and genetics. Circulation 1995; 91:2488-96.
9. Davies MJ. Atherosclerosis and ischemic heart disease. Diseases of the heart. 2nd Ed. Saunders 1996: 944-77.
10. Nilsson J. Cytokines and smooth muscle cells in atherosclerosis. Cardiovasc Res 1993; 27:1184-90.
11. Davies MJ. Pathology and morphology of atherosclerosis. Br J Cardiol 1997; 4 Suppl 1:S4-7.
12. Poston RN. Atherosclerosis goes to the wall. Lancet 1992; 339: 647-8.
13. Honda HM, Wortham CM, Derner LL. Disturbed flow alters endothelial cell shape and monocyte adhesion. Presented at the American Heart Association, 1995.
14. Stary HC, Chandler AB, Dinsmore RE et al. A definition of advanced types of atherosclerotic lesions and a histological classification of atherosclerosis. A report from the Committee on Vascular Lesions of the Council on Arteriosclerosis, American Heart Association. Arterioscl Thromb Vasc Biol 1995; 15:1512-31.
15. Scott J. Unravelling atherosclerosis. Nature 1989; 338:118-9.
16. Stryer L. Biosynthesis of membrane lipids and steroid hormones. Biochemistry 3rd ed. Freeman 1988: 560.
17. Poston RN, Hussain IF. The immunohistochemical heterogeneity of atheroma macrophages: Comparison with lymphoid tissue suggests that recently blood-derived macrophages can be distinguished from longer-resident cells. J Histochem Cytochem 1993; 41:1503-12.
18. Albrecht W, Schuler W. The effect of short term cholesterol feeding on the development of aortic atheromatosis in the rabbit. J Atheroscler 1965; 5:353-68.
19. Davies MJ. The role of plaque pathology in coronary thrombosis. Clin Cardiol 1997; 20 suppl I:I-2-I-7.
20. Fuster V, Stein B, Ambrose JA et al. Atherosclerosis plaque rupture and thrombosis-evolving concepts. Circulation 1990; 82 suppl II: II-47-II-59.
21. Frenette PS, Wagner DD. Adhesion molecules—part I. N Eng J Med 1996; 334: 1526-9.
22. Jang Y, Lincoff AM, Plow EF et al. Cell adhesion in coronary artery disease. JACC 1994; 24:1591-601.
23. Sakai A, Kume N, Nishi E et al. P-selectin and VCAM-1 are focally expressed in aortas of hypercholesterolemic rabbits before intimal accumulation of macrophages and T lymphocytes. Arterioscl Thromb Vasc Biol 1997; 17:310-6.
24. Weyrich AS, McIntyre TM, McEver RP et al. Monocyte tethering by P-selectin regulates monocyte chemotactic protein-1 and tumor necrosis factor—alpha secretion: Signal integration and NF$_\kappa\beta$ translocation. J Clin Invest 1995; 95:2297-303.
25. Pongratz G, Pohle CF, Bachmann K. Components from oxidatively modified low-density lipoproteins induce P-selectin expression on platelets. Eur J Cardiol 1997; 18 Abstr Suppl: ref P451.
26. Johnson-Tidey RR, McGregor JL, Taylor PR et al. Increase in the adhesion molecule P-selectin in endothelium overlying atherosclerotic plaques. Am J Pathol 1994; 144: 952-61.
27. Poston RN, Johnson-Tidey RR. Localized adhesion of monocytes to human atheroscle-

rotic plaques demonstrated in vitro. Am J Pathol 1996; 149:73-80.

28. Stiegler H, Fischer Y, Schoebel FC et al. P-selectin β-thromboglobulin plasma levels as indicator for platelet and leukocyte activation in acute myocardial infarction. Eur J Cardiol 1997; 18 Abstr Suppl: ref P537.

29. Frenette PS, Wagner DD. Insights into selectin function from knockout mice. Thrombosis and Haemostasis 1997; 78:60-4.

30. O'Brien KD, McDonald TO, Alpers C et al. E-selectin expression is highly specific for human coronary atherosclerosis. Presented at the American Heart Association, 1995.

31 Davies MJ, Gordon JL, Gearing AJH et al. The expression of the adhesion molecules ICAM-1, VCAM-1, PECAM and E-selectin in human atherosclerosis. J Pathol 1993; 171:223-9.

32. Economou E, Vasiliadou K, Trikas A et al. Plasma levels of soluble E-selectin are elevated while plasma levels of von Willebrand factor and plasminogen activator inhibitor-1 remain unaffected in patients with mild hypertension. Eur J Cardiol 1997; 18 Abstr Suppl: ref P1136.

33. Siminiak T, Kazmierczak M, Minczykowski A et al. Dipyridamole stress test decreases soluble E-selectin and soluble VCAM-1 levels in patients with ischemic heart disease. Eur J Cardiol 1997; 18 Abstr Suppl: ref P541.

34. Languino L.R, Plescia J, Duperray A et al. Fibrinogen mediates leukocyte adhesion to vascular endothelium through an ICAM-1 dependent pathway. Cell 1993; 73:1423-34.

35. Couffinhal T, Duplaa C, Labat L et al. Tumor necrosis factor-alpha stimulates ICAM-1 expression in human vascular smooth muscle cells. Arteriosclerosis & Thrombosis 1993; 13:407-14.

36. Kume N, Cybulsky M.I, Gimbrone MA. Lysophospatidylcholine, a component of atherogenic lipoproteins, induces mononuclear leukocyte adhesion molecules in cultured human and rabbit arterial endothelial cells. J Clin Invest 1992; 90:1138-44.

37. Palkama L, Majure M-L, Mattila P et al. Regulation of endothelial adhesion molecules by ligands binding to the scavenger receptor. Clin Exp Immunol 1993; 92:353-60.

38. Frostegard J, Wu R, Haegerstrand A et al. Mononuclear leukocytes exposed to oxidized low density lipoprotein secrete a factor that stimulates endothelial cells to express adhesion molecules. Atherosclerosis 1993; 103: 213-9.

39. Zünd G, Colgan S, Dzus A et al. Hypoxia enhances endotoxin-stimulated induction of endothelial ICAM-1 through accumulation of NF-κβ. Eur J Cardiol 1997; 18 Abstr Suppl: ref P444.

40. Duplàa C, Couffinhal T, Labat L et al. Monocyte/macrophage recruitment and expression of endothelial adhesion proteins in human atherosclerotic lesions. Atherosclerosis 1996; 121:253-66.

41. Wood KM, Cadogan MD, Ramshaw AL et al. The distribution of adhesion molecules in human atherosclerosis. Histopath 1993; 22:437-44.

42. Poston RN, Haskard DO, Coucher JR et al. Expression of ICAM-1 in atherosclerotic plaques. Am J Pathol 1992; 140:665-73.

43. O'Brien KD, McDonald TO, Chait A et al. Neovascular expression of E-selectin, ICAM-1 and VCAM-1 in human atherosclerosis and their relation to intimal leukocyte content. Circulation 1996; 93:672-82.

44. Siminiak T, Dye JF, Egdell RM et al. The release of soluble adhesion molecules ICAM-1 and E-selectin after acute myocardial infarction and following coronary angioplasty. Int J Cardiol 1997; 61:113-8.

45. Ridker PM, Hennekens CH, Roitman-Johnson B et al. Plasma concentration of soluble ICAM-1 and risks of future myocardial infarction in apparently healthy men. Lancet 1998; 351:88-92.

46. Libby P, Li H. VCAM-1 and smooth muscle cell activation during atherogenesis. J Clin Invest 1993; 92:538-9.

47. Medford RM, Offerman MK, Benett F. Inhibition of TNFalpha induced VCAM-1 gene expression in human vascular endothelial and smooth muscle cells using transcriptional factor decoys. Presented at the American Heart Association, 1995.

48. Walpola PL, Gotlieb AI, Cybulsky MI et al. Expression of ICAM-1 and VCAM-1 and monocyte adherence in arteries exposed to altered shear stress. Arterioscler Thromb Vasc Biol 1995; 15:2-10.

49. Marui N, Offermann MK, Swerlick R et al. VCAM-1 gene transcription and expression are regulated through an antioxidant-sensitive mechanism in human vascular endothelial cells. J Clin Invest 1993; 92:1866-74.

50. Richardson M, Hadcock SJ, DeReske M et al. Increased expression in vivo of VCAM-1 and E-selectin by the aortic endothelium of normolipaemic and hyperlipaemic diabetic rabbits. Arteriosclerosis & Thrombosis 1994; 14:760-9.

51. Li H, Cybulsky MI, Gimbrone MA et al. Inducible expression of vascular cell adhesion molecule-1 by vascular smooth muscle cells in vitro and within rabbit atheroma. Am J Pathol 1993; 143:1551-59.

52. O'Brien KD, Allen MD, McDonald TO et al. Vascular cell adhesion molecule-1 is ex-

pressed in human coronary atherosclerotic plaques. J Clin Invest 1993; 92:945-51.

53. Prado K, Ribeiro JP, Quadros A et al. Higher plasma levels of soluble VCAM-1 are associated with the presence of chest pain in coronary artery disease. Eur J Cardiol 1997; 18 Abstr Suppl: P448.

54. Calderon TM, Factor SM, Hatcher VB et al. An endothelial cell adhesion protein for monocytes recognized by monoclonal antibody IG9. Lab Invest 1994; 70:836-849.

55. Berman JW, Calderon TM. The role of endothelial cell adhesion molecules in the development of atherosclerosis. C V P 1992; 1:17-28.

56. Devitt A, Moffatt OD, Raykundalia C et al. Human CD14 mediates recognition and phagocytosis of apoptotic cells. Nature 1998; 392:505-9.

57. Piali L, Hammel P, Uherek C et al. CD31 / PECAM-1 is a ligand for alpha$_v$β$_3$ integrin involved in adhesion of leukocytes to endothelium. J Cell Biol 1995; 130:451-60.

58. Stern M, Savill J, Haslett C. Human monocyte-derived macrophage phagocytosis of senescent eosinophils undergoing apoptosis. Mediation by alpha$_v$β$_3$/CD36/thrombospondin recognition mechanism and lack of phlogistic response. Am J Pathol 1996; 149:911-21.

59. Beaudet AL, Nageh MF, Collins RG et al. Genetic analysis of leukocyte and endothelial cell adhesion molecules. Presented at Inflammation, Growth Factors and Atherosclerosis, Keystone, Jan. 1997.

60. Raines EW, Ross R. Multiple growth factors are associated with lesions of atherosclerosis: Specificity or redundancy. BioEssays 1996; 18:271-82.

61. Wilson VJ, Britten K, Thomas CR et al. Multiple growth factors are expressed in human atherosclerotic lesions: Most are macrophage associated. Presented at the International Symposium on the Etiology and Pathobiology of Transplant Vascular Sclerosis, Bermuda, Mar. 1995.

62. Ruben et al. Lancet 1988; 1353-

63. Kume N, Gimbrone MA. Lysophosphatidylcholine transcriptionally induces growth factor gene expression in cultured human endothelial cells. J Clin Invest 1994; 93: 907-11.

64. Zwijsen RML, Japenga SC, Heijen AMP et al. Induction of platelet-derived growth factor chain A gene expression in human smooth muscle cells by oxidized low density lipoproteins. Biochemical and Biophysical Research Communications 1992; 186: 1410-16.

65. Inaba T, Gotoda T, Harada K et al. Induction of sustained expression of proto-oncogene

c-fms by platelet-derived growth factor, epidermal growth factor and basic fibroblast growth factor and its suppression by interferon-gamma and macrophage colony-stimulating factor in human aortic medial smooth muscle cells. J Clin Invest 1995; 95:1133-39.

66. Cunningham LD, Brecher P, Cohen RA. Platelet-derived growth factor receptors on macrovascular endothelial cells mediate relaxation via nitric oxide in rat aorta. J Clin Invest 1992; 89:878-82.

67. Hughes SE, Crossman D, Hall PA. Expression of basic and acidic fibroblast growth factors and their receptor in normal and atherosclerotic human arteries. Cardiovasc Res 1993; 27:1214-19.

68. Klagsbrun M, Edelman ER. Biological and biochemical properties of fibroblast growth factors—implications for the pathogenesis of atherosclerosis. Arteriosclerosis 1989; 9: 269-78.

69. Olson NE, Chao S, Lindner V et al. Intimal smooth muscle cell proliferation after balloon catheter injury. Am J Pathol 1992; 140:1017-23.

70. Miyagawa J, Higashiyama S, Kawat S et al. Localization of heparin-binding EGF-like growth factor in the smooth muscle cells and macrophages of human atherosclerotic plaques. J Clin Invest 1995; 95:404-11.

71. Bahadori L, Milder J, Gold L, Botney M. Active macrophage-associated TGF-β colocalizes with type I procollagen gene expression in atherosclerotic human pulmonary arteries. Am J Pathol 1995; 146:1140-9.

72. Grainger DJ, Kemp PR, Metcalfe JC at al. The serum concentration of active transforming growth factor-beta is severely depressed in advanced atherosclerosis. Nature Med 1995; 1:74-9.

73. Dickson K, Philip A, Warshawsky H et al. Specific binding of endocrine transforming growth factor-beta to vascular endothelium. J Clin Invest 1995; 95:2539-54.

74. Grainger DJ, Byrne CD, Witchell CM et al. Transforming growth factor beta is sequestered in to an inactive pool by lipoproteins. J Lipid Research 1997; 38:2344-52.

75. Wang X, Feuerstein GZ, Gu JL et al. Interleukin-1 beta induces expession of adhesion molecules in human vascular smooth muscle cells and enhances adhesion of leukocytes to smooth muscle cells. Atherosclerosis 1995; 115:89-98.

76. Bevilacqua MP, Schleef RR, Gimbrone MA et al. Regulation of the fibrinolytic system of cultured human vascular endothelium by interleukin 1. J Clin Invest 1986; 78:587-91.

77. Kaplanski G, Porat R, Aiura K et al. Activated platelets induce endothelial secretion of interleukin-8 in vitro via an interleukin-1-mediated event. Blood 1993; 81:2492-5.
78. Barath P, Fishbein MC, Cao J et al. Detection and localisation of tumor necrosis factor in human atheroma. Am J Cardiol 1990; 65:297-302.
79. Rus HG, Niculescu F, Vlaicu R. Tumor necrosis factor-alpha in human arterial wall with atherosclerosis. Atherosclerosis 1991; 89:247-54.
80. Barath P, Fishbein MC, Cao J et al. Tumor necrosis factor gene expression in human vascular intimal smooth muscle cells detected by in situ hybridization. Am J Pathol 1990; 137:503-9.
81. Bazzoni F, Beutter B. The tumor necrosis factor ligand and receptor families. N Engl J Med 1996; 334:1717-25.
82. Schwimmbeck PL, Kaplan H, Schwimmbeck N et al. Elevated gene expression of tumor necrosis factor-alpha in human coronary atherectomy specimens. Eur J Cardiol 1997; 18 Abstr Suppl: ref P1581.
83. Yoshizumi M, Perrrella MA, Burnett JC et al. Tumor necrosis factor down-regulates an endothelial nitric oxide synthase mRNA by shortening its half-life. Circ Res 1993; 73:205-9.
84. Noble KE, Panayiotidis P, Collins PW et al. Monocytes induce E-selectin gene expression in endothelial cells: Role of CD11/CD18 and extracellular matrix proteins. Eur J Immunol 1996; 26:2944-51.
85. Simon AD, Yazdani S, Wang W et al. Inflammatory cytokines in stable versus unstable angina. Presented at 46th A.C.C. meeting 1997 Anaheim, California, USA.
86. Geng YJ, Hansson GK. Interferon-gamma inhibits scavenger receptor expression and foam cell formation in human monocyte-derived macrophages. J Clin Invest 1992; 89:1322-30.
87. Nelken NA, Coughlin SR, Gordon D et al. Monocyte chemoattractant protein-1 in human atheromatous plaques. J Clin Invest 1991; 88:1121-7.
88. Rollins BJ, Yoshimura T, Leonard EJ et al. Cytokine-activated human endothelial cells synthesize and secrete a monocyte chemoattractant, MCP-1/JE. Am J Pathol 1990; 136:1229-33.
89. Cushing SD, Berliner JA, Valente AJ et al. Minimally modified low density lipoprotein induces monocyte chemotactic protein 1 in human endothelial cells and smooth muscle cells. Proc Natl Acad Sci 1990; 87:5134-8.
90. Ylä-Herttuala S, Lipton B, Rosenfeld ME et al. Expression of monocyte chemoattractant protein-1 in macrophage rich areas of human and rabbit atherosclerotic lesions. Proc Natl Acad Sci 1991; 88:5252-6.
91. Takeya M, Yoshimura T, Leonard EJ et al. Detection of monocyte chemoattractant protein-1 in human atherosclerotic lesions by an antimonocyte chemoattractant protein-1 monoclonal antibody. Hum Pathol 1993; 24:534-9.
92. Witztum JL. The oxidation hypothesis of atherosclerosis. Lancet 1994; 344:793-5.
93. Haberland ME, Fogelman AM. The role of altered lipoproteins in the pathogenesis of atherosclerosis. Am Heart J 1987; 113:573-7.
94. Smith EB, Staples EM. Distribution of plasma proteins across the human aortic wall—barrier functions of endothelium and internal elastic lamina. Atherosclerosis 1980; 37:579-90.
95. Falcone DJ, Hajjar DP, Minick CR. Lipoprotein and albumin accumulation in reendothelialized and deendothelialized aorta. Am J Pathol 1984; 114:112-20.
96. Crawford DW, Blankenhorn DH. Arterial wall oxygenation, oxyradicals and atherosclerosis. Atherosclerosis 1991; 89:97-108.
97. Matsumoto A, Naito M, Itakura H et al. Human macrophage scavenger receptors: Primary structure, expression and localization in atherosclerotic lesions. Proc Natl Acad Sci 1990; 87:9133-7.
98. Naito M, Suzuki H, Mori T et al. Coexpression of type I and type II human macrophage scavenger receptors in macrophages of various organs and foam cells in atherosclerotic lesions. Am J Pathol 1992; 141:591-9.
99. Loughheed M, Lum CM, Ling W et al. High affinity saturable uptake of oxidized low density lipoprotein by macrophages from mice lacking the scavenger receptor class A type I/II. J Biol Chem 1997; 272:12938-44.
100. Yia-Herttuala S, Palinski W, Rosenfeld ME et al. Evidence for the presence of oxidatively modified low density lipoprotein in atherosclerotic lesions of rabbit and man. J Clin Invest 1989; 84:1086-95.
101. Orekhov AN, Tertov VV, Kabakov AE et al. Autoantibodies against modified low density lipoprotein—nonlipid factor of blood plasma that stimulates foam cell transformation. Arteriosclerosis and Thrombosis 1991; 11:316-26.
102. Palinski W, Yiä-Herttuala S, Rosenfeld ME et al. Autoantibodies prevalent in human sera bind to oxidized LDL in atherosclerotic lesions. Arteriosclerosis 1989; 9:898a.
103. Boyd HC, Gown AM, Wolfbauer G et al. Direct evidence for a protein recognized by a monoclonal antibody against oxidatively modified LDL in atherosclerotic lesion from

a Watanabe Heritable Hyperlipidemic Rabbit. Am J Pathol 1989; 135:815-25.

104. Regnström J, Nilsson J, Tornvall P et al. Susceptibility to low-density lipoprotein oxidation and coronary atherosclerosis in man. Lancet 1992; 339:1183-6.

105. Salonen J, Salonen R, Seppännen K et al. Interactions of serum copper, selenium and low density lipoprotein cholesterol in atherogenesis. BMJ 1991; 302:756-60.

106. Orekhov AN, Tertov VV, Mukhin DN. Desialylated low density lipoprotein—naturally occurring modified lipoprotein with atherogenic potency. Atherosclerosis 1991; 86:153-61.

107. Kume N, Arai H, Kawai C et al. Receptors for modified low-density lipoproteins on human endothelial cells: Different recognition for acetylated low-density lipoprotein and oxidized low-density lipoprotein. Biochemica et Biophysica Acta 1991; 1091:63-7.

108. Boulanger CM, Tanner FC, Béa M-L et al. Oxidized low density lipoproteins induce mRNA expression and release of endothelin from human and porcine endothelium. Circ Res 1992; 70:1191-7.

109. Parhami F, Fang ZT, Fogelman AM et al. Minimally modified low density lipoprotein-induced inflammatory responses in endothelial cells are mediated by cAMP. J Clin Invest 1993; 92:471-8.

110. Hamilton T, Guoping MA, Chisolm GM. Oxidized low density lipoprotein suppresses the expression of TNFalpha mRNA in stimulated murine peritoneal macrophages. J Immunol 1990; 144:2343-50.

111. Terkeltaub R, Banka CL, Solan J et al. Oxidized LDL induces monocytic cell expression of IL8, a chemokine with T-lymphocyte chemotactic activity. Arterioscler Thromb 1994; 14:47-53.

112. Berliner JA, Territo MC, Sevanien A et al. Minimally modified low density lipoprotein stimulates monocyte endothelial interactions. J Clin Invest 1990; 85:1260-6.

113. Endemann G, Pronzcuk A, Friedman G et al. Monocyte adherence to endothelial cells in vitro is increased by ß-VLDL. Am J Pathol 1987; 126:1-6.

114. Frostegård J, Haegerstrand A, Gidlund M et al. Biologically modified LDL increases the adhesive properties of endothelial cells. Atherosclerosis 1991; 90:119-26.

115. Liao L, Starzyk RM, Granger DN. Molecular determinants of oxidized low-density lipoprotein-induced leukocyte adhesion and microvascular dysfunction. Arterioscl Thromb Vasc Biol 1997; 17:437-44.

116. Gebuhrer V, Murphy J F, Bordet J-C et al. Oxidized low-density lipoprotein induces the expression of P-selectin (GMP140/PADGEM/CD62) on human endothelial cells. Biochemistry 1995; 306:293-8.

117. Mehta A, Yang B, Khan S et al. Oxidized low-density lipoproteins facilitate leukocyte adhesion to aortic intima without affecting endothelium-dependent relaxation. Role of P-selectin. Arteriosclerosis, Thrombosis & Vascular Biology 1995; 15:2076-83.

118. Pritchard KA, Wong PY-K, Stemerman MB. Atherogenic concentrations of low-density lipoprotein enhance endothelial cell generation of epoxyeicosatrienoic acid products. Am J Pathol 1990; 136:1383-91.

119. Thorin E, Hamilton CA, Dominiczak MH et al. Chronic exposure of cultured bovine endothelial cells to oxidized LDL abolishes prostacyclin release. Arterioscler Thromb 1994; 14:453-9.

120. Yokode M, Kita T, Kikawa Y et al. Stimulated arachidonate metabolism during foam cell transformation of mouse peritoneal macrophages with oxidized low density lipoprotein. J Clin Invest 1988; 81:720-9.

121. Lehr HA, Hübner C, Finckh B et al. Role of leukotrienes in leukocyte adhesion following systemic administration of oxidatively modified human low density lipoprotein in hamsters. J Clin Invest 1991; 88:9-14.

122. Bath PMW, Gladwin A-M, Martin JF. Human monocyte characteristics are altered in hypercholesterolemia. Atherosclerosis 1991; 90:175-81.

123. Fan JL, Yamada T, Tokunaga O et al. Alterations in the functional characteristics of macrophages induced by hypercholesterolemia. Virchows Archiv B Cell Pathology incl. Molecular Pathology 1991; 61:19-27.

124. Parhami F, Fang ZT, Fogelman AM et al. Minimally modified low density lipoprotein-induced inflammatory responses in endothelial cells are mediated by cAMP. J Clin Invest 1993; 92:471-8.

125. Syvänne M, Taskinen M-R. Lipids and lipoproteins as coronary risk factors in non-insulin dependent diabetes mellitus. Lancet 1997; 350 suppl I:20-3.

126. Gwynne JT. High density lipoprotein cholesterol levels as a marker of reverse cholesterol transport. Am J Cardiol 1989; 64:10G-17G.

127. Packard C.J. LDL subfractions and atherogenicity: An hypothesis from the University of Glasgow. Curr Med Res Opin 1996; 13:379-90.

128. Brewer HB. Clinical significance of plasma lipid levels. Am J Cardiol 1989; 64:3G-9G.

129. Cockerill GW, Rye KA, Gamble JR et al. High-density lipoproteins inhibit cytokine-induced expression of endothelial cell adhesion molecules. Arterioscl Thromb Vasc Biol 1995; 15:1987-94.

130. Watson AD, Navab M, Hama SY et al. Effect of platelet activating factor-acetylhydrolase on the formation and action of minimally oxidized low density lipoprotein. J Clin Invest 1995; 95:774-82.

131. Navab M, Berliner JA, Watson AD et al. The yin and yang of oxidation in the development of the fatty streak: A review based on the 1994 George Lyman Duff memorial lecture. Arterioscler Thromb Vasc Biol 1996; 16:831-42.

132. De Caterina R, Cybulsky MI, Clinton SK et al. The omega-3 fatty acid docosahexaenoate reduces cytokine-induced expression of proatherogenic and proinflammatory proteins in human endothelial cells. Arteriosclerosis & Thrombosis 1994; 14:1829-36.

133. Scandinavian Simvastatin Survival Study Group. Randomised trial of cholesterol lowering in 4444 patients with coronary heart disease: The Scandinavian Simvastatin Survival Study (4S). Lancet 1994; 344:1383.

134. Matz J, Andersson TL, Ferns GA et al. Dietary vitamin E increases the resistance to lipoprotein oxidation and attenuates endothelial dysfunction in the cholesterol-fed rabbit. Atherosclerosis 1994; 110:241-9.

135. Martin A, Foxall T, Blumberg JB et al. Vitamin E inhibits low-density lipoprotein-induced adhesion of monocytes to human aortic endothelial cells in vitro. Arteriosclerosis, Thrombosis & Vascular Biology 1997; 17:429-36.

136. Lehr H-A, Frei B, Arfors K-E. Vitamin C prevents cigarette smoke induced leukocyte aggregation and adhesion to endothelium in vivo. Proc Natl Acad Sci 1994; 91:7688-92.

137. Diaz MN, Frei B, Vita JA et al. Antioxidants and atherosclerotic heart disease. N Engl J Med 1997; 337:408-16.

138. Sharma RC, Crawford DW, Kramsch DM et al. Immunolocalization of native antioxidant scavenger enzymes in early hypertensive and atherosclerotic arteries. Arteriosclerosis & Thrombosis 1992; 12:430-15.

139. Libby P, Ganz P. Restenosis revisited—new targets, new therapies. N Engl J Med 1997; 337:418-19.

140. Ferns GAA, Forster L, Stewart-Lee A et al. Probucol inhibits neointimal thickening and macrophage accumulation after balloon injury in the cholesterol-fed rabbit. Proc Natl Acad Sci 1992; 89.

141. Tardif J-C, Côté G, Lespérance J et al. Multivitamins and Probucol study group. Probucol and multivitamins in the prevention of restenosis after coronary angioplasty. N Engl J Med 1997; 337:365-72.

142. Frankel EN, Kanner J, German JB et al. Inhibition of oxidation of human low-den-sity lipoprotein by phenolic substances in red wine. Lancet 1993; 341:454-7.

143. Vallance P, Collier J. Biology and clinical relevance of nitric oxide. BMJ 1994; 309: 453-7.

144. Labinjoh C, Webb D. Lipids, lipid lowering and endothelial function. Br J Cardiol 1997; 4—suppl.1:S12-S17.

145. Bath PMW, Hassall DG, Gladwin A-M et al. Nitric oxide and prostacyclin—divergence of inhibitory effects on monocyte chemotaxis and adhesion to endothelium in vitro. Arteriosclerosis & Thrombosis 1991; 11: 254-60.

146. Nabel EG. Biology of the impaired endothelium. Am J Cardiol 1991; 68:6C-8C.

147. Fösterman V, Closs EI, Pollock JS et al. Nitric oxide synthase isoenzymes. Characterisation, purification, molecular cloning and functions. Hypertension 1994; 23: 1121-31.

148. Leeson CPM, Whincup PH, Cook DG et al. Flow-mediated dilation in 9 to 11-year-old children: The influence of intrauterine and childhood factors. Circulation 1997; 96:2233-8.

149. Zeiher AM, Drexler H, Saurbier B et al. Endothelium-mediated coronary blood flow modulation in humans—effects of age, atherosclerosis, hypercholesterolemia and hypertension. J Clin Invest 1993; 92:652-62.

150. Deanfield JE. Endothelial dysfunction as a marker of vascular disease. ACCEL 1996; 28.

151. Henderson AH, Jones CJH. Reversible endothelial dysfunction in epicardial coronary arteries. Lancet 1993; 342:253.

152. Leung W-H, Lau C-P, Wong C-K. Beneficial effect of cholesterol lowering therapy on coronary endothelium-dependent relaxation in hypercholesterolaemic patients. Lancet 1993; 341:1496-500.

153. Wang XL, Sim AS, Badenhop RF et al. A smoking-dependent risk of coronary artery disease associated with a polymorphism of the endothelial nitric oxide synthase gene. Nature Med 1996; 2:41-5.

154. Lamas S, Michel T, Collins T et al. Effects of interferon-gamma on nitric oxide synthase activity and endothelin-1 production by vascular endothelial cells. J Clin Invest 1992; 90:879-87.

155. Ihling C, Göbel HR, Lippold A et al. Endothelin-1-like immunoreactivity in human atherosclerotic coronary tissue: A detailed analysis of the cellular distribution of endothelin-1. J Pathol 1996; 179:303-8.

156. Zeiher AM, Ihling C, Pistorius K et al. Increased tissue endothelin immunoreactivity in atherosclerotic lesions associated with acute

coronary syndromes. Lancet 1994; 344: 1405-6.

157. Golino P, Cappelli-Bigazzi M, Ambrosio G et al. Endothelium-derived relaxing factor modulates platelet aggregation in an in vivo model of recurrent platelet activation. Circ Res 1992; 71:1447-56.

158. Hunt BJ, Jurd KM. Endothelial function in inflammation, sepsis, reperfusion and the vasculitides. An Introduction to vascular biology 1st Ed. Cambridge University Press 1998: 225-247.

159. D'Souza D, Wu KK, Hellums JD et al. Platelet activation and arterial thrombosis. Lancet 1994; 344:991-4.

160. Parmentier S, Kaplan C, Catimel B et al. New families of adhesion molecules play a vital role in platelet functions. Immunol Today 1990; 11:225-7.

161. Roth GJ. Platelets and blood vessels: The adhesion event. Immunol Today 1992; 13:100-5.

162. Lefkovits J, Plow EF, Topol EJ. Platelet glycoprotein IIb/IIIa receptors in cardiovascular medicine. N Engl J Med 1995; 332: 1553-8.

163. Rush DS, Kerstein MD, Bellan JA et al. Prostacyclin, thromboxane A$_2$, and prostaglandin E$_2$ formation in atherosclerotic human carotid artery. Arteriosclerosis 1988; 8:73-8.

164. Gawaz M, Neumann F-J, Dickfield T et al. Vitronectin receptor alpha$_v$β$_3$ mediates platelet adhesion to the luminal aspect of endothelial cells—implications for reperfusion in acute myocardial infarction. Circulation 1997; 96:1809-18.

165. Narins CR, Topol EJ. Attention shifts to the white clot. Lancet 1997; 350-suppl III:sIII2.

166. van Zanten GH, de Graaf S, Slootweg PJ et al. Increased platelet deposition on atherosclerotic coronary arteries. J Clin Invest 1994; 93:615-32.

167. Henn V, Slupsky JR, Gräfe M et al. CD40 ligand on activated platelets triggers an inflammatory reaction of endothelial cells. Nature 1998; 391:591.

168. Rapaport SI, Rao LVM. Initiation and regulation of tissue factor-dependent blood coagulation. Arteriosclerosis and Thrombosis 1992; 12:1111-19.

169. Taubman MB, Fallon JT, Schechter AD et al. Tissue factor in the pathogenesis of atherosclerosis. Thrombosis and Haemostasis 1997; 78:200-4.

170. Bar-Shavit R, Eldor A, Vlodavsky I. Binding of thrombin to subendothelial extracellular matrix. J Clin Invest 1989; 84: 1096-104.

171. Wilcox JN, Smith KM, Schwartz SM et al. Localization of tissue factor in the normal vessel wall and in the atherosclerotic plaque. Proc Natl Acad Sci 1989; 86:2839-43.

172. Drake TA, Hannani K, Fei H et al. Minimally oxidized low-density lipoprotein induces tissue factor expression in cultured human endothelial cells. Am J Pathol 1991; 138:601-7.

173. Nelken NA, Solfer SJ, O'Keefe J et al. Thrombin receptor expression in normal and atherosclerotic human arteries. J Clin Invest 1992; 90:1614-21.

174. Selwyn AP, Kinlay S, Creager M. Cell dysfunction in atherosclerosis and the ischemic manifestations of coronary artery disease. Am J Cardiol 1997; 79:17-23.

175. Carter AM, Ossei-Gerning N, Grant PJ. Platelet glycoprotein IIIa Pl^A polymorphism in young men with myocardial infarction. Lancet 1996; 348:486-7.

176. Weiss EJ, Bray PF, Tayback M et al. A polymorphism of a platelet glycoprotein receptor as an inherited risk factor for coronary thrombosis. N Engl J Med 1996; 334: 1090-4.

177. Ardissino D, Merlini PA, Coppola R et al. Thrombogenic potential of human coronary atherosclerotic plaques. Presented at the 46th A.C.C. meeting 1997, Anaheim, California, USA.

178. Fyfe AI, Qiao JH, Lusis AJ. Immune-deficient mice develop typical atherosclerotic fatty streaks when fed an atherogenic diet. J Clin Invest 1994; 94:2516-20.

179. Hansson GK, Jonasson L, Seifert PS et al. Immune mechanisms in atherosclerosis. Arteriosclerosis 1989; 9:567-78.

180. Wissler RW, Group RD. Atheroarteritis: A combined immunological and lipid imbalance. Int J Cardiol; 1997; 54 suppl:S37-49.

181. Wissler RW, Robert L. Aging and cardiovascular disease. Circulation 1996; 93: 1608-12.

182. Frostegård J, Wu R, Giscombe R et al. Induction of T cell activation by oxidized low density lipoprotein. Arteriosclerosis and Thrombosis 1992; 12:461-7.

183. McMurray HF, Parthasarathy S, Steinberg D. Oxidatively modified low density lipoprotein is a chemoattractant for human T lymphocytes. J Clin Invest 1993; 92:1004-8.

184. Salonen JT, Yia-Hertuala S, Yamamoto R et al. Autoantibody against oxidised LDL and progression of carotid atherosclerosis. Lancet 1992; 339:883-7.

185. Lip GYH, Beevers DG. Can we treat coronary artery disease with antibiotics? Lancet 1997; 350:378-9.

186. Danesh J, Collins R, Peto R. Chronic infections and coronary heart disease: Is there a link?. Lancet 1997; 350:430-6.

187. Gupta S, Camm AJ. Chlamydia Pneumoniae and coronary heart disease. BMJ 1997; 314:1778-9.

188. Wissler RW. Significance of Chlamydia pneumoniae in atherosclerotic lesions. Circulation 1995; 92:3376.

189. Gurfinkel E, Bozovich G, Daroca A et al. Randomized trial of roxithromycin in nonQ-wave coronary syndromes: ROXIS pilot study. Lancet 1997; 350:404-7.

190. Gupta S, Leatham EW, Carrington D et al. The effect of azithromycin in postmyocardial infarction patients with elevated Chlamydia pneumoniae antibody titres. J Am Coll Cardiol 1997; 29-suppl A:209A.

191. Penco M, Sessa R, Di Pietro M et al. Chlamydia Pneumoniae infection and atherosclerotic coronary disease. Presented at the 46th A.C.C. meeting 1997 Anaheim, California, USA.

192. De Stefano F, Anda RF, Kahn HS et al. Dental disease and the risk of coronary artery disease and mortality. BMJ 1993; 306:688-91.

193. Roma P, Catapano AL. Stress proteins and atherosclerosis. Atherosclerosis 1996; 127: 147-54.

194. Xu Q, Dietrich H, Steiner HJ et al. Induction of arteriosclerosis in normocholesterolemic rabbits by immunization with heat shock protein 65. Arteriosclerosis & Thrombosis 1992; 12:789-99.

195. Xu Q, Kleindienst R, Waitz W et al. Increased expression of heat shock protein 65 coincides with a population of infiltrating T lymphocytes in atherosclerotic lesions of rabbits specifically responding to heat shock protein 65. J Clin Invest 1993; 91:2693-702.

196. Kleindienst R, Xu Q, Willeit J et al. Immunology of atherosclerosis. Demonstration of heat shock protein 60 expression and T lymphocytes bearing alpha/beta or gamma/delta receptor in human atherosclerotic lesions. Am J Pathol 1993; 142:1927-37.

197. Xu Q, Luef G, Weimann S et al. Staining of endothelial cells and macrophages in atherosclerotic lesions with human heat shock protein-reactive antisera. Arteriosclerosis & Thrombosis 1993; 13:1763-9.

198. Xu Q, Willeit J, Marosi M et al. Association of serum antibodies to heat-shock protein 65 with carotid atherosclerosis. Lancet 1993; 341:255-9.

199. Berberian PA, Myers W, Tytell M et al. Immunohistochemical localization of heat shock protein 70 in normal-appearing and atherosclerotic specimens of human arteries. Am J Pathol 1990; 136:71-80.

200. Caligiuri G, Van de Greef W, Summaria F et al. Antigenic stimuli may be responsible for instability of angina. Eur J Cardiol 1997; 18 Abstr suppl: ref P480.

201. Vallance P, Collier J, Bhagat K. Infection, inflammation and infarction: Does acute endothelial dysfunction provide a link. Lancet 1997; 349:1391-2.

202. Geng YJ, Libby P. Evidence for apoptosis in advanced human atheroma. Colocalization with interleukin-1 beta-converting enzyme. Am J Pathol 1995; 147:251-66.

203. Han DK, Haudenschild CC, Hong MK et al. Evidence for apoptosis in human atherogenesis and in a rat vascular injury model. Am J Pathol 1995; 147:267-77.

204. Hegyi L, Skepper JN, Cary NRB et al. Foam cell apoptosis and the development of the lipid core of human atherosclerosis. J Pathol 1996; 180:423-9.

205. Bauriedel G, Hutter R, Willert P et al. Increased apoptosis in human atherectomy specimens from patients with acute coronary syndrome versus stable angina. Eur J Cardiol 1997; 18 Abstr suppl:P236.

206. Hardwick SJ, Hegyi L, Clare K et al. Apoptosis in human monocyte-macrophages exposed to oxidized low density lipoprotein. J Pathol 1996; 179:294-302.

207. Reid VC, Mithcinson MJ, Skepper JN. Cytotoxicity of oxidized low density lipoprotein to mouse peritoneal macrophages: An ultrastructural study. J Pathol 1993; 171: 321-8.

208. Bjorkerud B, Bjorkerud S. Contrary effects of lightly and strongly oxidized LDL with potent promotion of growth versus apoptosis on arterial smooth muscle cells, macrophages and fibroblasts. Arterioscl Thromb Vasc Biol 1996; 16:416-24.

209. Nishio E, Arimura S, Watanabe Y. Oxidized LDL induces apoptosis in cultured smooth muscle cells: A possible role for 7-ketocholesterol. Biochemical & Biophysical Research Communications 1996; 223:413-8.

210. Bennett MR, Evan GI, Schwartz SM. Apoptosis of human vascular smooth muscle cells derived from normal vessels and coronary atherosclerotic plaques. J Clin Invest 1995; 95:2266-74.

211. Bennett MR, Evan GI, Newby AC. Deregulated expression of the c-myc oncogene abolishes inhibition of proliferation of rat vascular smooth muscle cells by serum reduction, interferon-gamma, heparin and cyclic nuclotide analogues and induces apoptosis. Circ Res 1994; 74:525-36.

212. Demer LL. A skeleton in the atherosclerosis closet. Circulation 1995; 92:2029-32.

213. Hirota S, Imakita M, Kohri K et al. Expression of osteopontin mRNA by macrophages in atherosclerotic plaques. Am J Pathol 1993; 143:1003-8.

214. Feener EP, King GL. Vascular dysfunction in diabetes mellitus. Lancet 1997; 350—suppl I:9-13.

215. Zimmet PZ, Alberti KGMM. The changing face of macrovascular disease in noninsulin dependent diabetes mellitus: An epidemic in progress. Lancet 1997; 350—suppl I:1-4.

216. Hadcock S, Richardson M, Winocour PD et al. Intimal alterations in rabbit aortas during the first 6 months of alloxan-induced diabetes. Arteriosclerosis & Thrombosis 1991; 11:517-29.

217. Schmidt AM, Yan SD, Stern DM. The dark side of glucose. Nature Med 1995; 1:1002-4.

218. Nakamura Y, Horii Y, Nishino T et al. Immunohistochemical localization of advanced glycosylation endproducts in coronary atheroma and cardiac tissue in diabetes mellitus. Am J Pathol 1993; 143: 1649-56.

219. Vlassara H, Fuh H, Donnelly T et al. Advanced glycation end-products promote adhesion molecule (VCAM-1, ICAM-1) expression and atheroma formation in normal rabbits. Molecular Med 1995; 1:447-56.

220. Nathan DM, Meigs J, Singer DE. The epidemiology of cardiovascular disease in type 2 diabetes mellitus: How sweet it is, or is it? Lancet 1997; 350 suppl I: 4-9.

221. Kim JA, Berliner JA, Natarajan RD et al. Evidence that glucose increases monocyte binding to human aortic endothelial cells. Diabetes 1994; 43:1103-7.

222. Gibbons GH. Endothelial function as a determinant of vascular function and structure: A new therapeutic target. Am J Cardiol 1997; 79 - 5A:3-8.

223. Keidar S, Kaplan M, Aviram M. Angiotensin II-modified LDL is taken up by macrophages via the scavenger receptor, leading to cellular cholesterol accumulation. Arterioscler Thromb Vasc Biol 1996; 16:97-105.

224. Sudhir K, Chou TM, Messina LM et al. Endothelial dysfunction in a man with disruptive mutation in estrogen-receptor gene. Lancet 1997; 349:1146-7.

225. Mazière C, Auclair M, Ronveaux M-F et al. Estrogens inhibit copper and cell-mediated modification of low density lipoprotein. Atherosclerosis 1991; 89:175-82.

226. Caulin-Glaser TL, Watson C, Pardi R. Effects of 17β-oestradiol on cytokine-mediated endothelial cell adhesion molecule expression. J Clin Invest 1996; 98:36-42.

227. Bellamy MF, McDowell IF. Putative mechanisms for vascular damage by homocysteine. J Inherited Metabolic Disease 1997; 29: 307-15.

228. Nygård O, Nordrehaug JE, Refsum H et al. Plasma homocysteine levels and mortality in patients with coronary artery disease. N Engl J Med 1997; 337:230-6.

229. van den Berg M, Stehouwer CDA, Bierdrager E et al. Plasma homocysteine and severity of atherosclerosis in young patients with lower-limb atherosclerotic disease. Arterioscler Thromb Vasc Biol 1996; 16: 165.

CHAPTER 3

The Immune Response to Endothelial Cells

Marlene L. Rose

Endothelial cells forming the interface between donor and recipient are the first donor cells to be recognized by the host's immune system. This fact plus the observation that they are highly responsive to cytokines and express numerous molecules which interact with ligands on lymphocytes (Tables 3.1, 3.2) has stimulated much research into their precise role in transplant rejection. It is our view that endothelial cells are pivotal both in controlling the egress of inflammatory cells into the allografted organ but also as specific antigen presenting cells, by presenting foreign molecules to the immune (Fig. 3.1).

It has been recognized for many years that cells of the immune system which bear class II antigens of the Major Histocompatibility Complex (MHC) are potent stimulators of the rejection process. In particular, bone-marrow derived monocytes or dendritic cells (also found in donor organs as passenger leukocytes) are potent stimulators of an antiallograft response. However, it is now appreciated that many human parenchymal cells either constitutively express MHC class II molecules (such as endothelial and epithelial cells) or can be induced to express MHC class II (smooth muscle cells and fibroblasts) after exposure to appropriate cytokines. In view of these observations there is considerable interest in the role of donor derived MHC class II positive parenchymal cells in stimulating alloreactive T cells of the host. Part of this review will compare the characteristics of endothelial cells and other parenchymally derived cells as antigen presenting cells. The other part will describe interactions between endothelial cells and the immune system which are implicated in chronic rejection.

Basic Mechanism of Rejection

The major stimulus for rejection of allografted organs is recognition of foreign antigens coded by the Major Histocompatibility Complex (MHC). Class I (HLA-ABC) and class II (HLA-DR, DP, DQ) antigens are highly polymorphic glycoproteins encoded by the MHC locus found on chromosome 6 in humans. The frequency of circulating T cells which recognize foreign MHC molecules is very large (estimated at an astounding 0.1-1% of circulating T cells)—a fact which almost certainly accounts for the vigor of the antiallograft response.

Rejection is initiated by the CD4+ T cell subset recognizing MHC class II antigens on antigen presenting cells (APC) within the graft (Fig. 3.2). Recognition of foreign MHC molecules results in CD4+ T cell activation and release of cytokines (IL-2, IL-4, IL-5, IL-6, IFNγ, TNFα, TNFβ) which allow maturation of the effector mechanisms of rejection, namely maturation of CD8+ cytotoxic T cells, infiltration of macrophages, maturation of NK cells and lymphokine activated killer cells (LAK) and antibody formation (Fig. 3.2).

Transplant-Associated Coronary Artery Vasculopathy, edited by Marlene L. Rose.
©2001 Eurekah.com.

Table 3.1. Interaction of endothelial cells with the immune system; expression of cell surface molecules

Constitutive Expression	Induced by Cytokines
MHC class I (HLA-ABC antigens)	VCAM-1
MHC class II (HLA-DR and DP)	E-selectin
ICAM-1, ICAM-2 (CD54)	CD40L
LFA-3 (CD58)	Fas-L (CD95L)
Inhibitors of complement activation CD59, DAF, MCP	
CD40	
Fas (CD95)	

Table 3.2. Interaction of endothelial cells with the immune system; secretion of cytokines, growth factors and chemokines

Cytokines and Growth Factors	Chemokines
IL-1	CXC, attract neutrophils (IL-8, IP-10)
IL-6	CC, attract mononuclear cells, MCP-1
PDGF, TGFα, TGFβ, bFGF, HB-EFG	Rantes
	Fractalkine

Abbreviations: PDGF, platelet-derived growth factor; TGF, transforming growth factor; bFGF, basic fibroblast growth factor; HB-ERG, heparin-binding epidermal growth factor.

These effector mechanisms have been listed for the sake of completeness, there is little evidence that NK or LAK cells are important in allograft rejection. Indeed, the precise effector mechanisms which cause graft dysfunction are unknown; although CD8+ cytotoxic T cells are invariably found in allografts experimental studies have shown that CD8+ T cells are not essential for rejection.[1] It is quite possible that a direct effect of cytokines, in particular TNFα and IFNγ may be toxic to allografted cells. For example, TNFα has a negative ionotropic effect on cardiac myocytes[2] and elevated levels of TNFα have been reported in the serum of patients in heart failure[3] Similarly, induction of inducible nitric oxide synthase by activated macrophages and endothelial cells may be an important effector mechanism and has been associated with contractile dysfunction after cardiac transplantation.[4]

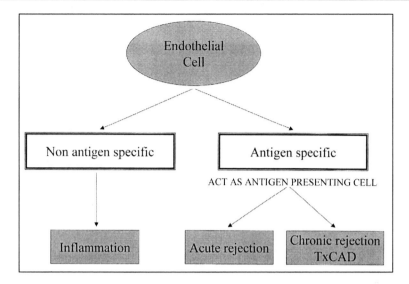

Fig. 3.1. Diagram to illustrate role of endothelial cells in transplant rejection.

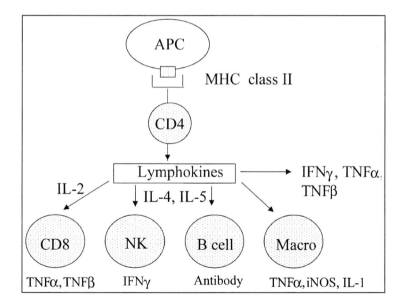

Fig. 3.2. Diagrammatic representation of T cell activation, illustrating the pivotal role of MHC class II antigens (presented by Antigen Presenting Cells within the graft) in initiating rejection. Activation of CD4⁺ T cells results in a cascade of lymphokines causing the maturation of a number of possible effector cells (in shaded circles). Note that the cytokines IFNγ, TNFβ and TNFα may be directly damaging to tissues.

Activation of CD4+ T cells is thus pivotal in initiating acute rejection (Fig. 3.2). In view of the fact that CD4+ T cells are activated by foreign MHC class II molecules, understanding the quantitative and qualitative distribution of these molecules on the allografted organ is of considerable importance. The advent of monoclonal antibodies, use of frozen sections and advances in immunocytochemical techniques have revolutionized knowledge about the normal distribution of MHC molecules in different tissues. Class II (HLA-DR and DP) antigens, originally thought to be restricted to macrophages, dendritic cells, monocytes and activated T cells have now been described on human endothelial and epithelial cells.[5,6] The expression of class II on human endothelial cells has been described in every organ[5,7] and it is particular striking on the microvessels i.e., capillaries, arterioles and venules. The large vessel endothelium (such as aorta, pulmonary artery, saphenous vein) are negative for MHC class II expression.[7]

Pathways of Antigen Presentation and Antigen Presenting Cells

The term Antigen Presenting Cell (APC) has a specific meaning to immunologists: it means the cell is able to present antigen to resting T cells i.e., is able to cause activation of resting T cells. Only specialized cells (traditionally recognized as B cells, dendritic cells and monocytes) can perform this task. T cells recognize nominal antigen as processed peptides presented by self MHC molecules. An important step in the understanding of alloreactivity came with the discovery that T cells can engage and respond to allogeneic MHC molecules directly (Fig. 3.3). This form of antigen recognition, termed direct presentation or the direct pathway is responsible for the strong proliferative response to alloantigens seen in vitro and quite possibly the early acute rejection seen in nonimmunosuppressed animals after transplantation of MHC mismatched organs. However, T cells can also recognize allogeneic peptides that have been processed and presented within self MHC molecules by recipient APC in the same

manner that T cells recognize nominal antigen (Fig. 3.3). This pathway is termed the indirect route or indirect pathway of T cell activation. Alloantigens shed from the graft are likely to be treated as exogenous antigen by recipient APC and will therefore be presented within MHC class II molecules to activate recipient CD4+ T cells. A number of experimental studies have exemplified this phenomenon.[8-10]

Any graft cell expressing class II antigens will be able to activate the indirect pathway—is likely that damaged endothelial cells are an important source of graft derived MHC class II antigens—since these are the only parenchymal cells expressing class II in the heart. The contribution indirect recognition of endothelial MHC class II makes to cellular rejection is currently not known. However, the question which has received much attention of a number of groups in recent years is whether endothelial cells can cause direct allostimulation of resting T lymphocytes (see below and ref. 11 for review). There are two reasons for this. First is that direct recognition of allo-MHC molecules results in a 'strong' response, the number of T cells recognizing MHC molecules directly is 10-100 higher than those recognizing nominal antigen, resulting in a strong in vitro proliferative response. Second is that it is known that expression of MHC class II is not sufficient to cause T cell activation.

One of the important concepts to emerge in recent years is the knowledge that T cells require two signals to become activated,[12] one is occupancy of the T cell receptor and the second is activation of one of the many 'accessory molecules' present on T cells (Fig. 3.4). Much attention has focused on the B7 family of receptors (13, CD80, CD86), known to be essential as second signals on APC of bone-marrow origin (e.g., monocytes, B cells and dendritic cells); B7 receptors interact with CD28 molecules on the surface of resting T cells and blockade of this pathway inhibits dendritic cell stimulated mixed lymphocyte responses in vitro.[14] It is thought that interaction of resting T cell receptor with antigen in the absence of

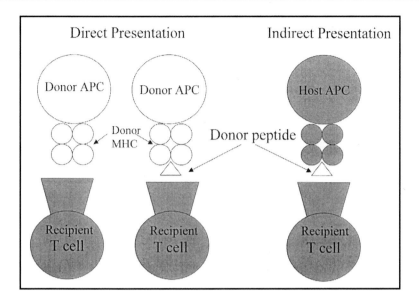

Fig. 3.3. Diagrammatic representation of mechanisms whereby recipient T cells recognize allo-class II determinants. Recipient T cells recognize donor MHC determinants on donor APC (direct presentation) or they recognize donor MHC peptides which have been released from donor cells and processed and presented by host APC within self MHC molecules (indirect presentation).

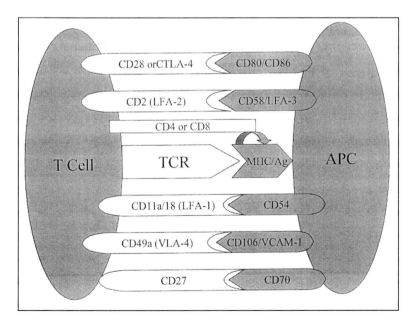

Fig. 3.4. Diagrammatic representation of possible interactions between receptors on T cells and their appropriate ligands on antigen presenting cells (APC).

costimulation results in T cells being rendered anergic[15] or depleted by apoptosis.[16] Another T cell surface antigen CTLA4, binds CD80 and CD86 with 10-20 times greater affinity than CD28.[17] Thus CTLA4-Ig acts as a strong competitive inhibitor of CD28 mediated T cell activation. Injection of this fusion protein into rodents results in long term survival of MHC mismatched cardiac allografts[18-20] and human islet xenografts.[21]

Acute Rejection

The consequences of T cell activation described above leads to infiltration of the graft with inflammatory cells (T cells and monocytes)—a process termed acute rejection. The majority of patients have one or two acute rejection episodes in the first 6 months following transplantation. Acute rejection may be suspected clinically but it is always confirmed by histological assessment of biopsies. In particular, endomyocardial biopsy is an essential part of the management of patients following cardiac transplantation. Acute rejection invariably responds to anti-T cell depletion therapy, steroids and cyclosporine.

A characteristic feature of biopsies showing histological signs of rejection is the upregulation of MHC and adhesion molecules (see ref. 22 for review). After cardiac transplantation there is massive upregulation of MHC class I antigens (normally found only on the interstitial cells) so that cardiac myocytes become MHC class I positive.[6] Upregulation of MHC class I antigens has also been described on hepatocytes after liver transplantation.[23] In kidneys, the majority of structures constitutively express MHC class I molecules[22] therefore studies have focused on changes in MHC class II expression[24,25] where increased expression is found on renal tubules and endothelial cells. There is also upregulation of adhesion molecules during acute rejection (see below). The upregulation of these molecules is almost certainly mediated by local production of cytokines by infiltrating cells; thus some cytokines (such as TNFα and IFNγ) have been directly visualized in graft biopsies using immunocytochemical methods,[26] others (IL-2, IL-1, IL-4, IL-6, IL-10) have been detected using polymerase chain reaction to amplify cytokine mRNA.[27]

Chronic Rejection

Chronic rejection, presenting as a rapidly progressing obliterative vascular disease occurring in the transplanted heart, is the major cause of late death and repeat transplantation after cardiac transplantation. This disease is variously termed cardiac allograft vasculopathy or transplant associated coronary artery disease (TxCAD). This same phenomenon is also present in renal, lung and liver allografts and has been designated chronic rejection, obliterative bronchiolitis and vanishing bile duct syndrome respectively. Chronic rejection does not respond to currently used immunosuppressive drugs (cytolytic anti-T cell agents, steroids and cyclosporine)—which primarily inhibit T cell mediated immune responses. The reported incidence of TxCAD, detected by routine angiography varies greatly between cardiac transplant centers. Incidences of 18% at one year progressing to 44% at three years have been reported.[28] Use of more accurate and sensitive methods of detecting intimal hyperplasia, such as intracoronary ultrasound,[29] is likely to increase the reported incidence of the disease.

Histology of TxCAD

There are a number of reviews which describe the histological differences between TxCAD and spontaneous CAD and the various risk factors, both immunological and nonimmunological have been described (ref. 30, and Chapter 1). Importantly the lesions are limited to the allograft and involve both arterial and venous structures—demonstrating the role of the alloimmune response in the etiology. The formation and structure of plaques characteristic of spontaneous CAD are described in Chapter 2; there are important differences between these eccentric plaques and the lesions typical of TxCAD. It is interesting that TxCAD is a much more diffuse disease than spontaneous CAD, affecting the entire length of the epicardial vessels;

Table 3.3. Risk factors associated with transplant associated coronary artery disease

Immunological	Nonimmunological
Numbers of acute rejection episodes	Hypercholesterolemia
Numbers of mismatches at the HLA locus	Hypertriglyceridemia
Anti-HLA antibodies	Hypertension
Antiendothelial antibodies	CMV infection
	Increased donor age
	Smoking
	Diabetes

Reproduced from[112], permission of Birkhauser Verlag, Basel, 1998

the intimal proliferation is concentric as opposed to the eccentric plaque found in spontaneous CAD. These differences suggest the whole endothelium is the target of damage in TxCAD. Because the epicardial branches, including the intramyocardial branches are affected by TxCAD, coronary artery bypass surgery for revascularization is usually precluded (but see Chapter 1).

Immunobiology of TxCAD

The occurrence of a vasculopathy, affecting the allografted organs, is almost certainly of multi-factorial etiology. It is highly likely that the obstructive vascular lesions progress through repetitive endothelial injury followed by repair, smooth muscle cell proliferation and hypertrophy, all of which gradually produce luminal obliteration. It is useful to think of the disease in term of the Ross hypothesis[31]—namely an initial damage to the endothelium resulting in release of growth factors and intimal proliferation. The latter process will be assisted by risk factors (Table 3.3) some of which are common to both TxCAD and spontaneous CAD (the nonimmunological risk factors) and some of which are only found in transplant recipients (immunological risk factors). Of the nonimmunological risk factors, the role of hypercholesterolemia is firmly established[32,33] and is covered in detail in Chapters 2 and 9. The role of cytomegalovirus, a common opportunistic infection in transplant recipients is also covered in Chapter 8.

Most investigators would acknowledge that the initial damage to the endothelium is mediated by the alloimmune response, although it can also be argued that nonimmunological damage such as ischemia, surgical manipulation, and perfusion/reperfusion injury could also initially damage the endothelial cells (see Chapter 5). It is highly likely that damaged endothelial cells are a better target for the alloimmune response resulting in upregulation of alloantigens or exposure of cryptic epitopes.

Endothelial Cells as Antigen Presenting Cells

Endothelial Expression of MHC Class II Molecules In Vivo

Since MHC class II antigens initiate allograft rejection, it is of interest to describe the distribution of these molecules on endothelial cells of different origins (Table 3.4). All endothelial cells constitutively express MHC class I molecules and many endothelial cells constitutively express MHC class II molecules. However, there is an interesting heterogeneity with regard to constitutive expression of class II antigens; the large vessels (aorta, pulmonary artery, endocardium, umbilical vein, umbilical artery) are negative but the capillaries within all organs examined are strongly positive (Table 3.4). Coronary artery endothelial cells, from unused donors, were always found to be MHC class II positive.[7] Thus foreign MHC class I and class II antigens are introduced into a host every time a vascularized organ is transplanted. Arterioles and venules within the heart show weak

Table 3.4. Distribution of adhesion molecules, MHC molecules and vWf in endothelial cells derived from microvessels and large vessels of the human cardiovascular system

	Myocardial Biopsies			Large Vessels		
	capillaries	arterioles	Venules	coronary	PA	Aorta
CD31	++	++	++	++	++	++
ICAM-1	++	+	+	++	+	+
VCAM-1	neg	+/-	+/-	+	Neg	Neg
E-selectin	neg	neg	+/-	+	+/-	+/-
VWf	+/-	++	++	++	++	++
Class I	++	+	+	++	+	+
Class II	++	+/-	+/-	++	Neg	Neg

Summarized from refs. 6,7, and 71.
++ strong, even expression; + strong but patchy expression; ± weak and patchy expression

or patchy basal expression of MHC class II antigens.

MHC Class II Expression on Cultured Endothelial Cells

The most common endothelial cells used in cell culture are those derived from human umbilical vein (HUVEC), these do not express MHC class II antigens in situ[7] and it is therefore not surprising that they are also negative in vitro. Treatment of HUVEC with interferon (IFN)α, IFNβ, tumor necrosis factor (TNF), lymphotoxin or CD40 ligand increases the level of MHC class I antigen without inducing MHC class II molecules.[34,35] An exception to this is that TNFα enhances MHC class II expression in porcine endothelial cells.[36] IFNγ is uniquely able to induce MHC class II antigens on HUVEC,[37] HLA-DR is the most strongly expressed class II locus, followed by DP antigens, DQ is induced by IFNγ but at very low levels. When used in combinations the IFNs can act inter-changeably with each other and synergistically with TNF and LT to increases class I. IFNα and IFNβ strongly inhibit the ability of IFNγ to induce class II antigens as does TGFβ[38,39] The experiments to date using TGFβ have used monocytes or astrocytes as target cells,[38,39] recent studies confirm that TGFβ inhibits IFNγ mediated upregulation of class II on

human endothelial cells (Moyes and Rose unpublished). The vast majority of studies on human endothelial cells have been performed using HUVEC, because of their plentiful supply. However, endothelial cells can be cultured from any human vascular bed and where experiments have been performed on aorta or pulmonary artery cells the results are the same as with HUVEC.[40] Interestingly, the exception to this rule appears to be microvascular endothelial cells derived from the adult human heart. These cells are MHC class II positive in situ, and they lose their class II after two weeks in culture.[41] This observation raises the interesting possibility that factors in normal serum act to maintain microvascular MHC class II expression in vivo.

Stimulation of Alloreactive T Cells

The fact that human microvascular endothelial cells constitutively express MHC class II molecules has stimulated a number of workers to ask whether endothelial cells directly cause allostimulation of resting T cells. We and others[42,43] have cultured stringently purified CD4+ T cells with pure passaged human umbilical vein endothelial cells and looked for T cell proliferation (measured by uptake of 3H-thymidine) at day 6. The endothelial cells are treated with mitomycin C to stop them proliferating. Any proliferation

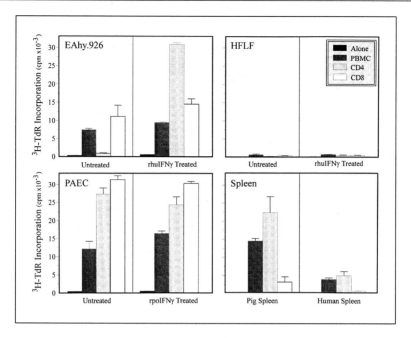

Fig. 3.5. Response of purified PBMC (peripheral blood mononuclear cells) CD4⁺ and CD8⁺ T cells to untreated and IFNγ treated human endothelial cells (EAhy.926), human fetal lung fibroblasts (HFLF), porcine aortic endothelial cells (PEAC) and porcine or human spleen cells. CD4⁺ and CD8⁺ T cells respond strongly to PAEC, regardless of cytokine treatment. CD4⁺ T cells respond well to human endothelial cells, providing they have been pretreated with IFNγ to upregulate MHC class II antigens. CD8⁺ T cells respond to human endothelial cells in the absence of cytokine treatment. There is no proliferative response of human lymphocytes to human fetal lung fibroblasts. Reproduced from ref. 112, permission of Birkhauser Verlag, Basel.

which is detected is thus due to responding T cells. The results in Figure 3.5 show the response of CD4⁺ T cell to human endothelial cells (EAhy.926), porcine aortic endothelial cells (PAEC) and fetal lung fibroblasts. It can be seen that provided IFNγ is used to upregulate MHC class II, there is a strong proliferative response to human EC, but not to fibroblasts. There is also a strong response to PAEC, which is independent of IFNγ treatment. The reason for this is that PAEC class II expression persists in culture. That the response was direct and not indirect was proven by the findings that responder T cells were free of contaminating APC[42,43] and could not be inhibited by CTAL-4-Ig.[44] The response of CD4⁺ T cells depends on pretreatment of the ECs with IFNγ and is inhibited by antibodies to MHC class II determinants[44] and not affected by antibodies to MHC class I. In contrast isolated CD8⁺ T cells proliferate to untreated EC and the response is inhibited by antibodies to MHC class I determinants.[43,45]

The majority of studies use endothelial cells of fetal origin (from the umbilical vein); we have confirmed these findings using endothelial cells from adult aorta, coronary artery and microvascular sources and in addition have shown that smooth muscle cells derived from the same vessel fail to stimulate allogeneic T cells.[46]

When restimulated, T cells respond to endothelial cells giving kinetics similar to that seen in the primary response.[47] There is no evidence for anergy induction.

It must be concluded therefore that donor endothelial cells can present alloantigen to recipient T cells. It is interesting to note that there is a species difference between rodents and humans. Rodents do not constitutively express MHC class II antigens on their endothelial cells. This is an important species difference and may explain why it is easier to suppress transplant rejection in rodents than it is in humans. Thus in rodents the only MHC class II positive cells of donor origin in the graft are cells (dendritic cells and macrophages) of bone marrow origin. These cells have a finite life span and in addition they migrate from the graft to lodge in other organs. Thus after about two weeks the rodent allografted organ is bereft of donor MHC class II presenting cells. In contrast, in humans, endothelial expression of donor MHC class II antigens ensures persistence of donor class II for long periods (if not indefinitely) after transplantation.[48] It follows that understanding the signals that allow human endothelial cells to stimulate T cells may lead to new strategies of preventing rejection.

Second Signal Requirements of Endothelial Cells

We have questioned whether endothelial cells utilize the B7 pathway to stimulate T cells, and our results[44] and those of others[49] demonstrate that endothelial cells do not express B7 receptors. Thus EC from umbilical vein, aorta, coronary artery and heart microvascular endothelial cells do not bind CTLA-4-Ig as assessed by flow cytometry and do not bind monoclonal antibodies against CD80/CD86.[44,46,49] Endothelial/T cell interactions are not inhibited by doses of CTLA-4-Ig which inhibit splenocyte/T cell interactions.[44,49] There has been a single report that HUVEC and dermal microvascular EC express CD86 antigens.[50] This study used commercially derived EC and has not, as yet, been corroborated. However, EC/T cell interactions are inhibited by antibodies to LFA/3 (CD58, ref. 51) the ligand for which, found on all T cells, is CD2. It is interesting that the affinity of CD2 for the human ligand CD58 is much greater than its affinity for the

mouse ligand CD48.[52] The possibility that second signals other than B7 could be important in allograft rejection is supported by the evidence of skin allograft rejection in CD28 deficient mice,[53] blocking the receptor for LFA3 prolongs cardiac allograft survival in primates[54] and use of antibodies against ICAM-1 induces tolerance in a murine model of allograft rejection.[55] Indeed, it has been shown that adhesion molecules such as VCAM-1, ICAM-1 and ICAM-2 can act a costimulatory signals inducing T cell proliferation in vitro.[56,57] Human endothelial cells constitutively express ICAM-1 and ICAM-2 each of which can bind to T cell LFA-1(CD11aCD18). ICAM-1 is constitutively expressed at quite low levels but unlike ICAM-2 it can be rapidly and markedly upregulated by in response to IL-1, TNFα, LT, CD40 ligand or IFNγ.[58] VCAM-1 is not normally expressed on ECs but is upregulated by IL-1, TNFα, LT, CD40 ligand and IL-4. VCAM-1 binds to T cell VLA4 (CD49/CD29). Several groups have demonstrated upregulation on VCAM-1 on microvascular EC within rejecting cardiac allografts.[59,60] Cytokines also induce ECs to express selectins (both E and P)[61,62] as well as ligands for L-selectin[63] but there is no evidence that these molecules can act as costimulators for T cell activation. However, the possibility that extracellular matrix molecules (fibronectin, collagen, laminin) can act as second signals via interaction with β1 integrins on lymphocytes should also be considered.[64] There remains much to learn about the nature of the second signals presented by endothelial cells.

Other Molecules Necessary for Antigen Presentation

The actual structure recognized by T cells is a complex of the MHC molecule and a bound peptide. Class I molecules acquire their peptides predominantly from proteins degraded in the cytoplasm by the proteosome; cytoplasmic peptides are translocated into the endoplasmic reticulum and loaded on to nascent class I molecules by the action of transporter in antigen processing (TAP) proteins.[65] Although peptide loading has not been examined directly in ECs, it has been shown that ECs express TAP proteins[66] and that alloreactive CD8[+] CTL lines can specifically lyse EC implying that the TAP system is functional.[67,68] One of the effects of IFNγ on endothelial cells is to upregulate TAP functional activity.[69] Class II MHC molecules associate with a non-MHC coded protein, the invariant chain, until they reach an intracellular compartment that fuses with late endosomes/lysosomes. Here, the invariant chain is degraded and peptides derived from endocytosed proteins are loaded. The loading of peptides into class II molecules depends on an MHC encoded molecule called DM.[70,71] Endothelial cells have been shown to express invariant chain after IFNγ treatment[37] and we have recently shown that DM is expressed coordinately with DR and DP molecules in human endothelial cells (Moyes and Rose unpublished results). These results strongly suggest that EC must be able to process antigen and present it within MHC class I and class II molecules. It has been shown that human EC are able to process exogenous proteins to produce a peptide-class II MHC molecule complex recognized by antigen specific CD4[+] T cells.[72,73] These results imply that ECs must be able to form specific peptide complexes with IFNγ induced class II MHC molecules.

Endothelial Cells Compared to other Parenchymal Cells

Fibroblasts,[43] epithelial cells from lung[74] or kidney[75] and smooth muscle cells from blood vessels[41,76] can all be induced to express MHC class II antigens by in vitro treatment with IFNγ, achieving levels of expression similar to that found on IFNγ treated HUVEC. Nevertheless these cells do not cause proliferation of highly purified CD4[+] T cells.[43,74,75,41,76] A direct comparison between microvascular endothelial cells and alveolar and small airway epithelial cells derived from the same source (lungs) showed proliferation of T cells to endothelial but not epithelial cells.[74] Similarly a direct comparison between endothelial cells or smooth muscle cells from saphenous vein[76] or

coronary artery[46] showed only the endothelial cells caused cell proliferation. The smooth muscle cells failed to stimulate T cell proliferation. They did not anergize the T cells. Some have reported exposure of CD4[+] T cells to thyroid and renal epithelial cells energizing both the CD45RO and CD45RA population of T cells.[75] The same authors have reported that HUVEC anergise the CD45RA[+] population,[77] but such a phenomenon has never been reported by any other group.[47] The reason why non-EC parenchymal cells fail to stimulate T cells is not known; but it might relate to density of expression of accessory molecules. The accessory molecules of endothelial cells (LFA-3 and possibly ICAM-1) are not as strongly expressed on lung epithelial cells compared to lung microvascular endothelial cells.[74]

Consequences of Endothelial Activation of T Cells

It view of the lack of involvement of CD28 stimulation, it is not surprising that EC stimulation of T cells produces less IL-2 than stimulation by professional APC. Studies[78,79] have shown some IL-2 is produced from HUVEC/T cell cultures, but far less than is produced from culturing porcine aortic EC with human T cells.[78] Porcine aortic EC express B7 receptors and therefore stimulate via CD28.[78,79] The response of human T cells to allogeneic EC is however highly sensitive to cyclosporine (Batten and Rose in preparation), and the sensitivity is reversed by addition of B7 transfectants to culture wells, a result which accords entirely with the reported cyclosporine resistance of CD28 mediated stimulation of T cells.[80] In terms of numbers of T cells which are activated by endothelial cells, far fewer are activated than by professional APC. Thus limiting dilution analysis, investigating numbers of CD4[+] T cells stimulated to produce IL-2 showed monocytes stimulate 1/232 T cells, whereas HUVEC treated with IFNγ to express MHC class II stimulated 1/1801 T cells.[81] The relatively small number of CD4[+] T cells which respond to allogeneic HUVEC is explained by the fact that only T cells with a memory phenotype (CD45RO[+]) respond to endothelial cells.[45] Thus numbers of circulating CD45RO[+] B7 independent T cells will be relatively small.

It is well established that in vitro activation of T cells by allogeneic professional APC results in a vigorous proliferative response accompanied by production of substantial amounts of IL-2 and IFNγ and small amounts of IL-4. As a result of a primary proliferative response T cells are generated with a memory phenotype that give a different cytokine profile on restimulation (more IL-2 and IFNγ). In addition, cytotoxic T cells are generated with specificity for class I MHC antigens present on the original stimulator cells. Pober[82] has reported that endothelial cell stimulation of allogeneic T cells results in IL-2, IL-4 and IFNγ production, but upon secondary stimulation there is no evidence for a maturation of the T cell response: the second proliferative response is no different from the first one in terms of cytokines produced. The model used by Pober consisted of comparing endothelial cells and monocytes as accessory cells in the presence of the polyclonal activator PHA. The results obtained using this model may not be the same as directly investigating endothelial stimulation of T cells and more work needs to be done in this area. That a virally induced murine endothelial cell line activates cloned Th2 cells and not cloned Th1 cells[83] also suggests endothelial cells may activate T cells to produce a different cytokine profile than produced by professional APC. It has been suggested that the deficient antigen presenting ability of endothelial cells leads to an aberrant immune response when T cells are present in vessel walls[84] and that cytokines are produced which result in intimal thickening as opposed to an immune response resulting in tissue destruction (as happens in acute rejection of the heart).

Role of Endothelial Cells in Chronic Rejection

Presentation of Antigens

It is likely that endothelial cells are involved in chronic rejection via a number of different mechanisms. This review has focussed on class II/CD4[+] T cell interactions, but it is clear from rodent models that MHC class II is not essential for chronic rejection. Thus severe vasculopathy occurs in the coronary arteries of mice transplanted with hearts from MHC class I or non-MHC mismatched donors (ref. 85 and see Chapter 7) and vasculopathy can occur (albeit to a lesser degree) in CD4[+] T cell deficient recipients.[86] There are thus a number of ways that the immune system of the recipient could be presented with molecules from allografted vessels:

1. Direct presentation of alloantigens by endothelial cells
2. Indirect presentation of donor HLA antigens leading to an alloantibody response
3. Indirect presentation of non-HLA or immunodominant minor antigens
4. Recognition of allogeneic endothelium by recipient CD8[+] T cells

Rodent and human endothelial cells constitutively express MHC class I antigens. Interestingly, it appears that very little MHC class II is induced on mouse arterial and capillary endothelial cells even during chronic rejection.[85] Thus rodents may be providing us with a model of rejection in the absence of endothelial class II. The expression of CD40 on normal EC and induced expression of CD40L on endothelial cells in CAD[87] suggests EC may be able to stimulate B cells and monocytes by a T independent mechanism. Functional CD40 has been demonstrated on EC by their ability to induce B7.2 on a B cell line.[87] Indeed it has been shown that anti-CD40L antibody inhibits allograft rejection[88] and not only by inhibiting CD80 expression on recipient monocytes.

The mechanisms listed above have not included direct recognition of HLA antigens by donor dendritic cells. The reason being, it is assumed such cells have left the graft early (i.e., after several weeks) after transplantation. Recent evidence has demonstrated the importance of indirect recognition in cardiac transplant patients with chronic rejection.[89] Thus Hornick et al detected a very low frequency of donor reactive helper and cytotoxic T cells in patients with chronic rejection suggesting they were hyporeactive for the direct pathway.[89] The same group then went on to demonstrate high frequencies of T cells in the indirect pathway in patients with chronic rejection.[90]

Mechanisims of Damage

The paradox is that whereas T cell damage to the myocardium is limited and controlled by immunosuppressive drugs, initial damage to the coronary endothelium, in some patients, progresses to TxCAD. It must be remembered that the mainstay immunosuppressive drug, cyclosporine, acts by inhibiting early events in T cell activation, namely transcription of IL-2.[91] T cell activation leads to maturation of a number of different effector pathways (Fig. 3.2) depending on the release of different cytokines. Thus production of cytokines IL-4 and IL-5 will lead to antibody production. Some studies report that production of these cytokines is less sensitive to cyclosporine than IL-2 production,[92] and proliferating B cells are known to be less sensitive to cyclosporine.[93] It is clear therefore that chronic rejection is driven by mechanisms which are relatively resistant to CsA; such mechanisms may include the T-independent mechanisms described above. One of the effector mechanisms which is little effected by immunosuppression is antibody production. Despite the heavy immunosuppression received by patients after solid organ transplantation, the majority make a vigorous antibody response against the allografted organ (reviewed in ref. 94).

It is our hypothesis that a sustained antibody response against HLA and non-HLA antigens on donor endothelial cells is one of the factors which leads to TxCAD. The most common way of detecting antibodies formed after transplantation is a complement-dependent cytotoxicity test against a panel of HLA typed leukocytes (termed Panel Reactive Antibodies or PRA test) or donor cells (termed a donor specific response). Many clinical studies have reported

an association between antibody producers and development of chronic rejection.[95] Thus Suciu-Foca et al reported a 90% 4 year actuarial survival in patients who had not made antibody following cardiac transplantation versus 38% 4 year survival in the antibody producers.[96,97] These authors looked for anti-HLA antibodies, but our own studies have shown a correlation between antiendothelial antibodies and chronic rejection.[98] This has recently been confirmed by Fredrich et al.[99] Using gel electrophoreses to separate endothelial peptides according to molecular weight followed by probing blots with patients' sera, we found that the majority of patients who had TxCAD had antibodies against endothelial peptides of 56/58kDa. We have subsequently confirmed this association in a separate study of new patients using both western blotting and flow cytometry (unpublished data). A similar association between antiendothelial antibodies, detected by flow cytometry and chronic rejection has been reported after renal transplantation.[100,101] Since this test[98] detected antibodies against unrelated HUVEC, it is clear that donor specific HLA antigens could not be involved. Use of SDS gel electrophoresis and amino acid sequencing revealed that the most immunogeneic endothelial peptide (at 56/58 kDa) was the intermediate filament vimentin and other immunoreactive peptides were identified as triose phosphate isomerase and glucose regulating protein—in all, 40 different proteins were identified which reacted with patients IgM.[102] Vimentin is the intermediate filament characteristic but not restricted to endothelial cells and fibroblasts. Whereas smooth muscle cells predominantly express desmin as their intermediate filament, they co-express desmin and vimentin when migrating or proliferating. Vimentin is diffusely expressed in the intima and media of normal and diseased coronary arteries. Our working hypothesis is that antibodies to vimentin reflect disease activity in the coronary arteries. We have developed an ELISA method to detect antivimentin antibodies, which will be used to sequentially follow patients.[103] The outstanding questions are how vimentin, a cytosolic protein is exposed to the immune system and whether and how the antibodies are damaging.

It is highly likely that endothelial cells are damaged early after transplantation (possibly by nonimmunological factors such as ischemia/reperfusion injury) and some vimentin is released into the circulation to be bound by host B cells. Use of the MHCPEP database of MHC binding peptides has revealed a sequence homology between epitopes of vimentin and class II presented peptides, these being an HLA-DRα peptide and as heat shock protein peptide (HSP65). Our hypothesis is that host T cells will recognize vimentin fragments, presented indirectly by host B cells and other APC as a consequence of cross reactivity with MHC class II presented peptides (such as DRα). The hypothesis assumes that vimentin is not normally exposed to the immune system and it will therefore be recognized as an autoantigen. Such cross reactions between DRα and infectious agents/normal components of tissues have been suggested as a mechanism for a number of autoimmune diseases.[104] It is likely that damaged endothelial cells are a source of many peptides which will be presented indirectly to recipient T cells. It is also possible that CD40 ligand expressing EC directly stimulate B cells to make antibody.

A number of experimental models support an important role for antibody in pathogenesis of TxCAD (see ref. 105 and Chapter 7 for review). Thus Russell and colleagues reported that transplantation of B10A hearts into B10.BR recipients (across a MHC class I mismatch) resulted in severe coronary lesions and the recipients made complement-dependent cytotoxic antibody against donor MHC class I antigens. That antibody could induce arterial lesions was shown by use of severe combined immunodeficient mice as recipients of B10.BR hearts; arterial lesions were induced in these hearts by injections of anti-B10.BR anti-serum.[106] Cardiac vasculopathy was significantly reduced and modified in Ig deficient allogeneic recipients.[107] Recently, elegant experiments have used transplantation of carotid arteries from mu-rine B10.A(2R) donors into mutant MHC mismatched recipients with various genetic defects of their immune system.[86] The neointimal thickening characteristic of TxCAD was dependent on presence of CD4+ T cells, antibody formation and macrophages. Cytotoxic CD8+ T cells and NK cells were not involved in the process.

One of the major drawbacks to ascribing a role for antibodies in pathogenesis is lack of understanding about the way antibodies interact with their cellular targets. The serum derived from our patients does not exhibit complement-dependent cytotoxicity against endothelial cells derived from HUVEC or aorta, nor does it exhibit antibody-dependent cellular cytotoxicity. This is not surprising with the exception of serum from patients with Kawasaki disease, where IgM antibodies are directly cytotoxic to endothelial cells in the presence of complement.[108] Antiendothelial antibodies have not been found to mediate complement mediated damage to endothelial cells. However, complement-mediated lysis is a severe and acute form of damage, usually associated with hyperacute rejection. It is more important to investigate whether antiendothelial antibodies can mediate more subtle forms of damage. Recently a number of reports have demonstrated that antibodies from patients with autoimmune disease[109] or transplant patients[110] can upregulate adhesion molecules on endothelial cells. Ligation of monomorphic regions of MHC class I molecules on endothelial cells results in tyrosine phosphorylation (Smith and Rose, in preparation), endothelial cell proliferation and upregulation of fibroblast growth factor receptor.[111] We believe the information that antibodies can activate endothelial cells is very promising and should be explored as a mechanism whereby antibodies could damage endothelial cells in both autoimmune disease and chronic rejection after solid organ transplantation.

References

1. Steinmuller D. Which T cells mediate allograft rejection? Transplantation 1985; 40:229-233.

2. Ungureanu-Longrois D, Balligand JL, Simmons WW et al. Induction of nitric oxide synthase activity by cytokines in ventricular myocytes is necessary but not sufficient to decrease contractile responses to beta-adrenergic agonists. Circulation Research 1995; 77:494-502.

3. Matsumori A, Yamada T, Suzuki H, Matoba Y et al. Increased circulating cytokines in patients with myocarditis and cardiomyopathy. British Heart J. 1994; 72:561-566.

4. Lewis NP, Tsao PS, Rickenbacher PR et al. Induction of nitric oxide synthase in the human cardiac allograft is associated with contractile dysfunction of the left ventricle. Circulation 1996; 93:720-729.

5. Daar AS, Fuggle SV, Fabre JW et al. The detailed distribution of MHC class II antigens in normal human organs. Transplantation 1984; 38:292-297.

6. Rose ML, Coles MI, Griffin RJ et al. Expression of class I and class II major histocompatibility antigens in normal and transplanted human heart. Transplantation 1986; 41:776-780.

7. Page CS, Rose ML, Piggott R et al. Heterogeneity of vascular endothelial cells. Am J Pathology 1992; 141:673-683.

8. Parker KE, Dalchau R, Fowler VJ et al. Stimulation of CD4+ T lymphocytes by allogeneic MHC peptides presented on autologous antigen presenting cells. Transplantation 1992; 53:918-24.

9. Terness P, Dufter C, Otto G et al. Allograft survival following immunization with membrane bound or soluble peptide MHC class I donor antigens: Factors relevant for the induction of rejection by indirect recognition. Transplant International 1996; 9:2-8.

10. Liu Z, Colovai AI, Tugulea S et al. Indirect recognition of donor HLA-DR peptides in organ allograft rejection. J Clin Invest 1996; 98:1150-1157.

11. Pober JS, Orosz CG, Rose ML et al. Can graft endothelial cells initiate a host antigraft immune response? Transplantation 1996; 61:343-349.

12. Janeway CAJ, Bottomly K. Signals and signs for lymphocyte responses. Cell 1994; 57:275-285.

13. June CH, Blustone JA, Nadler LM et al. The B7 and CD28 receptor families. Immunology Today 1994; 15:321-331.

14. Tan P, Anasetti C, Hansen JA et al. Induction of alloantigen specific hyporesponsiveness in human T lymphocytes by blocking interaction of CD28 with its natural ligand B7/BB1. J Exp Med 1993; 177:165-173.

15. Schwartz RH. A cell culture model for T lymphocyte clonal anergy. Science 1990; 248:1349-1356.

16. Liu Y, Janeway CAJ. Interferon γ plays a critical role in induced cell death of effector T cells: A possible third mechanism of self-tolerance. J Exp. Med 1990; 172:1735-.

17. Linsley PS, Brady W, Urnes M et al. CTLA-4 is a second receptor for the B cell activation antigen B7. J Exp Med 1991; 147: 561-569.

18. Turka LA, Linsley PS, Lin H et al T cell activation by the CD28 ligand B7 is required for cardiac allograft rejection in vivo. Proc Natl Acad Sci USA 1992; 89:11102-11105.

19. Lin H, Bolling SF, Linsley PS et al. Long term acceptance of major histocompatibility complex mismatched cardiac allografts induced by CTLA4-Ig plus donor specific transfusions. J Exp Med 1993; 178: 1801-1806.

20. Pearson TC, Alexandre DZ, Winn KJ et al. Transplantation tolerance induced by CTLA4-Ig. Transplantation 1994; 57: 1701-1706.

21. Lenschow D, Zeng Y, Thistlethwaite J et al. Long term survival of xenogeneic pancreatic islet grafts induced by CTLA4-Ig. Science 1992; 257:789-792.

22. Fuggle SV. MHC antigen expression in vascularized organ allografts: Clinical correlations and significance. Transplantation Reviews 1989; 3:81-102.

23. So SKS, Platt JL, Ascher NL et al. Increased expression of class I major histocompatibility complex antigens on hepatocytes in rejecting human liver allografts. Transplantation 1987; 43:79-84.

24. Fuggle SV, Errasti P, Daar AS et al. Localisation of major histocompatibility complex (HLA-ABC and DR) antigens in 46 kidneys: Differences in HLA-DR staining of tubules among kidneys. Transplantation 1983; 35:385-390.

25. Milton AD, Spencer SC, Fabre JW. Detailed analysis and demonstration of differences in the kinetics of induction of class I and class II major histocompatibility complex antigens in rejecting cardiac and kidney allografts in the rat. Transplantation 1986; 41:499-508.

26. Arbustini E, Grasso M, Diegoli M et al. Expression of tumor necrosis factor in human acute cardiac rejection. An immunohistochemical and immunoblotting study. American J Pathology 1991; 139:709-715.

27. Cunningham DA, Dunn M J, Yacoub MJ et al. Local production of cytokines in the human cardiac allograft. Transplantation 1994; 57:1333-1337.

28. Gao SJ, Schroeder JS, Alderman EL et al. Prevalence of accelerated coronary artery disease in heart transplant survivors. Comparison of cyclosporine and azathioprine regimens. Circulation 1989; 8:III100-105.

29. Johnson JA, Kobashigawa JA. Quantitative analysis of transplant coronary artery disease with use of intracoronary ultrasound. J Heart & Lung Transplantation 1995; 14:S198-202.

30. Hosenpud JD, Shipley GD, Wagner CR. Cardiac allograft vasculopathy: Current concepts, recent developments, and future directions. J Heart & Lung Transplantation 1992; 11:9-23.

31. Ross R. The pathogenesis of atherosclerosis: A perspective for the 1990s. Nature 1993; 362:801-809.

32. Barbir M, Banner N, Thompson GR et al. Relationship of immunosuppression and serum lipids to the development of coronary arterial disease in the transplanted heart. Int J. Cardiology 1991; 32:51-56.

33. Barbir M, Kushwaha S, Hunt B et al. Lipoprotein (a) and accelerated coronary artery disease in cardiac transplant recipients. Lancet 1992; 340:1500-1502.

34. Collins T, Lapierre LA, Fiers W et al. Recombinant human tumor necrosis factor increases mRNA levels and surface expression of HLA-A,B antigens in vascular endothelial cells and dermal fibroblasts in vitro. Proc Natl Acad Sci USA 1986; 83:446-450.

35. Lapierre LA, Fiers W, Pober JS. Three distinct classes of regulatory cytokines control endothelial cell MHC antigen expression: Interactions with immune (γ) interferon differentiate the effects of tumor necrosis factor and lymphotoxin form those of leukocyte (α) and fibroblast (β) interferons. J Exp Med 1988; 167:794-804.

36. Batten P, Yacoub MH, Rose ML. Effect of human cytokines (IFN-γ, TNF-α, IL-1β, IL-4) on porcine endothelial cells: Induction of MHC and adhesion molecules and functional significance of these changes. Immunol 1996; 87:127-133.

37. Collins T, Korman AJ, Wake CT et al. Immune interferon activates multiple class II major histocompatibility complex genes and the associated invariant chain gene in human endothelial cells and dermal fibroblasts. Proc Natl Acad Sci USA 1984; 81:4917-4921.

38. Panek RB, Lee Y-J, Benveniste EN. TGF-β suppression of IFN-γ induced class II MHC gene expression does not involve inhibition of phosphorylation of JAK1, JAK2 or signal transducers and activators of transcription, or modification of IFN-γ enhanced factor X expression. J Immunol 1995; 154:610-619.

39. Nanden D, Reiner NE. TGF-β attenuates the class II transactivator and reveals an accessory pathway of IFN-γ action. J Immunol 1997; 158:1095-1101.

40. McDouall RM, Batten P, McCormack A et al. MHC class II expression on human microvascular endothelial cells: Exquisite sensitivity to IFNγ and NK cells. Transplantation 1997; 64:1175-1180.

41. McDouall RM, Yacoub MH, Rose ML. Isolation, culture and characterisation of MHC class II positive microvascular endothelial cells from the human heart. Microvascular Research 1996; 51:137-152.

42. Savage COS, Hughes CCW, McIntyre BW et al. Human CD4+ T cells proliferate to HLA-DR+ allogeneic vascular endothelium: Identification of accessory interactions. Transplantation 1993; 56:128-.

43. Page CS, Holloway N, Smith H et al. Alloproliferative responses of purified CD4+ and CD8+ T cell subsets to human vascular endothelial cells in the absence of contaminating accessory cells. Transplantation 1994; 57:1628.

44. Page CS, Thompson C, Yacoub MH et al. Human endothelial cell stimulation of allogeneic T cells via a CTLA-4 independent pathway. Transplant Immunology 1994; 2:342-347.

45. Epperson DE, Pober JS. Antigen presenting function of human endothelial cells, direct activation of resting CD8+ T cells. J Immunol 1994; 153:5042-.

46. McDouall RM, Page CS, Hafizi H et al. Alloproliferation of purified CD4+ T cells to adult human heart endothelial cells and study of second signal requirements. Immunology 1996; 89:220-226.

47. Savage COS, Brooks CJ. Human vascular endothelial cells do not induce anergy in allogeneic CD4+ T cells unless costimulation is prevented. Transplantation 1995; 60: 734-740.

48. Rose ML, Navarette C, Yacoub MH et al. Persistence of donor specific class II antigens in allografted human heart two years after transplantation. Human Immunol 1988; 23:179-182.

49. Murray AG, Khodadoust MM, Pober JS et al. Porcine aortic endothelial cells activate human T cells: Direct presentation of MHC antigens and costimulation by ligands for human CD2 and CD28. Immunity 1994; 1:57-63.

50. Seino K, Azuma M, Bashuda H et al. CD86 (B70/B7-2) on endothelial cells co-stimulates allogeneic CD4+ T cells. Int Immunol 1995; 7:1331-1337.

51. Savage COS, Hughes CCW, Pepinsky RS et al. Endothelial cell lymphocyte function associated antigen-3 and an unidentified ligand act in concert to provide costimulation to human peripheral blood CD4+ T cells. Cell Immunol 1991; 137:150-163.

52. Arulanandam AR, Moingeon P, Concino MF et al. A soluble multimeric recombinant CD2 protein identifies CD48 as a low affinity

ligand for human CD2: Divergence of CD2 ligands during evolution of humans and mice. J Exp Med 1993; 177:1439-1450.

53. Kawai K, Shahinian A, Mak TW et al. Skin allograft rejection in CD28-deficient mice. Transplantation 1996; 61:352-355.

54. Kaplon AG, Hochman PS, Michler RE et al. Short course single agent therapy with an LFA-3-IgG1 fusion protein prolongs primate cardiac allograft survival. Transplantation 1996; 61:356-363.

55. Isobe IM, Suzuki J, Yamazaki S et al. Assessment of tolerance induction to cardiac allograft by anti-ICAM-1 and anti-LFA-1 monoclonal antibodies. J Heart & Lung Transplantation 1997; 16:1149-1156.

56. VanSeventer GA, Shimizy Y, Horgan KJ et al. The LFA-1 ligand ICAM-1 provides a costimulatory signal for T cell receptor mediated activation of resting T cells. J Immunol 1990; 144:4579-4586.

57. Damle N, Aruffo A. Vascular cell adhesion molecule induces T cell antigen receptor dependent activation by CD4+ T lymphocytes. Proc Natl Acad Sci 1991; 88:6403-.

58. Karman K, Hughes CCW, Schechner J et al. CD40 on human endothelial cells: Inducibility by cytokines and functional regulation of adhesion molecule expression. Proc Natl Acad Sci 1995; 92:4342-.

59. Taylor PM, Rose ML, Yacoub MH et al. Induction of vascular adhesion molecules during rejection of human cardiac allografts. Transplantation 1992; 54:451-457.

60. Briscoe DM, Schoen FJ, Rice GE et al. Induced expression of endothelial-leukocyte adhesion molecules in human cardiac allografts. Transplantation 1991; 51:537-539.

61. Pober JS, Bevilacqua MP, Mendrick DL et al. Two distinct monokines, interleukin 1 and tumor necrosis factor each independently induce biosynthesis and transient expression of the same antigen on the surface of cultured endothelial cells. J Immunol 1986; 136:1680-.

62. Luscinskas FW, Ding H, Lichtman AH. P-selectin and vascular adhesion molecule 1 mediate rolling and arrest, respectively, of CD4+ T lymphocytes on tumor necrosis factor α activated vascular endothelium under flow. J Exp Med 1995; 181:1179-.

63. Spertini O, Luscinskas FW, Kansas GS et al. Leukocyte adhesion molecule -1 (LAM-1, L-selectin) interacts with an inducible endothelial cell ligand to support leukocyte adhesion. J Immunol 1991; 147:2565-.

64. Yamada A, Nikaido T, Nojima Y et al. Activation of human CD4+ T lymphocytes. Interaction of fibronectin with VLA-5 receptor on CD4 cells induces the AP-1 transcription factor. J Immunology 1991; 146:53-56.

65. Germain RH, Margulies DH. The biochemistry and cell biology of antigen processing and presentation. Annual Revs Immunol 1993; 11:403-.

66. Epperson DE, Arnold D, Spies T et al. Cytokines increase transporter in antigen processing -1 (TAP-1) expression more rapidly than HLA class I expression in endothelial cells. J Immunol 1992; 149:3297.

67. Clayberger C, Uyehara T, Hardy B et al. Target specificity and cell surface structures involved in the human cytolytic T cell response to endothelial cells. J Immunol 1985; 135:12.

68. Pardi R, Bender JR. Signal requirements for the generation of CD4+ and CD8+ T cell responses to human allogeneic microvascular endothelium. Circ Res 1991; 69:1269-.

69. Ayalon O, Hughes EA, Creswell P et al. Induction of transporter associated with antigen processing by interferon γ confers endothelial cell cytoprotection against natural killer-mediated lysis. Proc Natl Acad Sci 1998; 95:2435-2440.

70. Fling SP, Arp B, Pious D. HLA-DMA and DMB genes are both required for the MHC class II /peptide complex formation in antigen presenting cells. Nature 1994; 368:554.

71. Ceman S, Rudersdorf RA, Peteran JM et al. DMA and DMB are the only genes in the class II region of the human MHC needed for class II associated antigen processing. J Immunol 1995; 154:2545.

72. Vora M, Yssel H, de Vries JE et al. Antigen presentation by human dermal microvascular endothelial cells. Immunoregulatory effects of IFNγ and IL-10. J Immunol 1994; 154:573-.

73. Savage COS, Brooks CJ, Harcourt GC et al. Human vascular endothelial cells process and present autoantigen to human T cell lines. Int Immunol 1995; 7:471-479.

74. Cunningham AC, Zhang J-G, Moy JV et al. A comparison of the antigen-presenting capabilities of class II MHC expressing human lung epithelial and endothelial cells. Immunol 1997; 91:458-463.

75. Marelli-Berg FM, Weetman A, Frasca L et al. Antigen presentation by epithelial cells induces anergic immunoregulatory CD45RO+ T cells and deletion of CD45RA+ T cells. J Immunol 1997; 159:5853-5861.

76. Murray AG, Libby P, Pober JS. Human vascular smooth muscle cells poorly stimulate and actively inhibit allogeneic CD4+ T cell proliferation in vitro. J Immunol 1995; 154:151-161.

77. Marelli-Berg FM, Hargreaves REG, Carmichael P et al. Major histocompatibility complex class II expressing endothelial cells induce

allospecific nonresponsiveness in naive T cells. J Exp Med 1996; 183: 1603-1612.

78. Bravery CA, Batten P, Yacoub MH et al. Direct recognition of SLA and HLA-like class II antigens on porcine endothelium by human T cells results in T cell activation and release of Interleukin-2. Transplantation 1995; 60:1024-1033.

79. Hughes CCW, Pober JS. Costimulation of peripheral blood T cell activation by human endothelial cells. Enhanced IL-2 transcription correlates with increased c-fos synthesis and increased fos content of AP-1. J Immunol 1993; 150:3148-3160.

80. June CH, Ledbetter JA, Gillespie MM et al. T cell proliferation involving the CD28 pathway is associated with cyclosporine resistant interleukin-2 gene expression. Mol Cell Biol 1987; 7:4472-.

81. Adams PW, Lee SH, Waldman WJ et al. Alloantigenecity of human endothelial cells. 1. Frequency and phenotype of human T helper lymphocytes that can react to allogeneic endothelial cells. J Immunol 1992; 148:3753-3760.

82. Ma W, Pober JS. Human endothelial cells effectively costimulate cytokine production by, but not differentiation of, naïve CD4+ T cells. J Immunol 1998; 161:2158-2167.

83. St. Louis JD, Lederer JA, Lichtman AH. Costimulator deficient antigen presentation by an endothelial cell line induces a nonproliferative T cell activation response without anergy. J Exp Med 1993; 178: 1597-1605.

84. Pober JS, Ma W, Biedermann B et al. Vascular cells have limited capacities to activate and differentiate T cells: Implications for transplant vascular sclerosis. Transplant Immunol 1997; 5:251-255.

85. Russel PS, Chase CM, Winn HJ et al. Coronary atherosclerosis in transplanted mouse hearts. 1. Time course and immunogenetics and immunopathological considerations. Am J Pathol 1994; 144:260-274.

86. Shi C, Lee WS, He Q et al. Immunologic basis of transplant associated arteriosclerosis. Proc Natl Acad Sci USA 1996; 93:4051-4056.

87. Mach F, Schonbeck U, Sukhova GK et al. Functional CD40 ligand is expressed on human vascular endothelial cells, smooth muscle cells, and macrophages: Implications for CD40-CD40 ligand signalling in atherosclerosis. Proc Natl Acad Sci USA 1997; 94:1931-1936.

88. Hancock WW, Sayegh MH, Zheng XG et al. Costimulatory function and expression of CD40 ligand, CD80, and CD86 in vascularized murine cardiac allograft rejection. Proc Natl Acad Sci USA 1996; 93: 13967-13972.

89. Hornick PI, Mason PD, Yacoub MH et al. Assessment of the contribution that direct allorecognition makes to the progression of chronic cardiac transplant rejection in humans. Circulation 1998; 97:1257-1263.

90. Hornick PI, Mason PD, Lombardi G et al. Significant frequencies of T cells with indirect antidonor specificity in heart graft recipients with chronic rejection. Circulation in press.

91. Clipstone NA, Crabtree GR. Identification of calcineurin as a key signalling enzyme in T lymphocyte activation. Nature 1992; 357:695-697.

92. Han CW, Imamura M, Hashino S et al. Differential effects of the immunosuppressants cyclosporin A, FK506, and KM2210 on cytokine expression. Bone Marrow Transplantation 1995; 15:733-739.

93. Morris RE. Rapamycin: FK506's fraternal twin or distant cousin? Immunology Today 1991; 12:137-140.

94. Rose ML. Antibody mediated rejection following cardiac transplantation. Transplantation Reviews 1993; 7:140-152.

95. Rose EA, Smith CR, Petrossian GA et al. Humoral immune responses after cardiac transplantation: Correlation with fatal rejection and graft atherosclerosis. Surgery 1989; 106:203-8.

96. Suciu-Foca N, Reed E, Marboe C et al. The role of anti-HLA antibodies in heart transplantation. Transplantation 1991; 51: 716-724.

97. Reed EF, Hong B, Ho E et al. Monitoring of soluble HLA alloantigens and anti-HLA antibodies identifies heart allograft recipients at risk of transplant associated coronary artery disease. Transplantation 1996; 61: 566-572.

98. Dunn MJ, Crisp SJ, Rose ML et al. Antiendothelial antibodies and coronary artery disease after cardiac transplantation. Lancet 1992; 339:1566-70.

99. Fredrich R, Toyoda M, Czer LS et al. The clinical significance of antibodies to human vascular endothelial cells after cardiac transplantation. Transplantation 1999; 67:385-91.

100. Al Hussein KA, Talbot D, Proud D et al. The clinical significance of posttransplantation non-HLA antibodies in renal transplantation. Transplant Int 1995; 8:214-220.

101. Ferry BL, Welsh KI, Dunn MJ et al. Anticell surface endothelial antibodies in sera from cardiac and kidney transplant recipients: Association with chronic rejection. Transplant Immunol 1997; 5:17-24.

102. Wheeler CH, Collins A, Dunn MJ et al. Characterisation of endothelial antigens associated with transplant associated coronary

artery disease. J Heart & Lung Transplantation 1995; 14:S188-97.

103. Jurcevic S, Dunn MJ, Crisp S et al. A new enzyme linked immunosorbent assay to measure antiendothelial antibodies after cardiac transplantation demonstrates greater inhibition of antibody formation by tacrolimus compared with cyclosporine. Transplantation 1998; 65:1197-1202.

104. Baum H, Davies H, Peakman M. Molecular mimicry in the MHC: Hidden clues to autoimmunity? Immunology Today 1996; 17:64-70.

105. Rose ML Role of antibody and indirect antigen presentation in transplant associated coronary artery disease. J Heart & Lung Transplantation 1996; 15:342-9.

106. Russell PS, Chase CM, Winn HJ et al. Coronary atherosclerosis in transplanted mouse hearts. II. Importance of humoral immunity. J Immunology 1994; 152:5134-41.

107. Russell PS, Chase CM, Colvin RB. Alloantibody and T cell mediated immunity in the pathogenesis of transplant arteriosclerosis: Lack of progression to sclerotic lesions in B cell-deficient mice. Transplantation 1997; 64:1531-6.

108. Leung DY, Collins MT, Lapierre LA et al. Immunoglobulin M antibodies present in the acute phase of Kawasaki syndrome lyse cultured vascular endothelial cells stimulated by gamma interferon. J Clinical Investigation 1986; 77:1428-1435.

109. Carvalho D, Savage COS, Black CM et al. IgG antiendothelial cell autoantibodies from scleroderma patients induce leukocyte adhesion to human vascular endothelial cells in vitro. J Clin Investigation 1996; 97:1-9.

110. Pidwell DW, Heller MJ, Gabler D et al. In vitro stimulation of human endothelial cells by sera from a subpopulation of high percentage panel reactive antibody patients. Transplantation 1995; 60:563-569.

111. Harris PE, Bian H, Reed EF. Induction of high affinity growth factor receptor and proliferation in human endothelial cells by anti-HLA antibodies: A possible mechanism for transplant atherosclerosis. J Immunol 1997; 159:5697-5704.

112. Rose ML. Endothelial cells as antigen-presenting cells: Role in human transplant rejection. Cell Mol Life Sci 1998; 54:965-978.

CHAPTER 4

Role of Hemostasis, Anticoagulation, Fibrinolysis and Endothelial Activation

Carlos A. Labarrere and David R. Nelson

The success or failure of any transplanted organ is directly associated with the recipient's response to the allografts and with changes within the allografts that promote such a response. Although it has been recognized that antibodies and complement as well as cellular infiltrates are key elements associated with the success or failure of the allografts, it has also been recognized that grafts lacking immunologic responses have similar degrees of success or failure.[1]

The survival of any allograft depends upon the maintenance of a patent microvasculature, and this is particularly evident in cardiac transplantation. Indeed, the long-term survival of cardiac allografts depends largely on the development of vascular disease, a particular type of atherosclerotic lesion defined as transplant-associated coronary artery disease (CAD). Transplant-associated CAD is of significant clinical importance since the angiographically detectable incidence of this disease increases approximately 10% a year and is 50% at 5 years.[2] Much higher incidences of the disease are reported using intravascular ultrasound.[3] Transplant-associated CAD involves the entire arterial vasculature from epicardial to endomyocardial arteries, and it is characterized by a diffuse and concentric narrowing of the vessels.[4] The compromise of the entire arterial tree and the additional involvement of the venous site of the microvasculature[5] suggest a panvascular disease.

Clinical interest in transplant-associated CAD is stimulated by several observations. First, there is a lack of correlation with known atherogenic risk factors.[4,6-10] Second, the disease always involves the entire vasculature, and in a significant number of allografts develops rapidly, often achieving angiographically evident epicardial disease within one year.[4,8-11] Third, some allografts do not develop the disease, even after 5 or 10 years, and these allograft recipients do not differ in traditional risk factors from allograft recipients that develop the disease. Fourth, cellular infiltrates are infrequently found within the myocardium in recipients that develop the disease,[1,12] and if these infiltrates are present, they are predominantly focal, whereas the vascular disease is always diffuse. Fifth, despite the inclusion of different new therapeutic approaches directed to impede cellular infiltrates, this approach has proven to diminish the number of histologically diagnosed rejection episodes, but does not impact the incidence of transplant-associated CAD.[13,14]

Our approach to the study of transplant-associated CAD was initiated by previous observations of increasing plasma concentrations of coagulation factors such as coagulation factor VII and fibrinogen in patients that subsequently died of cardiovascular disease, predominantly ischemic heart disease.[15,16] Although several studies have shown the relationship between plasma abnormalities in coagulation factors and ischemic heart dis-

Transplant-Associated Coronary Artery Vasculopathy, edited by Marlene L. Rose.
©2001 Eurekah.com.

ease,[15-19] there is no indication as to whether plasma abnormalities are represented in vascular tissues within the heart.

Using immunohistochemical techniques, we evaluated whether components of the hemostatic, fibrinolytic and anticoagulation pathways in endomyocardial biopsies were related to the development of transplant-associated CAD. In addition, markers of endothelial activation were evaluated and compared with the vascular findings. Results of these investigations have demonstrated that the vascular control of hemostasis and fibrinolysis is altered in grafts that subsequently develop transplant-associated CAD, and that grafts with this vascular dysfunction show early activation of arterial and arteriolar endothelial cells. These findings expand the immunopathological focus on transplant-associated CAD to include other systems besides the immunological system.

Patients and Methods

Patients

A total of 121 consecutive adult cardiac allograft recipients, who received transplants between 1988 and 1995 and were followed through 1996, were studied at Methodist Hospital of Indiana, Clarian Health (Methodist, Indiana University and Riley Hospitals). All recipients survived the first three months after transplantation, had pretransplantation and serial endomyocardial biopsy specimens for light microscopy and immunohistochemical studies and angiographic or histopathological evaluations of coronary arteries. Coronary angiography was performed at 1 year following transplantation and yearly thereafter, and recipients averaged (mean ± SE) 3.2 ± 0.2 angiograms. Evaluation of the coronary arteries in those recipients who died before the first coronary angiogram (n=7) was performed by histopathological examination of the allograft.[4] All recipients received triple drug immunosuppression with prednisone at an initial dose of 1 mg per kilogram of body weight per day, with the dose tapered to 0.1 mg per kilogram per day; azathioprine at

a dose of 1.5-2.0 mg per kilogram per day; and cyclosporine at an initial dose of 7-10 mg per kilogram per day, with the dose tapered to 3-5 mg per kilogram per day. Major episodes of rejection (grades 3 and 4)[20] were treated by augmenting the immunosuppressive therapy with high-dose corticosteroids and by using rabbit antithymocyte globulin or mouse monoclonal antibody OKT3 to human lymphocytes.

Functional classification was determined with the use of New York Heart Association criteria,[21] and the ejection fractions were measured by radionuclide ventriculography. Recipients with functional classifications of class III and IV and decreasing ejection fractions were classified as being unstable clinically, and recipients with clinical function scores of I and II and nondecreasing ejection fractions were classified as being clinically stable. Graft failure was defined as death due to cardiac allograft dysfunction or need for a second transplant.

Donor hearts were perfused with Stanford cardioplegia solution, and the mean ± SE ischemic time for donor hearts was 138.0 ± 5.0 min. Endomyocardial biopsies were obtained by right cardiac catheterization at 7-10 days; every 2 weeks during the first 2 months, and at 3, 4.5, 6, 9, and 12 months postoperatively; and every 6 months thereafter. The mean ± SE of biopsy specimens per patient obtained during the first three months posttransplant was 5.5 ± 0.1. A control biopsy from the right ventricle was obtained before transplantation from all donor hearts. Cellular infiltrates were graded according to the International Society for Heart Transplantation.[20] Cytomegalovirus infection was defined as a fourfold rise in antibody, cytomegalovirus inclusion bodies, or positive cultures.[22] Recipients with evidence of cytomegalovirus infection were treated with ganciclovir.

Criteria for Diagnosis of CAD

Diagnosis of CAD was based on evidence of coronary artery irregularities and/or any decrease in luminal diameter,[23,24] whether in proximal or distal branches. Annual arteriograms were compared with identical projections in serial studies and evaluated by side-by-side serial comparisons to detect development and progression of disease. These comparisons allowed the recognition of focal atherosclerotic-type lesions, identification of mid-to-distal concentric narrowing, and distal tapering or abrupt closure of vessels characteristic of transplant-associated CAD.[23,24] Side-by-side comparisons allowed the detection of minimal changes that otherwise could remain undetected. The presence or absence of disease and progression of lesions were determined by a consensus of two experienced angiographers. Coronary arteries were examined histopathologically[4] in those recipients who died before their first annual angiogram was performed, and the presence of intimal thickening exceeding the thickness of the media or the presence of plaques was considered as representing CAD.[25]

Antibodies and Control Experiments

Primary antibodies used were monoclonals to fibrin (Mab 350, American Diagnostica, Greenwich, CT), tissue plasminogen activator (ESP 4, American Diagnostica), HLA-DR (L-243, Becton Dickinson, San Jose, CA), intercellular adhesion molecule-1 or ICAM-1 (LB-2, Becton Dickinson), antithrombin (2333, American Diagnostica), antithrombin heparin binding site (2331, American Diagnostica), venous endothelium (PAL-E, Sera-Lab, Sussex, UK) and smooth-muscle-specific α-actin (1A4, Sigma, St. Louis, MO) and rabbit polyclonal antibodies to antithrombin (A 296, Dako, Carpinteria, CA) and von Willebrand factor (A 082, Dako). Secondary antibodies were affinity-purified, fluorochrome-labeled $F(ab')_2$ to mouse immunoglobulins (Protos ImmunoResearch, San Francisco, CA), and fluoroblue-labeled goat antirabbit IgG (Biomeda, Foster City, CA). All experiments included antibody and conjugate controls and isotype-matched irrelevant antibodies. Irrelevant antibodies were IgG1 anti-cytomegalovirus early antigen (Chemicon,

Temecula, CA) and IgG2a anti-Epstein-Barr virus capsid antigen (Chemicon). The dilution for each antibody was two serial-dilutions before end-point titration.[26] Antibodies were ultracentrifuged 100,000xg at 4°C for 1 hour to remove complexes and aggregates, and monthly titrations and ultracentrifugations were done for each antibody.

Immunohistochemistry

Biopsies were snap frozen in liquid nitrogen and stored at -20°C. Cryostat sections (4 μm) were air dried overnight without chemical fixation. Antibody experiments and their evaluations were performed as described.[12,23,24,27] Tissue sections were washed 30 minutes in 0.01M pH 7.2 phosphate-buffered saline (PBS), incubated 15 min. in a moisture box with 25 μl of primary antibody, rinsed in PBS and washed 3 times for 10 minutes in a glass reservoir (7.5 x 10 x 12.5 cm) filled with PBS agitated with a magnetic stirrer. After the last wash, sections were incubated for 15 min. with 25 μl of fluorescein isothiocyanate (FITC), rhodamine isothiocyanate (RITC) or fluoroblue labeled antibody to the species-specific isotype of the first antibody. Double antibody experiments were done by using primary antibodies from different species and appropriately matched antispecies FITC, RITC or fluoroblue conjugates or fluorochrome labeled primary antibodies.[24,26] Triple antibody experiments were done by using primary antibodies from different species and appropriately matched antispecies RITC and fluoroblue conjugates and FITC labeled primary antibodies.[24,26] After the final wash, tissue sections were coverslipped with pH 8.0 PBS-buffered glycerol. Serial biopsies obtained during the first three months posttransplant (5.5 ± 0.1 biopsies/patient) for each recipient were graded independently for each one of the immunohistochemical variables studied and were considered for the evaluation of subsequent outcome (i.e., development and progression of CAD and allograft failure). The concomitant presence of fibrin within the microvasculature and myocardial cells in one or more of the serial biopsies obtained during the first three months posttransplant was considered as representing myocardial fibrin, whereas the lack of fibrin in all biopsies was considered as absence of myocardial fibrin. The presence of arterial/arteriolar endothelial ICAM-1/HLA-DR reactivity was considered abnormal since it is not found in normal hearts,[24,28] and the percentage of biopsies with such reactivity during the first three months posttransplant was calculated.[24] Since vascular antithrombin is found in arterial/arteriolar smooth muscle cells, arterial intima and venous endothelium, but not in capillaries, of normal donor hearts,[1,12] the persistence of such reactivity in serial biopsies obtained during the first three months posttransplant was considered normal. The persistent loss of vascular antithrombin in serial biopsies obtained during the first three months posttransplant was considered as absence of vascular antithrombin. The recovery of vascular antithrombin following initial loss of antithrombin immunoreactivity with concomitant development of novel capillary antithrombin binding during the first three months posttransplant was considered as capillary antithrombin reactivity. The presence of tissue plasminogen activator in all arteriolar smooth muscle cells of serial biopsies obtained during the first three months posttransplant was considered normal.[23] The persistent loss of tissue plasminogen activator from arteriolar smooth muscle cells in serial biopsies obtained during the first three months posttransplant was considered as depleted tissue plasminogen activator.[23]

Statistical Analysis

Analyses were performed to prospectively evaluate each of the four immunohistochemical variables studied. Initially, significant relationships between demographic/clinical-laboratory variables and immunohistochemical variables were assessed with Fisher's exact tests, Wilcoxon rank-sum tests, and Spearman correlations, where appropriate. Demographic/clinical-laboratory variables that were significantly related to the immunohistochemical variables were included in subsequent logistic regression (progression of CAD)

and Cox regression (interval to CAD and graft failure) analyses. Kaplan-Meier methods were used to evaluate time to CAD and graft failure. The correlation between serum troponin T and myocardial fibrin was performed with generalized linear models, and troponin T was also used as a time varying covariate in Cox regression to examine CAD. Lastly, comparisons between allografts that remained normal for the immunohistochemical variables studied during the first three months posttransplant and those showing immunohistochemical abnormalities during the first three months posttransplant were performed. Statistical significance was assessed by using two-tailed tests at the 0.05 level and calculated using PC-SAS version 6.12. Summary statistics are reported as mean ± standard error.

Myocardial Fibrin and Transplant-Associated CAD

Normal hearts do not contain fibrin, since endomyocardial biopsies obtained from normal donor hearts before transplantation studied with a monoclonal antibody to fibrin do not show any immunohistochemical evidence of fibrin deposits.[1,12,27,29-31] Following transplantation, the allografts that remained clinically stable did not contain deposits of fibrin.[1,12,27,29,30] However, those allografts that became clinically unstable and subsequently failed showed the deposition of fibrin not only within the microvasculature of the heart, but also within the cardiomyocytes,[27] suggesting that those cardiomyocytes were damaged or dead. To determine if the deposition of fibrin within the microvasculature and cardiomyocytes was associated with myocardial damage, we studied the relationship between immunohistochemically detected fibrin deposition within the myocardium and serum concentrations of cardiac troponin T, which is a contractile protein found specifically in myocardial cells.[32] Indeed, troponin T concentrations increase with higher degrees of myocardial damage,[33] and serum concentrations of cardiac troponin T have been used to diagnose the presence and extension of myocardial infarctions.[34] Our studies showed

Table 4.1. Relationship between serum troponin T (TnT) concentrations, myocardial fibrin and subsequent development of coronary artery disease (CAD) in heart transplant recipients

	TnT (ng/ml)*		
	< 0.10 (n = 43)	≥ 0.10 (n = 8)	p-value
Myocardial Fibrin (% Biopsies)	36.4	65.6	< 0.001
Time Free of CAD(Mo.)+	23.8 ± 2.8	7.1 ± 2.0	< 0.001

* Calculations were performed on a total of 103 serial serum samples (2.0 ± 0.2/patient) from 51 cardiac transplant recipients without initial CAD by angiography.
+ Mean ± SE

significant associations between the presence of fibrin within the endomyocardial biopsies and serum concentrations of cardiac troponin T (Table 4.1).[30]

We also asked if the presence of endomyocardial fibrin and myocardial cell damage was associated with allograft outcome. A significant association between the amount of fibrin deposition within the myocardium and development of angiographically detected transplant-associated CAD was found.[30] Also, allografts that showed increasing amounts of fibrin deposits failed more than allografts that did not have any deposition of fibrin.[27] Interestingly, allografts with increasing serum concentrations of cardiac troponin T which were associated with increasing fibrin deposition within endomyocardial biopsies developed transplant-associated CAD sooner than allografts with lower serum troponin T concentrations (Table 4.1),[35] suggesting that microvascular fibrin and myocardial cell damage preceded subsequent development of CAD in large epicardial arteries. In light of these observations, we investigated whether early deposition of myocardial fibrin following transplantation was associated with subsequent allograft outcome.

We found that the deposition of fibrin within the microvasculature and myocardial cells during the first three months following transplantation was a common event in human cardiac transplantation. Indeed, from

a total of 121 cardiac allografts, 71 (58.7%) allografts showed fibrin deposits within the first three months posttransplant (Fig. 4.1). The remaining 50 (41.3%) allografts remained without fibrin during the first three months posttransplant, showing similar immunohistochemical characteristics to normal donor hearts before transplantation (Fig. 4.1). Analysis of demographic and clinical-laboratory data for allografts with and without fibrin in the first three months posttransplant revealed significant differences for sex of recipients and clinical instability during follow-up (Table 4.2). However, analyses of these variables with development of transplant-associated CAD and/or graft failure indicated fibrin was the only significant risk factor (Table 4.3).

The presence of fibrin within the microvasculature of the heart and within cardiomyocytes during the first three months following transplantation was associated with the subsequent development of transplant-associated CAD (Table 4.4). Allografts without fibrin in the first three months following transplantation developed significantly less transplant-associated CAD than allografts that showed fibrin deposition. Allografts with myocardial fibrin developed the disease significantly sooner (p < 0.001) than allografts that remained without deposition of myocardial fibrin during the first three months following transplantation (Table 4.4).

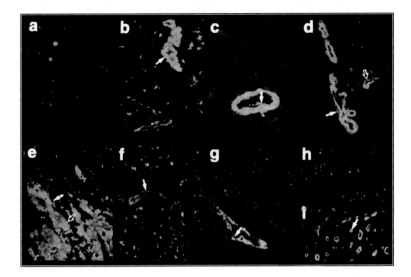

Fig. 4.1. Heart biopsies obtained 10 days following transplantation and reacted with monoclonal antibody to fibrin (a and e); monoclonal antibody to tissue plasminogen activator (b and f); double antibody technique with monoclonal antibodies to HLA-DR (red) and smooth-muscle-specific α-actin (green) to localize arteries and arterioles (c and g); and rabbit polyclonal antibody to antithrombin (d, h, and i). Note absence of fibrin (a), presence of tissue plasminogen activator in all arteriolar smooth muscle cells (arrow in b), absence of HLA-DR on arterial endothelium (arrow in c), and antithrombin in arterial smooth muscle cells and intima (arrow in d) and venous endothelium (open arrow in d) as found in normal hearts, in an allograft that had not developed coronary artery disease when last studied 5 years after transplantation. Note presence of fibrin within microvasculature (arrow in e) and myocardial cells (open arrow in e), depletion of arteriolar tissue plasminogen activator from arteriolar smooth muscle cells (arrow in f), presence of arterial endothelial HLA-DR (arrow in g), and absence of vascular antithrombin (h) in an allograft that developed coronary artery disease in the first posttransplant year. A group of allografts developed unusual capillary antithrombin reactivity during the first three months posttransplant (arrow in i), which is not found in normal hearts (d). Original magnification x 400.

Table 4.2. Correlation of demographic and clinical-laboratory data with myocardial fibrin during the first three months posttransplant

Characteristics*	Myocardial Fibrin		
	No fibrin (n = 50)	Fibrin (n = 71)	p-value
Sex of Recipient (% males)	80	61	0.03
Clinical Status (% unstable)	10	51	< 0.001

* Variables not found to be significant were age of donor and recipient, sex of donor, reason for transplantation, ischemic times, percentage of smokers, blood pressure, serum triglycerides and cholesterol, number of HLA-A, B, or DR mismatches, number of recipients with positive cell panel percent reactive antibodies or donor-specific cytotoxic antibodies, number of rejection episodes, incidence of cytomegalovirus infection or ejection fractions.

Table 4.3. Adjusted relative risk of coronary artery disease (CAD) and graft failure based on status of myocardial fibrin during the first three months posttransplantation*

	Interval to CAD		Interval to Graft Failure	
	RR (95% CI)	p-value	RR (95% CI)	p-value
Myocardial Fibrin	11.7 (5.4-25.5)	< 0.001	10.9 (2.5-47.0)	0.001
Sex of Recipient	0.9 (0.5-1.5)	0.87	0.7 (0.3-1.7)	0.40

* Values presented are multivariate relative risks (RRs) and 95% confidence intervals (CIs). The RRs are for myocardial fibrin and male recipients.

Table 4.4. Relationship between myocardial fibrin during the first three months posttransplant and transplant-associated coronary artery disease (CAD) or graft failure

	Myocardial Fibrin		
Outcome Variables	No Fibrin (n = 50)	Fibrin (n = 71)	p-value*
Coronary Artery Disease			
No CAD, No. (%)	42 (84.0)	19 (26.8)	< 0.001
CAD, No. (%)	8 (16.0)	52 (73.2)	
No Progression	5 (62.5)	24 (46.2)	< 0.001
Progression	3 (37.5)	28 (53.8)	
Time Free of CAD (Mo.)[+]	54.0 ± 2.2	25.9 ± 2.4	< 0.001
Graft Failure			
No Graft Failure, No. (%)	49 (98.0)	49 (69.0)	< 0.001
Graft Failure, No. (%)	1 (2.0)	22 (31.0)	
Interval to Graft Failure (Mo.)[+]	59.0 ± 1.7	44.6 ± 2.8	< 0.001

* Logistic regression (percent with CAD or graft failure and progression of CAD) and Cox regression (time free of CAD or interval to graft failure) after adjusting for demographic findings.
[+] Mean ± SE

Allografts with myocardial fibrin also showed more progression of transplant-associated CAD than allografts that remained without fibrin deposition during the first three months posttransplant.

Allografts with myocardial fibrin in the first three months following transplantation had significantly more graft failure than allografts without myocardial fibrin (Table 4.4). Only 2.0% of allografts without myocardial fibrin failed during follow-up compared with 31.0% of allografts with myocardial fibrin (p < 0.001). Failure also occurred significantly sooner (p < 0.001) in allografts with myocardial fibrin than those

without myocardial fibrin (Table 4.4). These data demonstrate that fibrin within the microvasculature and cardiomyocytes during the first three months posttransplant is associated with subsequent development and progression of transplant-associated CAD and with allograft failure.

The presence of microvascular fibrin as early as 7-10 days after transplantation (time at which the first biopsy is performed) in 54.9% (39 out of 71) of the allografts with myocardial fibrin during the first three months posttransplant suggests that endothelial damage occurs very early following transplantation. This is supported by reports of early microvascular damage as determined by endothelial activation and depletion of fibrinolytic and anticoagulant pathways in biopsies obtained from hearts that subsequently developed CAD or failed.[1,12,23,24,27,28,36] These microvascular changes are associated with increased endothelial cell thrombogenicity.[1,37] Mechanisms underlying these changes are not known, but allogeneic recognition and rejection reactions often are suggested,[14,38] and previous immunological data support this hypothesis.[14,38,39] However, vascular damage in allografts often occurs without the presence of cellular infiltrates,[1,12] and changes in immunosuppression have failed to affect the natural history of CAD in heart transplant recipients.[13,14,38]

A relevant observation in allograft vascular disease is that the initiating event for development of both CAD and graft failure occurs very soon following transplantation.[31] This is supported by the common finding of fibrin deposits within the first posttransplant month. Indeed, there is a significant association between myocardial fibrin in the first month posttransplant and subsequent development of CAD and graft failure in cardiac transplant recipients.[31] The concept of an early precipitating event also is supported by recent reports of arterial and arteriolar endothelial activation and depletion of arteriolar tissue plasminogen activator and vascular antithrombin within weeks of transplantation that associate significantly with the subsequent development of CAD and/or graft failure.[23,24,28,36]

Studies performed in transplanted hearts[23,24,28,36] and kidneys[40-42] suggest that the long-term allograft outcome may be determined by events occurring during the peritransplant period. Although the status of the coronary arteries cannot be monitored continuously to precisely establish the relationship between the deposition of myocardial fibrin and the development of CAD, and coronary angiography is less sensitive than intracoronary ultrasonography or histopathological examination of the coronary arteries for the identification of CAD,

our data suggest that deposition of myocardial fibrin is a very early occurring event which precedes the development of transplant-associated CAD.

A potential early event mediating vascular damage is the reaction of recipient antibodies with donor endothelial cells. Antibodies to histocompatibility antigens,[39] phospholipids[43] and solubilized endothelial antigens[44] have been described in cardiac allograft recipients and antibodies to solubilized endothelial antigens have been associated with CAD.[44,45] Since antibodies to endothelial cells promote thrombogenic responses resulting in fibrin deposition,[46] the role of these antibodies in the development of CAD is being explored. In this regard, it was recently found that identification of antibodies to solubilized endothelial antigens detect a subpopulation of recipients at risk of developing CAD.[47]

Whatever the cause(s) of microvascular damage early after transplantation, the result is ischemia and hypoxia, which is associated with endothelial cell damage,[48] up-regulation of histocompatibility antigens[49] and increased endothelial cell procoagulant activity.[50,51] Damage of myocardial cells is associated with ischemia and reperfusion injury,[52,53] and histological changes associated with ischemia identify recipients at risk of developing transplant-associated CAD.[52] Experimentally, less prolonged ischemic times significantly reduce development of vascular arteriosclerosis,[54] and attenuation of free radicals with a single intravenous injection of recombinant superoxide dismutase at the time of surgery significantly reduces chronic rejection in renal transplantation.[55] Although the cause(s) of early deposition of myocardial fibrin is(are) still unknown, our findings demonstrate that fibrin deposits significantly associate with subsequent CAD and/or graft failure. These data also suggest that perhaps fewer biopsies will be required to follow heart transplant recipients, particularly if the predictive value of an early biopsy could be amplified by similarly predictive blood determinations of cardiac troponin T[35] and/or antiendothelial cell antibodies.[47]

Vascular Antithrombin and Transplant-Associated CAD

The vasculature of the normal heart is thromboresistant, mainly because the antithrombin component of the heparan sulfate proteoglycan-antithrombin natural anticoagulant pathway is found within the microcirculation of normal donor hearts.[1,12,29,56-60] Antithrombin is normally present in arterial and arteriolar smooth muscle cells, arterial intima, and venous endothelium and normal hearts do not show any capillary antithrombin binding.[1,12,29,56-60] The localization of antithrombin in arterial and arteriolar smooth muscle cells, arterial intima and venous endothelium was confirmed by double and triple antibody experiments with antibodies to antithrombin, to smooth muscle-specific α-actin to identify smooth muscle cells and to von Willebrand factor to identify endothelial cells. The antithrombin molecule is bound to the plasma membrane of smooth muscle and endothelial cells by the heparan sulfate moiety of the heparan sulfate proteoglycan molecules.[61-63] Endothelial cells synthesize biologically active heparan sulfate proteoglycan molecules, and biochemical studies have shown that these heparan sulfate proteoglycan molecules contain the critical oligosaccharide domains that bind antithrombin.[62] Mediation of antithrombin binding by heparan sulfate has been demonstrated by biochemical, immunohistochemical and enzymatic experiments.[61,62,64-68] Indeed, our own immunohistochemical studies have shown that the heparin binding site of antithrombin in arterial and arteriolar smooth muscle cells as well as in venous endothelium is occupied.[36] Vascular antithrombin binding can be removed from smooth muscle cells and endothelium by incubation in excess heparin or heparan sulfate or by enzymatic hydrolysis with heparinase.[61] The heparan sulfate proteoglycan core protein that mediates vascular antithrombin binding is still unknown. Many types of cell surface heparan sulfate proteoglycans are capable of binding antithrombin.[69] Possible candidates are the phosphatidyl inositol-linked glypican or the members of the syndecan family of transmembrane proteoglycans.

The loss of the antithrombin component of the heparan sulfate proteoglycan-antithrombin natural anticoagulant pathway has been associated with unstable and failing allografts.[1,12,29,58] The loss of vascular antithrombin early after transplantation is associated with the subsequent development of transplant-associated CAD and failure of the allografts.[36,70] Indeed, the persistent loss of antithrombin binding from arterial and arteriolar smooth muscle cells, arterial intima and venous endothelium during the first three months following transplantation was significantly associated with the risk of future development of CAD. Allografts with loss of vascular antithrombin showed more CAD progression and developed the disease earlier than allografts that remained with normal vascular antithrombin reactivity. The loss of vascular antithrombin during the first three months posttransplant also was associated with the risk of subsequent allograft failure.[70] Allografts with early loss of vascular antithrombin failed more often and failed earlier than allografts that maintained a normal pattern of vascular antithrombin.

Since the early loss of vascular antithrombin from arterial and arteriolar smooth muscle cells, arterial intima and venous endothelium is associated with the outcome of human cardiac allografts, understanding of the mechanism(s) involved in the loss of vascular antithrombin binding is essential to introduce new therapies for maintaining the anticoagulant properties of the cardiac microvasculature.

One possible mechanism for the loss of vascular antithrombin could involve macrophage-mediated growth factor or cytokine release or heparinase release by T lymphocytes or macrophages,[71-74] but this seems improbable since no differences in the number of cellular rejection episodes between allografts with normal or negative vascular antithrombin were found.

The loss of vascular antithrombin binding could be associated with changes in the expression or availability of heparan sulfate proteoglycan molecules, such as glypican or syndecans. It has been recently demonstrated that thrombin accelerates endothelial cell shedding of syndecans 1 and 4.[75] Since much of the heparan sulfate at the cell surface is derived from syndecans,[76] increased shedding of these molecules by thrombin could be associated with less membrane-bound syndecan and decreased antithrombin binding. Other potential mechanisms for the loss of antithrombin involve alterations of the heparan sulfate proteoglycan core protein with or without changes in the length of the polysaccharide chains,[68,77] inadequate sulfation of heparan[78] or antibody-mediated complement activation,[79,80] which induces release of heparan sulfate proteoglycan molecules from endothelial cells.

The loss of vascular antithrombin occurs early after transplantation and always is associated with the deposition of microvascular and intramyocytic fibrin, suggesting a mechanism involving ischemia and reperfusion injury.[31] The recent finding that the use of antithrombin limits the damage caused by ischemia/reperfusion in an animal model suggests that thrombin participates in ischemia/reperfusion injury.[81] Interestingly, thrombin plays a critical role in reperfusion-induced leukocyte recruitment as well as increased microvascular permeability alterations.[81] Neutrophils can subsequently cleave heparan sulfate proteoglycan molecules from the cell surface through enzymatic degradation by neutrophil elastase or heparanase,[82] promoting the depletion of vascular heparan sulfate proteoglycans and subsequent intravascular thrombosis.

Although some cardiac allografts lose and never recover antithrombin binding and have a poor outcome, most allografts recover antithrombin and develop novel capillary antithrombin binding, which improves outcome.[70] Capillary antithrombin binding is mediated by heparan sulfate proteoglycan molecules, since it can be removed by incubation in excess heparin or heparan sulfate or by enzymatic hydrolysis with heparinase, but still it is not clear whether the heparan sulfate proteoglycan involved in the capillary

antithrombin binding is glypican or a member of the syndecan family of transmembrane proteoglycans.

Analysis of demographic and clinical-laboratory findings for cardiac allografts with normal, negative or capillary antithrombin binding (Fig. 4.1) during the first three months posttransplant revealed significant differences for sex of recipient, cardiomyopathy as a reason for transplantation, diastolic blood pressure, left ventricular ejection fraction and clinical status during follow-up (Table 4.5). However, only changes in vascular antithrombin associated significantly with development of CAD and/or cardiac allograft failure (Table 4.6).

Cardiac allografts that, following initial loss of vascular antithrombin, recover the antithrombin immunoreactivity and concomitantly showed unusual capillary antithrombin binding within the first three months posttransplant had a lower rate of CAD, less progression of the disease and developed the disease later than allografts that lost vascular antithrombin (Table 4.7). Graft failure also occurred less frequently and occurred later in allografts with capillary antithrombin than those allografts that lost vascular antithrombin (Table 4.7).

The capillary-sized vessels becoming antithrombin-reactive also react with antibodies to smooth muscle cell specific α-actin and PAL-E, which is characteristic of venous endothelium (Figs. 4.2 and 4.3).[60,83] These small α-actin-reactive, PAL-E reactive vessels are not identified in control heart biopsies obtained before transplantation. Interestingly, studies performed in animal models of myocardial ischemia have shown that the first vessels to appear as a response to gradual ischemia are vein-like structures with thin walls that show increased endothelial cell activation and become surrounded by vascular smooth muscle cells.[84] This sequence of events is remarkably similar to the findings previously described in human cardiac allografts showing microvascular and intramyocytic deposition of fibrin. Following the deposition of fibrin, capillary-sized vessels of cardiac allografts show increased expression

Table 4.5. Relationship of demographic and clinical-laboratory data with vascular antithrombin (AT) during the first three months posttransplant

| Characteristics* | Vascular AT | | | p-value |
	Normal AT (n = 37)	Capillary AT (n = 67)	Negative AT (n = 17)	
Sex of Recipient (% males)	84	60	76	0.03
Pretransplant Cardiomyopathy (%)	12	38	4	0.01
Diastolic Blood Pressure (mm Hg)+	92.3 ± 1.3	88.8 ± 1.1	92.8 ± 4.5	0.03
Ejection Fraction (%)+	60.6 ± 1.7	54.3 ± 1.4	51.2 ± 1.9	0.003
Clinical Status (% unstable)	5	39	76	< 0.001

* Variables not found to be significant were age of donor and recipient, sex of donor, coronary artery disease as a reason for transplantation, ischemic times, percentage of smokers, systolic blood pressure, serum triglycerides and cholesterol, number of HLA-A, B, or DR mismatches, number of recipients with positive cell panel percent reactive antibodies or donor-specific cytotoxic antibodies, number of rejection episodes or incidence of cytomegalovirus infection.
+ Mean ± SE

Table 4.6. Adjusted relative risk of coronary artery disease (CAD) and graft failure based on antithrombin (AT) status during the first three months posttransplant*

| | Interval to CAD | | Interval to Graft Failure | |
	RR* (95% CI)	p-value	RR* (95% CI)	p-value
Normal AT	1.00		1.00	
Capillary AT	4.76 (2.12-10.66)	< 0.001	9.13 (1.14-72.8)	0.04
Negative AT	7.63 (3.14-18.55)	< 0.001	20.32 (2.44-169.15)	0.005
Sex of Recipient	1.27 (0.63-2.52)	0.51	0.55 (0.19-1.60)	0.27
Cardiomyopathy	0.60 (0.31-1.17)	0.13	1.24 (0.45-3.39)	0.68
Diastolic Blood Pressure	1.00 (0.96-1.03)	0.87	0.98 (0.93-1.03)	0.35
Ejection Fraction	0.98 (0.95-1.01)	0.11	0.96 (0.92-1.01)	0.09

* Values presented are multivariate relative risks (RRs) and 95% confidence intervals (CIs). The RRs are for capillary AT, negative AT, male recipients, cardiomyopathy as reason for transplantation, 1 mm Hg increase in diastolic blood pressure, and 1% increase in left ventricular ejection fraction.

of endothelial activation markers and the presence of venous-specific antigens. These capillary-sized vessels become surrounded by smooth-muscle-specific α-actin positive cells and subsequently develop novel capillary antithrombin binding. All these capillary changes are suggestive of vascular growth and/or vascular remodeling. It was recently proposed that these changes could be the result of ischemia-induced production of vascular endothelial growth factor (VEGF).[85] Indeed, it has been clearly demonstrated that hypoxia induces VEGF expression in cultured cells,[86-89] and that ischemia induces VEGF expression in vivo.[85,90-93] These, and our previous findings showing increased VEGF expression in transplanted human hearts with deposition of fibrin and microinfarctions,[85]

Table 4.7. Relationship between vascular antithrombin (AT) during the first three months posttransplant and coronary artery disease (CAD) and graft failure

Outcome Variables	Vascular AT			p-value*
	Normal AT (n = 37)	Capillary AT (n = 67)	Negative AT (n = 17)	
Coronary Artery Disease				
No CAD, No. (%)	27 (73.0)	32 (47.8)	2 (11.8)	< 0.001
CAD, No. (%)	10 (27.0)	35 (52.2)	15 (88.2)	
No progression	6 (16.2)	19 (28.4)	4 (23.5)	< 0.001
Progression	4 (10.8)	16 (23.9)	11 (64.7)	
Time free of CAD (Mo.)+	53.7 ± 2.4	32.6 ± 3.0	20.1 ± 4.5	< 0.001
Graft Failure				
No graft failure, No. (%)	36 (97.3)	54 (80.6)	8 (47.1)	0.003
Graft failure, No. (%)	1 (2.7)	13 (19.4)	9 (52.9)	
Interval to graft failure (Mo.)+	68.2 ± NA	49.0 ± 2.7	35.9 ± 6.1	< 0.001

* Logistic regression (percent with CAD or graft failure and progression of CAD) and Cox regression (time free of CAD or interval to graft failure) after adjusting for demographic/clinical findings.
+ Mean ± SE
NA = Not Applicable

Fig. 4.2. Double antibody technique using antithrombin (purple) and smooth-muscle-specific α-actin (red) antibodies. Note the presence of smooth-muscle-specific α-actin reactive cells in capillaries that are antithrombin-reactive (arrows). Original magnification x 400.

suggest VEGF may play a key role in vascular growth and/or vascular remodeling in human cardiac allografts. The phenotypic changes culminating with novel capillary antithrombin binding could represent a wound healing process perhaps involving increased expression of heparan sulfate proteoglycan molecules. Syndecans were predominantly found to be associated with fibroblast-like cells and immature fetal endothelium, and it has been suggested that syndecan 4 may participate in fibroblast growth, fetal angiogenesis and wound healing.[94,95] These studies support the possibility of an increased antithrombin binding ability of new and/or remodeled endothelium.

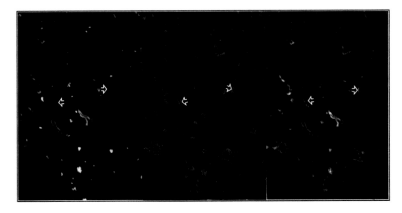

Fig. 4.3. Double-antibody technique using antithrombin (purple) and PAL-E endothelial antigen (red). Note that capillaries that are antithrombin-reactive show the presence of PAL-E endothelial antigen (arrows). Original magnification x 400.

Arterial and Arteriolar Endothelial Activation and Transplant-Associated CAD

The endothelial cell response to cytokines such as interleukin-1 and tumor necrosis factor, which is characterized by the synthesis of altered levels of proteins, is defined as endothelial activation, and activated endothelial cells then acquire new functional capacities.[96] Markers such as intercellular adhesion molecule-1 (ICAM-1) and class II major histocompatibility antigen HLA-DR are commonly used to detect endothelial activation.[97] The activation of endothelial cells is associated not only with the increased expression of those markers but also with an altered functional state of those cells. Indeed, the plasma membranes of resting (nonactivated) endothelial cells are thromboresistant, whereas the plasma membranes of activated endothelial cells are thrombogenic.[98]

Recent reports revealed that ICAM-1 and HLA-DR antigens normally are not expressed by arterial and arteriolar endothelium,[24,28,99,100] suggesting that arterial/arteriolar endothelium in vivo normally remains not activated. Contrarily, venous and capillary endothelial cells constitutively express ICAM-1 and HLA-DR antigens.[97,99,101] Following transplantation, the ICAM-1 and HLA-DR reactivity of venous and capillary endothelium was comparable with the reactivity observed before transplantation, but striking differences were found for the arterial and arteriolar endothelial ICAM-1 and HLA-DR reactivity.[24] Analyses of the immunohistochemical data have shown that both ICAM-1 and HLA-DR antigens were expressed concomitantly. Thus, when arterial and arteriolar endothelial cells became reactive, both ICAM-1 and HLA-DR antigens were expressed. Following cardiac transplantation, arterial and arteriolar endothelial cells remained negative for ICAM-1 and HLA-DR in a proportion of the allografts (35.5%), whereas most allografts (64.5%) became ICAM-1 and HLA-DR positive during the first three months posttransplant (Fig. 4.1). These allografts becoming arterial and arteriolar endothelial ICAM-1 and HLA-DR reactive showed a different proportion of biopsies ICAM-1 and HLA-DR positive during the first three months posttransplant. A mean ± SE of 53.6 ± 3.0% of all biopsy specimens per patient showed arterial and arteriolar endothelial ICAM-1 and HLA-DR during the first three months posttransplant in this population, and when arteries and arterioles showed endothelial ICAM-1 and HLA-DR reactivity, all arteries and arterioles in each biopsy section were reactive.[24] Immunohistochemical techniques using simultaneously two or three antibodies to

Table 4.8. Correlation of demographic and clinical-laboratory data with proportion of ICAM-1 and HLA-DR positive biopsies during the first three months posttransplant*

	Proportion (%) of Positive Biopsies			
	0	1-49	≥ 50	
Characteristics[++]	(n = 43)	(n = 28)	(n = 50)	p-value[†]
Sex of Recipient (% males)	86	75	52	.001
Sex of Donor (% males)	88	86	68	.04
Diastolic Blood Pressure (mm Hg)**	92.4 ± 1.2	89.9 ± 1.3	89.0 ± 1.4	.04
Clinical Status (% unstable)	12	50	44	.005

* ICAM-1 indicates intercellular adhesion molecule-1.

++ Variables not found to be significant were age of donor and recipient, reason for transplantation, ischemic times, percentage of smokers, systolic blood pressure, serum triglycerides and cholesterol, number of HLA-A, B, or DR mismatches, number of recipients with positive cell panel percent reactive antibodies or donor-specific cytotoxic antibodies, number of rejection episodes, incidence of cytomegalovirus infection or ejection fractions.

† Significance of the correlation between each characteristic with the continuous measure of proportion of ICAM-1 and HLA-DR positive biopsies.

** Mean ± SE

localize ICAM-1 or HLA-DR, vascular endothelium, and arterial or arteriolar smooth muscle cells have clearly demonstrated the activation of endothelial cells within arteries and arterioles.[24] Analyses of demographic and clinical-laboratory findings for cardiac allografts with different proportions of arterial and arteriolar endothelial ICAM-1 and HLA-DR positive biopsies during the first three months posttransplant revealed significant differences for sex of both donor and recipient, diastolic blood pressure and clinical instability during follow-up (Table 4.8). However, the association of these variables with development of transplant-associated CAD and/or graft failure was significant only for arterial and arteriolar ICAM-1 and HLA-DR (Table 4.9).

We found that the expression of arterial and arteriolar endothelial ICAM-1 and HLA-DR following cardiac transplantation is associated with the outcome of the allografts.[24] A significant relationship (p < 0.001) was described between the proportion of biopsies positive for arterial and arteriolar endothelial ICAM-1 and HLA-DR during the first three months posttransplant and the development

of transplant-associated CAD (Table 4.10). Most allografts (72.1%) without positive biopsies for arterial and arteriolar endothelial ICAM-1 and HLA-DR remained without developing CAD, whereas most allografts (64.0%) with 50% or more positive biopsies developed the disease during the outcome period. Progression of disease also occurred significantly more frequently in allografts with increasing proportion of biopsies positive for arterial and arteriolar endothelial ICAM-1 and HLA-DR during the first three months posttransplant than in allografts without reactivity (Table 4.10). Allografts without arterial and arteriolar endothelial ICAM-1 and HLA-DR reactivity remained free of CAD significantly longer than those allografts having an increasing proportion of positive biopsies (Table 4.10). The time interval for the appearance of the first positive biopsy following transplantation was also related (p < 0.001) to the development, progression and timing of CAD.[24]

The proportion of biopsies positive for arterial and arteriolar endothelial ICAM-1 and HLA-DR during the first three months posttransplant was associated with allograft

Table 4.9. Adjusted relative risks of coronary artery disease (CAD) and graft failure based on proportion of biopsies positive for arterial and arteriolar endothelial ICAM-1 and HLA-DR during the first three months posttransplant*

	Interval to CAD		Interval to Graft Failure	
	RR (95% CI)	p-value	RR (95% CI)	p-value
ICAM-1/DR	13.6 (5.9-31.4)	< 0.001	11.4 (3.1-42.2)	< 0.001
Sex of Recipient	0.8 (0.5-1.5)	0.58	0.6 (0.2-1.6)	0.34
Sex of Donor	0.7 (0.4-1.4)	0.29	.08 (0.3-2.3)	0.73
Diastolic Blood Pressure	1.0 (1.0-1.0)	0.93	1.0 (0.9-1.0)	0.61

* Values presented are multivariate relative risks (RRs) and 95% confidence intervals (CIs). The RRs are for each 1% increase in arterial/arteriolar endothelial ICAM-1/HLA-DR reactivity, for male donors and recipients and for 1 mm Hg increase in diastolic blood pressure.

Table 4.10. Relationship between proportion of biopsies positive for arterial and arteriolar ICAM-1 and HLA-DR during the first three months posttransplant and coronary artery disease (CAD) or graft failure*

	Proportion (%) of Biopsies with Arterial and Arteriolar ICAM-1 and HLA-DR			
Outcome Variables	0 (n = 43)	1-49 (n = 28)	≥ 50 (n = 50)	p-value[††]
<u>Coronary Artery Disease</u>				
No CAD, No. (%)	31(72.1)	12 (42.9)	18 (36.0)	< 0.001
CAD, No. (%)	12 (27.9)	16 (57.1)	32 (64.0)	
No Progression	7 (16.3)	8 (28.6)	14 (28.0)	< 0.001
Progression	5 (11.6)	8 (28.6)	18 (36.0)	
Time Free of CAD (Mo.)[†]	52.7 ± 2.5	37.9 ± 4.2	22.4 ± 2.9	< 0.001
<u>Graft Failure</u>				
No Graft Failure, No. (%)	40 (93.0)	21 (75.0)	37 (74.0)	< 0.001
Graft Failure, No. (%)	3 (7.0)	7 (25.0)	13 (26.0)	
Interval to Graft Failure (Mo.)[†]	57.0 ± 2.0	49.6 ± 4.1	45.3 ± 3.5	< 0.001

* ICAM-1 indicates intercellular adhesion molecule-1.
[††] Logistic regression (percent with CAD or graft failure and progression of CAD) and Cox regression (time free of CAD or interval to graft failure) after adjusting for demographic/clinical findings.
[†] Mean ± SE

failure.[24] Indeed, 26.0% of the allografts with 50% or more positive biopsies failed when compared with only 7.0% of the allografts without arterial and arteriolar endothelial ICAM-1 and HLA-DR reactivity (Table 4.10). The time interval to allograft failure was significantly shorter in allografts with increasing proportion of posi-

tive biopsies when compared with allografts that did not show any reactivity during the first three months following transplantation (Table 4.10).

The lack of expression of endothelial activation markers on endothelium of normal arteries and arterioles and the association between early expression of these markers following cardiac transplantation and subsequent development and progression of CAD or graft failure suggest that the expression of activation markers on arterial and arteriolar endothelium could be used as a risk factor in cardiac transplantation.

The importance of endothelial activation in the development of transplant-associated CAD has been clearly established in humans as well as in animal models. Indeed, our own experience has shown that the early expression of endothelial adhesion molecules such as ICAM-1 and major histocompatibility class II antigens such as HLA-DR on arterial and arteriolar endothelium is clearly associated with subsequent CAD and allograft failure.[24] The expression of ICAM-1 and other adhesion molecules such as P-selectin and VCAM-1 on arterial and arteriolar endothelium is associated with the development of transplant-associated CAD in animal models.[102-106] The inhibition of the development of transplant-associated CAD using monoclonal antibodies to ICAM-1 and to lymphocyte function-associated antigen-1 supports the idea of the involvement of adhesion molecules in the development of transplant-associated CAD.[106] This also seems to be the case for spontaneous CAD since it has been recently demonstrated that increased soluble concentrations of ICAM-1 were found in healthy individuals who later suffer acute myocardial infarctions probably due to development of spontaneous atherosclerotic disease.[107] All these data provide a clinical basis for the use of antiadhesion therapies to prevent transplant-associated CAD.

Several possibilities could be involved in the activation of arterial and arteriolar endothelium in cardiac allograft recipients. Cytokine release by activated immune cells could activate endothelium,[37] but this seems to be improbable since the number of rejection episodes early after transplantation was not related to the development of CAD.[24] The lack of association between cellular rejection and subsequent development of transplant-associated CAD has been demonstrated in humans and in animal models of cardiac transplantation.[29,108,109] The endothelial activation of the whole arterial and arteriolar vasculature also suggests that the mechanism of endothelial activation is diffuse more than focal.

Although bacterial products like endotoxin or the presence of viruses could activate endothelium and be related with subsequent development of CAD, the presence of bacterial or viral infections was not significantly increased in allografts that subsequently developed the disease when compared with those that did not develop CAD.[24]

The presence of antibodies to endothelial antigens has been suggested as a possible trigger for the development of CAD in cardiac allograft recipients.[39,44,45] The possibility of antibodies to endothelial cells being involved in the development of transplant-associated CAD also has been studied in an animal model of cardiac transplantation.[109,110] Indeed, T- and B-cell deficient mice that were injected with antibodies to transplantation antigens of the donor strain developed CAD, and those that did not receive those antibodies did not develop the disease. Interestingly, cardiac allografts in normal mice that did not develop antibodies to the donor strain did not develop CAD whereas mice that developed donor-specific antibodies developed the disease.[110]

The findings that (a) human cardiac allografts without arterial and arteriolar endothelial activation during the first three months posttransplant have very low risk of developing CAD, (b) allografts with a high proportion of biopsies positive for arterial and arteriolar endothelial ICAM-1 and HLA-DR during the first three months posttransplant have a significantly higher risk of developing the disease, and (c) allografts with positive arterial and arteriolar endothelial activation markers during the first month following transplantation have a significantly higher risk of developing CAD, show that early activation of arterial and arteriolar endothelium is a predictor of the development of transplant-associated CAD, which is associated with a poor prognosis.[11]

Arteriolar Tissue Plasminogen Activator and Transplant-Associated CAD

Tissue plasminogen activator (tPA), the principal activator of the fibrinolytic pathway, is normally found in arteriolar smooth muscle cells of endomyocardial biopsies from donor hearts obtained before transplantation.[23,27,111,112] Arteriolar smooth muscle cell tPA reactivity can become depleted following transplantation,[23,27,111] and tPA depletion is associated with the deposition of fibrin within the microvasculature.[27] Interestingly, when arteriolar smooth muscle cells become depleted of tPA, the endothelium of arterioles, capillaries and venules becomes tPA-reactive, and this reactivity is associated with reactivity for plasminogen activator inhibitor-1 (PAI-1), and complexes of PAI-1 with tPA, indicating that the microvascular endothelial tPA probably is not functional.[23]

In light of the previous observations, we analyzed the relationship between arteriolar tPA reactivity and the subsequent outcome of cardiac allograft recipients. We found that depletion of arteriolar tPA within the first three months following transplantation was associated with the subsequent development of transplant-associated CAD.[23] Analyses of demographic and clinical-laboratory data for allografts with normal or depleted arteriolar tPA (Fig. 4.1) in the first three months posttransplant revealed significant differences for sex of donor and recipient, diastolic blood pressure, number of rejection episodes during the first three months posttransplant, ejection fraction, and clinical instability during follow-up (Table 4.11). However, only arteriolar tPA and ejection fraction were significantly associated with development of transplant-associated CAD and/or allograft failure (Table 4.12).

Allografts with depletion of arteriolar tPA during the first three months posttransplant developed significantly more (p < 0.001) CAD than allografts that remained with normal arteriolar smooth muscle cell tPA. Allografts with depleted arteriolar tPA developed the disease sooner (p < 0.001) and showed more disease progression (p < 0.001) than allografts that maintained normal arteriolar tPA during the first three months posttransplant (Table 4.13).

Allografts with depletion of arteriolar tPA failed significantly more than allografts maintaining a normal arteriolar smooth muscle cell tPA reactivity (Table 4.13).[23] Only 3.9% of allografts with normal arteriolar tPA during the first three months posttransplant subsequently failed, when compared with 30.0% of allografts with depleted arteriolar tPA reactivity (p < 0.001). Allografts with depleted arteriolar tPA also failed significantly sooner (p < 0.001) than allografts that maintained normal arteriolar tPA reactivity (Table 4.13).

Interestingly, all allografts with depletion of arteriolar smooth muscle cell tPA showed tPA reactivity on the endothelium of arterioles, capillaries and venules, and this tPA was complexed to its inhibitor, PAI-1.[23] However, allografts that maintained normal arteriolar tPA reactivity following transplantation consistently showed absence of tPA-PAI-1 complexes within the microvasculature of the heart.[23]

All these data demonstrate the importance of the early tPA status within the cardiac microvasculature for the subsequent outcome of cardiac transplant recipients. The depletion of arteriolar tPA from smooth muscle cells and the identification of tPA-PAI-1 complexes on microvascular endothelium within transplanted hearts during the first three months posttransplant were associated with subsequent development and progression of CAD and allograft failure.[23] It has been recently demonstrated that cardiac transplant recipients with transplant-associated CAD have significantly elevated plasma levels of tPA and PAI-1 antigens when compared to patients without transplant-associated CAD.[113] Levels of tPA and PAI-1 antigens increased as angiographic severity of transplant-associated CAD worsened.[113] These findings seem to be not limited to transplant-associated CAD. Indeed, increased concentrations of immunologically identifiable tPA were found in blood from patients with ath-

Table 4.11. Relationship of demographic and clinical-laboratory data with arteriolar tissue plasminogen activator (tPA) during the first three months posttransplant*

| | Arteriolar tPA | | |
| | Normal | Depleted | |
Characteristics	(n = 51)	(n = 70)	p-value
Sex of Recipient (% males)	83	59	0.006
Sex of Donor (% males)	88	72	0.03
Diastolic Blood Pressure (mmHg)[+]	92.4 ± 1.0	88.9 ± 1.1	0.02
Rejection Episodes			0.02
Grade 0 (no.)	22	18	
Grade 1 or 2 (no.)	25	35	
Grade 3 or 4 (no.)	5	16	
Ejection Fraction (%)[+]	58.6 ± 1.4	53.6 ± 1.4	0.01
Clinical Status (% unstable)	6	55	< 0.001

* Variables not found to be significant were age of donor and recipient, reason for transplantation, ischemic times, percentage of smokers, systolic blood pressure, serum triglycerides and cholesterol, number of HLA-A, B, or DR mismatches, number of recipients with positive cell panel percent reactive antibodies or donor-specific cytotoxic antibodies, or incidence of cytomegalovirus infection.
[+] Mean ± SE.

Table 4.12. Adjusted relative risks of coronary artery disease (CAD) and graft failure based on status of arteriolar tissue plasminogen activator (tPA) during the first three months posttransplant*

| | Interval to CAD | | Interval to Graft Failure | |
	RR (95% CI)	p-value	RR (95% CI)	p-value
Arteriolar tPA	5.5 (2.8-10.8)	< 0.001	6.5 (1.8-23.5)	0.004
Sex of Recipient	1.0 (0.5-1.8)	0.92	0.6 (0.2-1.6)	0.31
Sex of Donor	0.7 (0.4-1.4)	0.37	0.8 (0.3-2.5)	0.74
Diastolic Blood Pressure	1.0 (1.0-1.0)	0.71	1.0 (0.9-1.0)	0.52
Rejection Episodes	1.1 (0.8-1.6)	0.59	1.1 (0.6-2.0)	0.86
Ejection Fraction	1.0 (1.0-1.0)	0.04	1.0 (0.9-1.0)	< 0.05

* Values presented are multivariate relative risks (RRs) and 95% confidence intervals (CIs). The RRs are for depleted arteriolar tPA, male donors and recipients, 1 mmHg increase in diastolic blood pressure, one grade increase in rejection severity, and 1% increase in left ventricular ejection fraction.

erosclerotic disease,[114] and these studies have revealed that circulating tPA in these patients was complexed with its inhibitor, PAI-1.[114,115]

The reason for the depletion of tPA from arteriolar smooth muscle cells in allografts prone to develop transplant-associated CAD is still not known. A common finding associ-

ated with depletion of arteriolar smooth muscle cell tPA is the presence of fibrin within the cardiac microvasculature and the identification of tPA-PAI-1 complexes on the microvascular endothelium.[23,27] These findings suggest that the microvasculature of these cardiac allografts favors thrombogenicity

Table 4.13. Relationship between arteriolar tissue plasminogen activator (tPA) during the first three months posttransplant and transplant-associated coronary artery disease (CAD) or graft failure

Outcome Variables	Arteriolar tPA		p-value*
	Normal (n = 51)	Depleted (n = 70)	
Coronary Artery Disease			
No CAD, No. (%)	37 (72.5)	24 (34.3)	< 0.001
CAD, No. (%)	14 (27.5)	46 (65.7)	
No Progression	9 (64.3)	20 (43.5)	< 0.001
Progression	5 (35.7)	26 (56.5)	
Time Free of CAD (Mo.)+	52.5 ± 2.1	26.2 ± 2.7	< 0.001
Graft Failure			
No Graft Failure, No. (%)	49 (96.1)	49 (70.0)	0.003
Graft Failure, No. (%)	2 (3.9)	21 (30.0)	
Interval to Graft Failure (Mo.)+	59.2 ± 1.1	43.8 ± 2.9	0.004

*Logistic regression (percent with CAD or graft failure and progression of CAD) and Cox regression (time free of CAD or interval to graft failure) after adjusting for demographic/clinical-laboratory findings.
+ Mean ± SE

instead of thromboresistance.[1] A metabolic consequence of these changes is the generation of thrombin, and thrombin generation results in deposition of fibrin[19,116] and secretion of tPA from endothelial and smooth muscle cells.[23,117] Interestingly, depletion of arteriolar tPA in cardiac allograft recipients is associated with identification of tPA complexed to PAI-1,[23,27] suggesting that endothelial secretion of these complexes could account for the increased concentration of tPA and PAI-1 found in patients developing transplant-associated CAD.[113]

It has also been recently found that cardiac transplant recipients with a 2/2 PAI-1 genotype are at increased risk for the development of transplant-associated CAD.[118] This suggests that the genotype-specific overexpression of PAI-1 could lead to depletion of tPA in arteriolar smooth muscle cells with the subsequent development of transplant-associated CAD. Allografts showing this genotype could be predisposed to maintaining a persistent depletion of arteriolar tPA, which is found in allograft recipients who subsequently develop transplant-associated CAD, and the presence of a 2/2 PAI-1 genotype in the donor heart could be another risk factor for developing the disease.

Relationship of all Immuno-histochemical Measures with Transplant-Associated CAD

The most striking observation from all these investigations is the association of components of the hemostasis (i.e., myocardial fibrin), anticoagulation (i.e., loss of vascular antithrombin), fibrinolysis (i.e., depletion of arteriolar tissue plasminogen activator) and markers of endothelial activation (i.e., arterial and arteriolar endothelial ICAM-1 and HLA-DR antigens) with the subsequent development of transplant-associated CAD or allograft failure. In light of these observations, we analyzed the relationship between all these immunohistochemical measures combined and the subsequent outcome of cardiac allograft recipients. We found that the presence of at least one abnormal measure (pres-

Table 4.14. Relationship of demographic and clinical-laboratory data with all immunohistochemical measures studied during the first three months posttransplant*

| Characteristics | Immunohistochemical Measures[tt] | | p-value* |
	All Normal (n = 29)	One or More Abnormal (n = 92)	
Sex of Recipient (% males)	90	63	< 0.001
Ejection Fraction (%)[+]	62 ± 1.8	53.8 ± 1.1	< 0.001
Clinical Status (% unstable)	0	45	< 0.001

* Variables not found to be significant were age of donor and recipient, sex of donor, reason for transplantation, ischemic times, percentage of smokers, blood pressure, serum triglycerides and cholesterol, number of HLA-A, B, or DR mismatches, number of recipients with positive cell panel percent reactive antibodies or donor-specific cytotoxic antibodies, number of rejection episodes or incidence of cytomegalovirus infection.

[tt] Immunohistochemical measures evaluated were myocardial fibrin, vascular antithrombin, arterial/arteriolar endothelial ICAM-1 / HLA-DR (expressed as percentage of reactive biopsies), and arteriolar tissue plasminogen activator, as described in Methods.

[+] Mean ± SE.

ence of myocardial fibrin, loss of vascular antithrombin, depletion of arteriolar tissue plasminogen activator or presence of arterial/arteriolar endothelial ICAM-1/HLA-DR) during the first three months posttransplant was associated with the subsequent development of transplant-associated CAD or allograft failure. Analysis of demographic and clinical-laboratory data for allografts with all normal or at least one abnormal immunohistochemical measure (Fig. 4.1) during the first three months posttransplant revealed significant differences for sex of recipient, ejection fraction, and clinical instability during follow-up (Table 4.14). However, only the presence of at least one abnormal immunohistochemical measure and ejection fraction were significantly associated with development of transplant-associated CAD, and only one or more abnormal immunohistochemical measures associated with subsequent graft failure (Table 4.15).

Allografts with at least one abnormal immunohistochemical measure during the first three months posttransplant developed significantly more CAD (p < 0.001), developed the disease sooner (p < 0.001), and showed more disease progression (p = 0.004) than allografts with all normal immunohis-

tochemical measures during the first three months posttransplant (Table 4.16). Allografts with at least one abnormal immunohistochemical measure during the first three months posttransplant failed more (p < 0.001) and failed sooner (p < 0.001) than allografts maintaining all normal immunohistochemical measures during the first three months posttransplant (Table 4.16).

Understanding of these microvascular events and of the sequences in which they occur will cast light on the pathophysiological mechanism(s) involved in the development of transplant-associated CAD and perhaps spontaneous CAD. The absence of fibrin within the microvasculature of the normal heart and the association of early myocardial fibrin deposits within the first month posttransplant with the subsequent outcome of cardiac allografts suggests that understanding why and how fibrin is deposited may be a key to understanding why some allografts fail and others succeed. The deposition of fibrin within the microvasculature and myocardial cells could be explained as an ischemia and reperfusion phenomenon occurring during the peritransplant period.[31] Ischemia and hypoxia can make available tissue factor on the endothelial plasma

Table 4.15. Adjusted relative risks of coronary artery disease (CAD) and graft failure based on status of all immunohistochemical measures studied during the first three months posttransplant*

	Interval to CAD		Interval to Graft Failure	
	RR (95% CI)	p-value	RR (95% CI)	p-value
One or More Abnormal Immunohistochemical Measure[††]	6.4 (2.5-16.7)	< 0.001	> 100	< 0.001
Ejection Fraction	0.97 (0.95-1.00)	0.03	0.96 (0.92-1.00)	0.052
Sex of Recipient	0.85 (0.49-1.5)	0.58	0.58 (0.23-1.46)	0.25

* Values presented are multivariate relative risks (RRs) and 95% confidence intervals (CIs). The RRs are for one or more abnormal immunohistochemical measure, 1% increase in left ventricular ejection fraction and male recipients.
[††] Immunohistochemical measures evaluated were myocardial fibrin, vascular antithrombin, arterial/arteriolar endothelial ICAM-1 / HLA-DR (expressed as percentage of reactive biopsies), and arteriolar tissue plasminogen activator, as described in Methods.

Table 4.16. Relationship of immunohistochemical measures studied during the first three months posttransplant and transplant-associated coronary artery disease (CAD) or graft failure

	Immunohistochemical Measures*		
	All Normal	One or More Abnormal	
Outcome Variables	(n = 29)	(n = 92)	p-value[+]
Coronary Artery Disease			
No CAD, No. (%)	24 (82.8)	37 (40.2)	< 0.001
CAD, No. (%)	5 (17.2)	55 (59.8)	
No Progression	3 (60.0)	26 (47.3)	0.004
Progression	2 (40.0)	29 (52.7)	
Time Free of CAD (Mo.)[††]	56.1 ± 2.4	31.7 ± 2.4	< 0.001
Graft Failure			
No Graft Failure, No. (%)	29 (100)	69 (75.0)	< 0.001
Graft Failure, No. (%)	0 (0)	23 (25.0)	
Interval to Graft Failure (Mo.)[††]	60.0 ± NA	47.4 ± 2.3	< 0.001

* Immunohistochemical measures evaluated were myocardial fibrin, vascular antithrombin, arterial/arteriolar endothelial ICAM-1/HLA-DR (expressed as percentage of reactive biopsies), and arteriolar tissue plasminogen activator, as described in Methods.
[+] Logistic regression (percent with CAD or graft failure and progression of CAD) and Cox regression (time free of CAD or interval to graft failure) after adjusting for demographic/clinical findings.
[††] Mean ± SE

membranes and the endothelial cells would become thrombogenic and not thrombo-resistant.[50,51] Myocardial cell damage following ischemia and reperfusion could allow also the deposition of fibrin within cardiomyocytes since cardiomyocytes are rich in tissue factor.[119,120] It has also been demonstrated that ischemia and hypoxia up-regulate histocompatibility antigens,[49] and then ischemia and hypoxia could promote endothelial activation which is commonly found in allografts having early deposition of myocardial fibrin following transplantation.[30] Another mechanism of endothelial cell activation involves antibody to endothelium or endothelial associated antigens which can activate endothelium cells.[45] The possibility of antibody-mediated endothelial activation being involved in the pathophysiology of transplant-associated CAD is a promising area of future research.

The deposition of fibrin within the microvasculature could be associated with a defect in the heparan sulfate proteoglycan-antithrombin natural anticoagulant pathway. A well-known pathophysiological process for the thrombin generation is the activation of factor VII through the availability of tissue factor on endothelial plasma membranes,[121] and the antithrombin component of the heparan sulfate proteoglycan-antithrombin natural anticoagulant pathway normally neutralizes enzymatically active thrombin generating inactive thrombin-antithrombin complexes.[59] Thus, if antithrombin is not present, thrombin will remain uninhibited and is free to convert fibrinogen to fibrin. The antithrombin component of the heparan sulfate proteoglycan-antithrombin pathway is absent from the vasculature of allografts that subsequently develop coronary artery disease.[1,36,60] The cause(s) of antithrombin loss is(are) not known, but the generation of thrombin could promote shedding of heparan sulfate proteoglycan molecules from the plasma membranes,[75] allowing further generation of thrombin within the microvasculature. Another consequence of the presence of enzymatically active thrombin within the cardiac microvasculature is related to the depletion of vascular tissue plasminogen activator.[117] It has been demonstrated that thrombin can induce the release of tissue plasminogen activator from endothelial and smooth muscle cells,[117] accounting for the depletion of arteriolar tissue plasminogen activator which precedes the development of coronary artery disease in heart transplant recipients.[23] The previously described relationship between arteriolar endothelial activation and depletion of arteriolar tissue plasminogen activator on the development of CAD[23,24] suggests a functional relationship between endothelium and smooth muscle cells on arteries and arterioles. The activated endothelium through the availability of tissue factor could promote thrombin generation and subsequent depletion of smooth muscle cell tissue plasminogen activator from the arterioles, which compromises fibrinolysis and allows persistent thrombin-mediated fibrin deposition within the microvasculature.[27]

The persistence of fibrin within the microvasculature results in the development of microinfarctions,[1,27] and areas of microinfarctions always develop in areas of myocardium with antithrombin-negative vessels and vessels with depleted tissue plasminogen activator.[27,58-60] Allografts that remain with these immunohistochemical characteristics have a poor prognosis, whereas allografts that, following the deposition of microvascular fibrin, subsequently recover vascular antithrombin and develop novel capillary antithrombin binding improve prognosis.[122] As previously mentioned, these vessels are not identified in normal hearts without fibrin and these capillary-sized vessels show particular phenotypic characteristics such as endothelial PAL-E antigen expression (venous related antigen) and they are surrounded by numerous smooth muscle-specific α-actin reactive cells. Since the development of these vessels is associated with improved prognosis, identification of their structure and function merits further study.

Although the role of immunopathological reactions such as the participation of cellular infiltrates, antibodies and complement in the development of transplant-associated CAD has been previously described,[1] we have focused in this Chapter on the importance of microvascular fibrin, alterations of the anticoagulation and fibrinolytic pathways and endothelial activation, all of which suggest that the status of the microvasculature is essential for the subsequent development of transplant-associated epicardial vascular disease. These findings further suggest that transplant-associated CAD is a panvascular disease and by studying the microvessels we could acquire further understanding about the pathophysiology of epicardial CAD.

Conclusions and Future Directions

We focused in this Chapter on the importance of microvascular changes occurring very early following heart transplantation and how these early changes are important for the subsequent outcome of the allografts. We identified that fibrin deposition within the microvasculature is a very early occurring event perhaps associated with ischemia and reperfusion injury,[31] that microvascular fibrin is associated with expression of arterial and arteriolar endothelial activation markers which are not normally expressed,[24,30] and that these changes are associated with loss of vascular antithrombin and depletion of arteriolar tissue plasminogen activator,[30,60] suggesting that the early status of the microvasculature within the allografts affects subsequent outcome. The identification of these changes opens new opportunities for the introduction of new therapeutic approaches soon after transplantation to improve outcome. The presence of microvascular fibrin suggests that anticoagulation therapies could impede further fibrin deposition, and the possibility of ischemia and reperfusion injury being involved suggests that attenuations of free radicals possibly with use of recombinant superoxide dismutase at the time of surgery could improve outcome, reducing the development of transplant-associated CAD. The presence of fibrin certainly favors the presence of thrombin generation. Since thrombin has also been involved in the generation of lesions of ischemia and reperfusion through the recruitment of neutrophils within the lesions,[81] the use of serine protease inhibitors such as antithrombin could also affect outcome by diminishing the generation of those lesions. The use of antithrombin and anticoagulation therapies may also affect the status of vascular antithrombin within the allografts, perhaps favoring the

antithrombin binding within the microvasculature, which has been demonstrated to improve outcome.[70,122] The expression of endothelial activation markers soon after transplantation in allografts that subsequently developed CAD[24] suggests the use of monoclonal antibodies to cell adhesion molecules, especially since these antibodies have proven to be useful to prolong graft survival and prevent rejection, and antibodies to adhesion molecules have been used to prolong function of renal allografts in humans and prevent CAD in experimental animal models.[106,123,124] The introduction of new therapeutic modalities such as the intracoronary transfer of human tissue plasminogen activator gene to in vivo over express tissue plasminogen activator[125] could prevent subsequent development of CAD. This is particularly promising based on the recent findings of successful delivery of human tissue plasminogen activator gene in a rabbit experimental animal model of heterotopic cardiac transplantation and the inhibition of transplant-associated CAD in the grafts showing early transgene expression following ex vivo intracoronary gene transfer at the moment of transplantation.[125] Regardless of the therapeutic intervention used, the previous data considered in this Chapter suggest that the treatment needs to be introduced very early following the transplant procedure in order to prevent the development of transplant-associated CAD, which, once developed, seriously affects the survival of human cardiac allografts.

References

1. Faulk WP, Labarrere CA. Fibrinolytic and anticoagulant control of hemostasis in human cardiac and renal allografts. Major Probl Pathol 1994; 30:49-65.
2. Gao S-Z, Hunt SA, Schroeder JS et al. Early development of accelerated graft coronary artery disease: Risk factors and course. J Am Coll Cardiol 1996; 28:673-679.
3. Miller LW, Wolford TL, Donohue TJ et al. Cardiac allograft vasculopathy: New insights from intravascular ultrasound and coronary flow measurements. Transplantation Rev 1995; 9:77-96.
4. Johnson DE, Gao S-Z, Schroeder JS et al. The spectrum of coronary artery pathologic

findings in human cardiac allografts. J Heart Transplant 1989; 8:349-359.

5. Oni AA, Ray J, Hosenpud JD. Coronary venous intimal thickening in explanted cardiac allografts. Evidence demonstrating that transplant coronary artery disease is a manifestation of a diffuse allograft vasculopathy. Transplantation 1992; 53:1247-1251.

6. Billingham ME. Cardiac transplant atherosclerosis. Transplant Proc 1987; 19:(suppl 5) 19-25.

7. Gao S-Z, Alderman EL, Schroeder JS et al. Accelerated coronary vascular disease in the heart transplant patient: Coronary arteriographic findings. J Am Coll Cardiol 1988; 12:334-340.

8. Schroeder JS, Hunt SA. Chest pain in heart-transplant recipients. N Engl J Med 1991; 324:1805-7.

9. Bajaj S, Shah A, Crandall C et al. Coronary collateral circulation in the transplanted heart. Circulation 1993; 88:II-263—II-269.

10. Ip JH, Fuster V, Badimon L et al. Syndromes of accelerated atherosclerosis: Role of vascular injury and smooth muscle cell proliferation. J Am Coll Cardiol 1990; 15: 1667-87.

11. Gao SZ, Hunt SA, Schroeder JS et al. Does rapidity of development of transplant coronary artery disease portend a worse diagnosis? J Heart Lung Transplant 1994; 13: 1119-1124.

12. Faulk WP, Labarrere CA, Pitts D et al. Vascular lesions in biopsy specimens devoid of cellular infiltrates: Qualitative and quantitative immunocytochemical studies of human cardiac allografts. J Heart Lung Transplant 1993; 12:219-229.

13. Gao SZ, Schroeder JS, Alderman EL et al. Prevalence of accelerated coronary artery disease in heart transplant survivors: Comparison of cyclosporin and azathioprine regimens. Circulation 1989; 80:III-100—III-105.

14. Marboe CC. Cardiac transplant vasculopathy. Major Probl Pathol 1994; 30: 111-32.

15. Meade TW, North WRS, Chakrabarti R et al. Haemostatic function and cardiovascular death: Early results of a prospective study. Lancet 1980; 1:1050-1054.

16. Meade TW, Mellows S, Brozovic M et al. Haemostatic function and ischaemic heart disease: Principal results of the Northwick Park Heart Study. Lancet 1986; 2:533-537.

17. Wilhelmsen L, Svardsudd K, Korsan-Bengtsen K et al. Fibrinogen as a risk factor for stroke and myocardial infarction. N Engl J Med 1984; 311:501-505.

18. Yarnell JWG, Baker IA, Sweetnam PM et al. Fibrinogen, viscosity, and white blood cell count are major risk factors for ischemic

heart disease. The Caerphilly and Speedwell Collaborative Heart Disease Studies. Circulation 1991; 83:836-844.

19. Meade TW, Miller GJ, Rosenberg RD. Characteristics associated with the risk of arterial thrombosis and the prethrombotic state. In: Fuster V, Verstraete M, eds. Thrombosis in Cardiovascular Disorders. Philadelphia: WB Saunders, 1992:79-97.

20. Billingham ME, Cary NRB, Hammond ME et al. A working formulation for the standardization of nomenclature in the diagnosis of heart and lung rejection: Heart Rejection Study Group. The International Society for Heart Transplantation. J Heart Transplant 1990; 9:587-593.

21. Braunwald E. The history. In: Braunwald E, ed. Heart Disease: A Textbook of Cardiovascular Medicine 5th Edition. Philadelphia: WB Saunders, 1997:1-14.

22. Grattan MT, Moreno-Cabral CE, Starnes VA et al. Cytomegalovirus infection is associated with cardiac allograft rejection and atherosclerosis. JAMA 1989; 261:3561-3566.

23. Labarrere CA, Pitts D, Nelson DR et al. Vascular tissue plasminogen activator and the development of coronary artery disease in heart transplant recipients. N Engl J Med 1995; 333:1111-1116.

24. Labarrere CA, Nelson DR, Faulk WP. Endothelial activation and development of coronary artery disease in transplanted human hearts. JAMA 1997; 278:1169-1175.

25. Schoen JF. Blood vessels. In: Cotran RS, Kumar V, Robbins SL, eds. Robbins Pathologic Basis of Disease. Philadelphia: WB Saunders, 1994:467-516.

26. Labarrere CA, Faulk WP. Antigenic identification of cells in spiral artery trophoblastic invasion: Validation of histological studies by triple-antibody immunocytochemistry. Am J Obstet Gynecol 1994; 171:165-171.

27. Labarrere CA, Pitts D, Halbrook H et al. Tissue plasminogen activator, plasminogen activatory inhibitor-1 and fibrin as indexes of clinical course in cardiac allograft recipients: An immunocytochemical study. Circulation 1994; 89:1599-1608.

28. Labarrere CA, Pitts D, Nelson DR et al. Coronary artery disease in cardiac allografts: Association with arteriolar endothelial HLA-DR and ICAM-1 antigens. Transplant Proc 1995; 27:1939-1940.

29. Faulk WP, Labarrere CA, Pitts D et al. Laboratory-clinical correlates of time-associated lesions in the vascular immunopathology of human cardiac allografts. J Heart Lung Transplant 1993; 12:S125-S134.

30. Faulk WP, Labarrere CA, Nelson DR et al. Hemostasis, fibrinolysis and natural antico-

agulation in transplant vascular sclerosis. J Heart Lung Transplant 1995; 14:S158-S164.

31. Labarrere CA, Nelson DR, Faulk WP. Myocardial fibrin deposits in first month after transplantation predict subsequent coronary artery disease and graft failure in cardiac allograft recipients. Am J Med 1998; 105:207-213.

32. Katus HA, Scheffold T, Remppis A et al. Proteins of the troponin complex. Lab Med 1992; 23:311-317.

33. Katus HA, Remppis A, Neumann FJ et al. Diagnostic efficiency of troponin T measurements in acute myocardial infarction. Circulation 1991; 83:902-912.

34. Katus HA, Schoeppenthau M, Tanzeem A et al. Non-invasive assessment of perioperative myocardial cell damage by circulating cardiac troponin T. Br Heart J 1991; 65:259-264.

35. Faulk WP, Labarrere CA, Torry RJ et al. Serum cardiac troponin-T concentrations predict development of coronary artery disease in heart transplant patients. Transplantation 1998; 66:1335-1339.

36. Faulk WP, Labarrere CA, Nelson DR et al. Coronary artery disease in cardiac allografts: Association with arterial antithrombin. Transplant Proc 1995: 27;1944-1946.

37. Pober JS. Warner-Lambert/Parke-Davis Award Lecture. Cytokine-mediated activation of vascular endothelium: Physiology and pathology. Am J Pathol 1988; 133:426-433.

38. Hosenpud JD, Shipley GD, Wagner CR. Cardiac allograft vasculopathy: Current concepts, recent developments, and future directions. J Heart Lung Transplant 1992; 11:9-23.

39. Rose EA, Pepino P, Barr ML et al. Relation of HLA antibodies and graft atherosclerosis in human cardiac allograft recipients. J Heart Lung Transplant 1992; 11:S120-S123.

40. Najarian JS, Gillingham KJ, Sutherland DER et al. The impact of the quality of initial graft function on cadaver kidney transplants. Transplantation 1994; 57:812-816.

41. Shoskes DA, Halloran PF. Delayed graft function in renal transplantation: Etiology, management and long-term significance. J Urol 1996; 155:1831-1840.

42. McLean AG, Hughes D, Welsh KI et al. Patterns of graft infiltration and cytokine gene expression during the first 10 days of kidney transplantation. Transplantation 1997; 63:374-380.

43. McIntyre JA, Wagenknecht DR, Faulk WP. Antiphospholipid antibodies in heart transplant recipients. Clin Cardiol 1995; 18: 575-580.

44. Dunn MJ, Crisp SJ, Rose ML et al. Antiendothelial antibodies and coronary artery disease after cardiac transplantation. Lancet 1992; 339:1566-1570.

45. Crisp SJ, Dunn MJ, Rose ML et al. Antiendothelial antibodies after heart transplantation: The accelerating factor in transplant-associated coronary artery disease? J Heart Lung Transplant 1994; 13:81-92.

46. Tannenbaum SH, Finko R, Cines DB. Antibody and immune complexes induce tissue factor production by human endothelial cells. J Immunol 1986; 137:1532-1537.

47. Faulk WP, Rose M, Meroni PL et al. Antibodies to endothelial cells identify myocardial damage and predict development of coronary artery disease in patients with transplanted hearts. Human Immunol 1999; 60: 826-832.

48. Billingham ME, Baumgartner WA, Watson DC et al. Distant heart procurement for human transplantation. Ultrastructural studies. Circulation 1980; 62(suppl 1):I11-I19.

49. Shoskes DA, Parfrey NA, Halloran PF. Increased major histocompatibility complex antigen expression in unilateral ischemic acute tubular necrosis in the mouse. Transplantation 1990; 49:201-207.

50. Thomas WS, Mori E, Copeland BR et al. Tissue factor contributes to microvascular defects after focal cerebral ischemia. Stroke 1993; 24:847-854.

51. Golino P, Ragni M, Cirillo P et al. Effects of tissue factor induced by oxygen free radicals on coronary flow during reperfusion. Nat Med 1996; 2:35-40.

52. Gaudin PB, Rayburn BK, Hutchins GM et al. Peritransplant injury to the myocardium associated with the development of accelerated arteriosclerosis in heart transplant recipients. Am J Surg Pathol 1994; 18:338-346.

53. Day JD, Rayburn BK, Gaudin PB et al. Cardiac allograft vasculopathy: The central pathogenetic role of ischemia-induced endothelial cell injury. J Heart Lung Transplant 1995; 14:S142-S149.

54. Yilmaz S, Paavonen T, Hayry P. Chronic rejection of rat renal allografts. II. The impact of prolonged ischemia time on transplant histology. Transplantation 1992; 53:823-827.

55. Land W, Schneeberger H, Schleibner S et al. The beneficial effect of human recombinant superoxide dismutase on acute and chronic rejection events in recipients of cadaveric renal transplants. Transplantation 1994; 57:211-217.

56. Labarrere CA, Pitts D, Halbrook H et al. Natural anticoagulation pathways in normal and transplanted human hearts. J Heart Lung Transplant 1992; 11:342-347.

57. Faulk WP, Labarrere CA. Vascular immunopathology and atheroma development in

human allografted organs. Arch Pathol Lab Med 1992; 116:1337-1344.

58. Faulk WP, Labarrere CA. Modulation of endothelial antithrombin III in human cardiac allografts. Haemostasis 1993; 23: 194-201.

59. Faulk WP, Labarrere CA. Antithrombin III in normal and transplanted human hearts: Indications of vascular disease. Semin Hematol 1994; 31:26-34.

60. Labarrere CA, Faulk WP. Antithrombin determinants of coronary artery disease in transplanted human hearts. Semin Hematol 1995; 32:61-66.

61. Absher E, Labarrere CA, Carter C et al. The endothelial heparan sulfate-antithrombin III natural anticoagulant pathway in normal and transplanted human kidneys. Transplantation 1992; 53;828-834.

62. Marcum JA, Rosenberg RD. Role of endothelial cell surface heparin-like polysaccharides. Ann NY Acad Sci 1989; 556:81-94.

63. Justus AC, Roussev R, Norcross JL et al. Antithrombin binding by human umbilical vein endothelial cells: Effects of exogenous heparan. Thromb Res 1995; 79:175-186.

64. Hook M, Kjellen L, Johansson S. Cell surface glycosaminoglycans. Ann Rev Biochem 1984; 53:847-869.

65. Marcum JA, Atha DH, Fritze LM et al. Cloned bovine aortic endothelial cells synthesize anticoagulantly active heparan sulfate proteoglycan. J Biol Chem 1986; 261: 7507-7517.

66. Marcum JA, McKenney JB, Rosenberg RD. Acceleration of thrombin-antithrombin complex formation in rat hindquarters via heparin-like molecules bound to the endothelium. J Clin Invest 1984; 74:341-350.

67. Marcum JA, Rosenberg RD. Anticoagulantly active heparin-like molecules from vascular tissue. Biochemistry 1984; 23:1730-1737.

68. Wight TN. Cell biology of arterial proteoglycans. Arteriosclerosis 1989; 9:1-20.

69. Mertens G, Cassiman JJ, Van den Berghe H et al. Cell surface heparan sulfate proteoglycans from human vascular endothelial cells. Core protein characterization and antithrombin III binding properties. J Biol Chem 1992; 267:20435-20443.

70. Labarrere CA, Torry RJ, Nelson DR et al. Vascular antithrombin and clinical outcome in heart transplant patients. Am J Cardiol (in press).

71. Dodge GR, Kovalszky I, Hassell JR et al. Transforming growth factor β alters the expression of heparan sulfate proteoglycan in human colon carcinoma cells. J Biol Chem 1990; 265:18023-18029.

72. Klein NJ, Shennan GI, Heyderman RS et al. Alteration in glycosaminoglycan metabolism and surface change on human umbilical vein endothelial cells induced by cytokines, endotoxin and neutrophils. J Cell Sci 1992; 102:821-832.

73. Fridman R, Lider O, Naparstek Y et al. Soluble antigen induces T lymphocytes to secrete an endoglycosidease that degrades the heparan sulfate moiety of subendothelial extracellular matrix. J Cell Physiol 1987; 130:85-92.

74. Savion N, Vlodavsky I, Fuks Z. Interaction of T lymphocytes and macrophages with cultured vascular endothelial cells: Attachment, invasion and subsequent degradation of the subendothelial extracellular matrix. J Cell Physiol 1984; 118:169-178.

75. Subramanian SV, Fitzgerald ML, Bernfield M. Regulated shedding of syndecan-1 and -4 ectodomains by thrombin and growth factor receptor activation. J Biol Chem 1997; 272:14713-14720.

76. Bernfield M, Kokenyesi R, Kato M et al. Biology of the syndecans: A family of transmembrane heparan sulfate proteoglycans. Ann Rev Cell Biol 1992; 8:365-393.

77. Kinsella MG, Wight TN. Structural characterization of heparan sulfate proteoglycan subclasses isolated from bovine aortic endothelial cell cultures. Biochemistry 1988; 27:2136-2144.

78. Pejler G, David G. Basement-membrane heparan sulphate with high affinity for antithrombin synthesized by normal and transformed mouse mammary epithelial cells. Biochem J 1987; 248:69-77.

79. Platt JL, Vercellotti GM, Lindman BJ et al. Release of heparan sulfate from endothelial cells. Implications for pathogenesis of hyperacute rejection. J Exp Med 1990; 171: 1363-1368.

80. Platt JL, Dalmasso AP, Lindman BJ et al. The role of C5a and antibody in the release of heparan sulfate from endothelial cells. Eur J Immunol 1991; 21:2887-2890.

81. Ostrovsky L, Woodman RC, Payne D et al. Antithrombin III prevents and rapidly reverses leukocyte recruitment in ischemia/reperfusion. Circulation 1997; 96: 2302-2310.

82. Key NS, Platt JL, Vercellotti GM. Vascular endothelial cell proteoglycans are susceptible to cleavage by neutrophils. Arterioscl Thromb 1992; 12:836-842.

83. Schlingemann RO, Dingjan GM, Emeis JJ et al. Monoclonal antibody PAL-E specific for endothelium. Lab Invest 1985; 52:71-76.

84. Schaper W, Sharma H, Quinkler W et al. Molecular biologic concepts of coronary anastomoses. J Am Coll Cardiol 1990; 15:513-518.

85. Torry RJ, Labarrere CA, Torry DS et al. Vascular endothelial growth factor expression in transplanted human hearts. Transplantation 1995; 60:1451-1457.

86. Shweiki D, Itin A, Soffer D et al. Vascular endothelial growth factor induced by hypoxia may mediate hypoxia-initiated angiogenesis. Nature 1992; 359:843-845.

87. Ladoux A, Frelin C. Hypoxia is a strong inducer of vascular endothelial growth factor mRNA expression in the heart. Biochem Biophys Res Comm 1993; 195:1005-1010.

88. Brogi E, Wu T, Namiki A et al. Indirect angiogenic cytokines upregulate VEGF and bFGF gene expression in vascular smooth muscle cells, whereas hypoxia upregulates VEGF expression only. Circulation 1994; 90;649-652.

89. Shore VH, Wang TH, Wang CL et al. Vascular endothelial growth factor, placenta growth factor and their receptors in isolated human trophoblast. Placenta 1997; 18:657-665.

90. Senger DR, Van de Water L, Brown LF et al. Vascular permeability factor (VPF, VEGF) in tumor biology. Cancer Metastasis Rev 1993; 12:303-324.

91. Li J, Brown LF, Hibberd MG et al. VEGF, flk-1 and flt-1 expression in a rat myocardial infarction model of angiogenesis. Am J Physiol 1996; 270:H1803-H1811.

92. Shinohara K, Shinohara T, Mochizuki N et al. Expression of vascular endothelial growth factor in human myocardial infarction. Heart Vessels 1996; 11:113-122.

93. Miller JW, Adamis AP, Shima DT et al. Vascular endothelial growth factor/vascular permeability factor is temporally and spatially correlated with ocular angiogenesis in a primate model. Am J Pathol 1994; 145:574-584.

94. Kojima T, Katsumi A, Yamazaki T et al. Human ryudocan from endothelium-like cells binds basic fibroblast growth factor, midkine and tissue factor pathway inhibitor. J Biol Chem 1996; 271:5914-5920.

95. Elenius K, Vainio S, Laato M et al. Induced expression of syndecan in healing wounds. J Cell Biol 1991; 114:585-595.

96. Pober JS, Cotran RS. The role of endothelial cells in inflammation. Transplantation 1990; 50:537-544.

97. Taylor PM, Rose MS, Yacoub MH et al. Induction of vascular adhesion molecules during rejection of human cardiac allografts. Transplantation 1992; 54:451-457.

98. Moore KL, Andreoli SP, Esmon NL et al. Endotoxin enhances tissue factor and suppresses thrombomodulin expression of human vascular endothelium in vitro. J Clin Invest 1987; 79:124-130.

99. Salomon RN, Hughes CC, Schoen FJ et al. Human coronary transplantation-associated arteriosclerosis. Evidence for a chronic immune reaction to activated graft endothelial cells. Am J Pathol 1991; 138:791-798.

100. Poston RN, Haskard DO, Coucher JR et al. Expression of intercellular adhesion molecule-1 in atherosclerotic plaques. Am J Pathol 1992; 140:665-673.

101. Tanio JW, Basu CB, Albelda SM et al. Differential expression of the cell adhesion molecules ICAM-1, VCAM-1, and E-selectin in normal posttransplantation myocardium. Cell adhesion molecule expression in human cardiac allografts. Circulation 1994; 89: 1760-1768.

102. Fuster V, Poon M, Willerson JT. Learning from the transgenic mouse: Endothelium, adhesive molecules, and neointimal formation. Circulation 1998; 97:16-18.

103. Adams DH, Russell ME, Hancock WW et al. Chronic rejection in experimental cardiac transplantation: Studies in the Lewis-F344 model. Immunol Rev 1993; 134:5-19.

104. Karnovsky MJ, Russell ME, Hancock W et al. Chronic rejection in experimental cardiac transplantation in a rat model. Clin Transplantation 1994; 8:308-312.

105. Koskinen PK, Lemstrom KB. Adhesion molecule P-selectin and vascular cell adhesion molecule-1 in enhanced heart allograft arteriosclerosis in the rat. Circulation 1997; 95:191-196.

106. Suzuki J, Isobe M, Yamazaki S et al. Inhibition of accelerated coronary atherosclerosis with short-term blockade of intercellular adhesion molecule-1 and lymphocyte function-associated antigen-1 in a heterotopic murine model of heart transplantation. J Heart Lung Transplant 1997; 16:1141-1148.

107. Ridker PM, Hennekens CH, Roitman-Johnson B et al. Plasma concentration of soluble intercellular adhesion molecule 1 and risks of future myocardial infarction in apparently healthy men. Lancet 1998; 351: 88-92.

108. Stovin PG, Sharples LD, Schofield PM et al. Lack of association between endomyocardial evidence of rejection in the first six months and the later development of transplant-related coronary artery disease. J Heart Lung Transplant 1993; 12:110-116.

109. Russell PS, Chase CM, Winn HJ et al. Coronary atherosclerosis in transplanted mouse hearts. I. Time course and immunogenetic and immunopathological considerations. Am J Pathol 1994; 144:260-274.

110. Russell PS, Chase CM, Winn HJ et al. Coronary atherosclerosis in transplanted mouse hearts. II. Importance of humoral immunity. Am J Pathol 1994; 152: 5135-5141.

111. Labarrere CA, Pitts D, Nelson DR et al. Coronary artery disease in cardiac allografts: Association with depleted arteriolar tissue plasminogen activator. Transplant Proc 1995; 27:1941-1943.

112. Labarrere CA, Pitts D, Halbrook H et al. Tissue plasminogen activator in human cardiac allografts. Transplantation 1993; 55: 1056-1060.

113. Warshofsky MK, Wasserman HS, Wang W et al. Plasma levels of tissue plasminogen activator and plasminogen activator inhibitor-1 are correlated with the presence of transplant coronary artery disease in cardiac transplant recipients. Am J Cardiol 1997; 80:145-149.

114. Olofsson BO, Dahlen G, Nilsson TK. Evidence for increased levels of plasminogen activator inhibitor and tissue plasminogen activator in plasma of patients with angiographically verified coronary artery disease. Eur Heart J 1989; 10:77-82.

115. Badimon L, Badimon JJ, Fuster V. Pathogenesis of thrombosis. In Fuster V, Verstraete M, eds. Thrombosis in Cardiovascular Diseases. Philadelphia: WB Saunders, 1992: 17-39.

116. Smith EB, Thompson WD. Fibrin as a factor in atherogenesis. Thromb Res 1994; 73;1-19.

117. Wojta J, Gallicchio M, Zoellner H et al. Thrombin stimulates expression of tissue-type plasminogen activator and plasminogen activator inhibitor type 1 in cultured human vascular smooth muscle cells. Thromb Haemost 1993; 70:469-474.

118. Benza RL, Grenett HE, Bourge RC et al. Gene polymorphisms for plasminogen activator inhibitor-1/tissue plasminogen activator and development of allograft coronary artery disease. Circulation. 1998; 98:2248-2254.

119. Drake TA, Morrissey JH, Edgington TS. Selective cellular expression of tissue factor in human tissues. Implications for disorders of hemostasis and thrombosis. Am J Pathol 1989; 134:1087-1097.

120. Fleck RA, Rao LV, Rapaport SI et al. Localization of human tissue factor antigen by immunostaining with monospecific, polyclonal antihuman tissue factor antibody. Thromb Res 1990; 59:421-437.

121. Hirsh J, Salzman EW, Marder VJ et al. Overview of the thrombotic process and its therapy. In: Colman RW, Hirsch J, Marker VJ, Salzman EW, eds. Hemostasis and Thrombosis. Philadelphia: JB Lippincott, 1994:1151-1163.

122. Labarrere CA, Torry RJ, Nelson DR et al. The status of microvascular antithrombin after early fibrin deposition is associated with outcome in cardiac transplant recipients. Transplantation 1998;65:102.

123. Isobe M, Yagita H, Okumura K et al. Specific acceptance of cardiac allograft after treatment with antibodies to ICAM-1 and LFA-1. Science 1992; 255:1125-1127.

124. Haug CE, Colvin RB, Delmonico FL et al. A phase I trial of immunosuppression with anti-ICAM-1 (CD54) mAb in allograft recipients. Transplantation 1993; 55:766-772.

125. Scholl FG, Sen L, Laks H et al. Effects of human tissue plasminogen gene transfer on allograft coronary atherosclerosis. J Heart Lung Transplant 1998; 17:74.

CHAPTER 5

Nonimmune Factors in the Etiology of Graft Coronary Arteriosclerosis

Hiroaki Nagano and Nicholas L. Tilney

Chronic rejection implies progressive and irreversible functional and structural deterioration of an organ graft in the months or years after transplantation. The process affects all solid grafts, including heart, kidney, lung, and probably, to a lesser extent, liver. In cardiac transplantation, the development of graft coronary arteriosclerosis, thought to be the characteristic feature of chronic rejection, has become the major limitation to long-term survival; up to 50% of heart grafts develop angiographically detectable coronary arteriosclerosis within 5 years, leading to a 25% mortality rate.[1,2] The etiology of this progressive luminal obliterative process remains unclear; assessment of the state of host alloimmunity (assumed to be the major contributor) as part of the diagnostic work-up has not been helpful. The possibility that at least some of its manifestations may result from alloantigen-independent factors is receiving increasing attention and will be the subject of this Chapter. Indeed, the term chronic rejection may be a misnomer and chronic graft dysfunction or graft vascular disease may be a more appropriate description as it is less specific and leaves room for a multi-factorial etiology.

The less steep rate of attrition of donor organs that are well matched for recipient HLA antigens compared with that of poorly matched grafts supports the notion that immunological mechanisms are important determinations of long-term outcome.[3,4] On the other hand, in kidney transplantation at least, the remarkable similarity between the excellent success rates of grafts from living unrelated and from related sources suggests that various nonspecific factors associated with procurement and storage of organs from brain dead cadavers may influence their less satisfactory success rate. The putative role of alloantigen-independent events in chronic graft dysfunction is also illustrated by the observation that renal transplants from identical human twins as well as long-term kidney isografts in rats may develop, albeit at a slower rate, progressive functional and morphologic changes that resemble the lesions of chronically rejecting allografts.[5-7] The eventual failure (10 years) of 40% of the human isografts was originally thought to result from recurrent nephritis, although the original investigators also suggested the effect of other undetermined causes. In further experimental studies, retransplantation of organ allografts into the original donor strain prevents the development of structural changes commensurate with chronic rejection if performed within a critical time period; late retransplantation does not alter the continuous progression of the chronic process despite removal of the host immunological drive.[8,9] This latter finding emphasizes the notion that tissue injury beyond a certain point may lead to autonomous and programmed propagation and advancement of the lesions. Thus, the early stages of chronic rejection appear alloantigen dependent and reversible, whereas the

Transplant-Associated Coronary Artery Vasculopathy, edited by Marlene L. Rose.
©2001 Eurekah.com.

later, presumably antigen-independent stages, are irreversible and progressive.

This Chapter focuses on several non-immunologic parameters (Table 5.1) implicated in the chronic graft arterial disease evolving in cardiac allografts, a phenomenon which often prevents clinical heart transplantation from being but a relatively short-term answer to an ultimately fatal condition.

The Cadaver Donor and Early Organ Injury

That the success of kidney allografts from cadavers is consistently inferior not only to those from living-related donors but to organs from living-unrelated sources as well is a compelling argument in favor of the importance of antigen-independent factors over time. As there is no genetic advantage between living-unrelated and cadaver groups and their recipients, this clear difference must relate in part to donor variables as well as to changes associated with organ removal and preservation. The living donor has been shown to be normal by a complete medical evaluation; the organ is removed under optimal circumstances and transplanted to the recipient promptly. In contrast, the use of tissues from cadavers implies profound physiological derangements that may occur long before transplantation. The organs are removed from individuals who have sustained massive irreversible central nervous system injury. In addition, the demographics of the donor population has changed substantially from a cohort of predominantly young persons suffering acute head injury to older patients, often with co-existing conditions, who have sustained an intracranial hemorrhage. The blood pressure of the brain dead individual is chaotic or labile and must often be sustained by powerful pharmacological agents. The organs may be taken under less than optimal conditions, perfused with electrolyte solutions and stored in the cold for many hours before eventual engraftment. Important changes associated with reperfusion may then occur.

Table 5.1. Antigen-independent risk factors in graft arteriosclerosis

| **Early Injury** |
| donor factors |
| brain death |
| ischemia/reperfusion |
| **Late Influences** |
| hyperlipidemia |
| immunosuppressive agents |
| cytomegalovirus |

Recipient Age, Donor Age and Gender

The contribution of these variables to the development of graft coronary arteriosclerosis is suggestive but still not clear. The relationship between the development of transplant coronary disease and recipient age has been examined in several studies which compared the incidence of risk factors in patients who did and who did not develop the condition, diagnosed either angiographically or at postmortem.[10,11] Some showed no correlation;[12] conversely in one study, the incidence of coronary disease in patients less than 20 years at the time of transplantation was found to be 100% by 3 years, suggesting that the incidence actually decreased as recipient age increased.[13] In another large series, recipient age less than 30 years or more than 40 years was associated with increased graft loss from coronary disease.[14] In an additional analysis, patients with coronary disease diagnosed more than 2 years after transplantation were older than patients in whom transplant coronary disease did not develop.[11]

Some studies have implicated the age of the donor with the development of the condition. Because the incidence is high even amongst children who receive a cardiac graft, existing coronary involvement in the donor should be considered carefully before a heart is used.[15] That the risk of transplant coronary disease was increased in hearts from donors greater than 35 years compared to those younger has been suggested.[16] A relationship between donor age and transplant coronary disease was noted in another analysis in which donor age over 40 years was associated with an increased risk of transplant coronary disease compared with younger donors.[14] However, in some series, no relationship has been found between donor age and transplant coronary disease.[15]

Only a few analyses have examined the relationship between donor/recipient gender mismatch and graft arteriosclerosis.[11,17] Compared with transplantation performed between donors and recipients of same gender, female recipients of male hearts had a relative low risk of graft loss; and male recipients of a female donor heart had a relatively higher risk. This difference may be based upon size of the organ in relationship to that of the recipient. In kidney transplantation, a high risk of chronic rejection in male recipients of smaller female kidneys was noted in one large series.[18] In an analysis of parent-to child living related transplants, the figures from the UCLA/UNOS Registry showed a lower graft half-life in sons who received kidneys from their mothers as compared to those from their fathers; the same trend was observed for daughters but was not statistically significant.[19] This point is still controversial.

Brain Death

Although brain death has been defined neurologically, philosophically and by social criteria, knowledge of the systemic changes occurring during this critical event is minimal. It is known that central regulatory processes are profoundly compromised; hypothalamus directed hormone systems may continue to function and release of neuroendocrine products may be increased.[20] Elevated circulating catecholamine levels may produce structural myocardial damage. It has been recently hypothesized that brain death "activates" surface molecules on peripheral organs via circulating factors.[21] The potentially increased immunogenicity of these organs, as exemplified by upregulated MHC class II expression, may explain, at least in part, the

striking divergence in clinical results between kidney grafts from cadavers and living sources. Release of macrophage and T-cell associated products into the blood stream may influence host cellular defense mechanisms, as yet undefined. Although little information is available on the systemic effects of brain death, a massive release of cytokines and adhesion molecules into the circulation from this irreparably injured central organ may influence initial and ultimate behavior of peripheral donor organs.

Ischemia/Reperfusion Injury

Graft arteriosclerosis develops selectively in the transplanted heart without comparable effect on host arteries. The immunological basis of such preferential involvement has been a subject of intense investigation throughout the history of cardiac transplantation. However, non-antigen-dependent factors may also explain these differences.[22,23] Ischemia occurring during organ removal or storage and reperfusion injury after re-establishment of blood flow may produce endothelial changes which may induce subsequent graft coronary obliterative disease.

Ischemia/reperfusion damage may cause desquamation and retraction of endothelial cells, increasing their permeability.[24] These damaged cells may augment adhesive interactions with circulating leukocytes that contribute to a cycle of pathological changes resembling those evoked by antigenic stimulation. Hypoxia can induce the expression of certain genes by endothelial cells, including VEGF (vascular endothelial growth factor) or vascular permeability factor.[25-27] In addition, b-FGF and PDGF, upregulated following vascular injury, may induce smooth muscle cell migration and proliferation in the damaged vessel wall.[28,29] VEGF not only stimulates replication of endothelial cells but can augment their permeability. Local production of TGF-β is associated with fibrogenesis, stimulates production of collagen, fibronectin, and proteoglycans, the synthesis of plasminogen activator inhibitor, and tissue inhibitor of metalloproteinase which results in decreased breakdown of extracellular matrix proteins.[30] Angiotensin II, a growth promoter for vascular smooth muscle cells, may participate in myointimal vascular proliferation following injury, in part by increasing production of TGF-β. As the activities and mRNA levels of angiotensin converting enzyme (ACE) are increased in the injured vessels, ACE inhibitors may ameliorate the arteriosclerotic process in the vessel wall.[31,32]

Ischemia may also increase tissue levels of proinflammatory mediators such as cytokines and products of arachidonic acid metabolism which in turn upregulate expression of the CD11/CD18 adhesion molecules on leukocytes, increase ICAM-1 and MHC expression on endothelial cells, and promote CD11/CD18 avidity for ICAM-1.[33-35] As a result, leukocytes adhere to endothelial cells, become activated, and upregulate additional inflammatory mediators. Whereas perfusion of mildly ischemic kidneys with normal neutrophils or perfusion of nonischemic kidneys with activated neutrophils does not produce renal insult, perfusion of mildly ischemic kidneys with primed neutrophils causes severe injury.[36] Anti-ICAM-1 monoclonal antibodies limit the myocardial damage following ischemia/reperfusion injury.[37]

The involvement of early ischemia in the later development of clinical graft coronary arteriosclerosis remains controversial. The results of a multicenter intravascular ultrasound study show no significant difference in the progression of intimal thickening based on perioperative ischemic time.[10] On the other hand, data from the International Registry for Heart and Lung Transplantation show that ischemia effects mortality at 1 year and persists as long as 3 years after injury.[11] Endothelial injury secondary to this initial insult may lead subsequently to the development of transplant coronary disease within 2 years after heart transplantation; although the mechanism of this injury is not clear, potential factors include hypoxia, cold injury, and the effect of oxygen-free radicals.[38]

Late Influences on Graft Arteriosclerosis

Hyperlipidemia

Hyperlipidemia is an important potential risk factor for recipients of heart grafts. Over half the prospective recipients in most cardiac transplant programs have developed ischemic cardiomyopathy as their primary condition; many with prior myocardial infarction have had premorbid dyslipidemia. By the time they have sustained end-stage heart failure and have become candidates for engraftment, lipid levels tend not to be elevated. However, hyperlipidemia with hypercholesterolemia, increased low-density lipoproteins and hypertriglycemia often develop 3-18 months after transplantation, as shown in several clinical studies.[39,40]

The cause of post-transplant lipid abnormalities is not certain, although immunosuppressive agents, particularly steroids, probably play an important role. Prednisone may increase hepatic apolipoprotein B production as well as contribute to the obesity that often develops post-operatively.[41,42] Cholesterol levels may decline when patients are tapered to lower maintenance levels of corticosteroids.[43] Cyclosporine has also been implicated; because it inhibits predonisolone clearance by the liver or because of interactions with the cytochrome P-450 system, steroid-induced effects may be enhanced.[44] A recent study has also suggested that this agent increases hepatic lipase activity and decreases lipoprotein lipase activity.[45] This could result in impaired very low-density (VLDL) and low-density lipoprotein (LDL) clearance, leading to hypertriglyceridemia and increased endothelial contact with small low-density lipoprotein particles.

Although hypercholesterolemia and combined (mixed) hyperlipidemia are established risk factors for nontransplant related atherosclerosis and coronary artery disease, their effects on graft arteriosclerosis remain unresolved. Low-density lipoproteins are directly toxic to endothelial cells in vitro; elevated cholesterol concentrations and hyperlipidemia foster vascular injury, arteriosclerosis, and glomerulosclerosis.[46] Several investigators have shown accelerated allograft arteriosclerosis in cholesterol fed rabbits.[47,48] Similar experiments in rats have yielded controversial results; graft arteriosclerosis was enhanced in cholesterol-fed rats in some studies but not in others.[49-51] Although hypercholesterolemia produced marked lipid deposition in areas of intimal thickening, careful analysis of the lesions failed to show a difference in the smooth muscle cell proliferative response compared to that of recipients on a standard diet.[51] However, a recent study showed that a diet high in cholesterol, cholic acid, and glycerol increased the rate of accelerated graft arteriosclerosis.[52] Upregulation of the adhesion molecules VCAM-1 and ICAM-1 and the expression of class II MHC antigens on coronary endothelium were also shown in heart transplants in hyperlipidemic rabbits.

Clinical evidence has also been inconclusive in renal allografts. Pretransplant hyperlipidemia has been found to constitute a risk factor for late loss by some but not others.[53,54] VLDL, LDL, total cholesterol and triglyceride levels are increased and the lipoproteins distributed in more atherogenic patterns in kidney allografts exhibiting chronic vascular rejection than in controls.[55] Although lipid derangements may result from impaired renal function, the presence of similar abnormalities in cardiac transplant patients with graft arteriosclerosis justifies the premise that hyperlipoproteinemia contributes to or is at least associated with its development.[56] This prospect is further supported by the presence of the apolipoproteins, A1, A2, B1, in the vessel walls of human renal allografts experiencing chronic rejection.[55] Taken together, some clinical and experimental evidence supports the hypothesis that hyperlipidemia plays a role in the pathophysiology of chronic graft arterial changes, although most clinical studies fail to show the expected correlation.

It is possible that the intensity of vessel wall damage by immunological mechanisms obscures the impact of the conventional atherogenic risk factors that usually take many years to become clinically detectable. That the allogeneic state may augment the arterial response to hypercholesterolemia has been shown by greater intimal thickening, increased angiogenesis in the expanded intima and accumulation of T cells in transplanted vessels compared with the native vessels of the same animal exposed to the same level of hyperlipidemia.[51,52] The clinical implication of these findings is that physicians caring for transplant patients should have a low threshold controlling dyslipidemia. Indeed, there is evidence that treatment with an inhibitor of the rate-limiting enzyme in cholesterol biosynthesis, hydroxymethylglutaryl coenzyme A (HMG co-A) reductase, can retard the development of intimal thickening in coronary arteries of allograft recipients.[57] The effects of antilipid lowering agents are described in more detail in Chapter 9.

Immunosuppressive Drugs

The impact of immunosuppressive agents on long-term graft structure and function has not been completely defined or even well examined. Although recipient immunosuppression with azathioprine and steroids does not evoke morphologic changes characteristics of cyclosporine nephrotoxicity in kidney grafts, their long-term attrition rate parallels that of kidney transplants in cyclosporine-treated hosts, suggesting that neither drug regimen is effective enough to avoid long-term graft loss.[3] The nephrotoxic effects of cyclosporine have been particularly problematic and difficult to differentiate from chronic allograft rejection. Indeed, patients with cardiac transplants who have been heavily immunosuppressed with this agent (as well as the profoundly nephrotoxic drug, FK506) not infrequently experience renal dysfunction and occasionally become dialysis-dependent. The associated structural renal lesions include interstitial inflammation, fibrosis, and focal glomerulosclerosis commensurate with reduced glomerular filtration rate and renal plasma flow; the interstitial inflammatory infiltrates consist of cells of the helper but not the cytotoxic T cell phenotype, B cells, and macrophages.[58]

Accelerated arteriosclerosis in heart grafts has also been associated with the use of cyclosporine. Some studies of rat cardiac grafts have shown that low doses of the drug reduce the frequency and severity of graft arteriosclerosis,[59,60] a condition that has been accelerated in other studies of both allografts and isografts.[61,62] Vasculitis and arteriosclerosis are increased by the agent in a rat aortic allograft model.[62] In contrast, although virtually all heart transplant patients receive cyclosporine as part of their standard postoperative immunosuppressive regimen, uncontrolled clinical data suggest that the incidence of graft coronary arteriosclerosis is not altered compared with prior treatment regimens.[63] Other clinical studies have confirmed the lack of effect of particular immunosuppressive regimens on the development of the process.[64,65] Similarly, no decrease in transplant coronary disease has been noted with the institution of triple versus double immunosuppressive therapy or with the use of cyclosporine/prednisone;[63,65] no difference in the incidence of coronary disease has been found in relationship to dose of prednisone or blood level of dose of cyclosporine.[66] Recently, rapamycin and HMG co-A reductase inhibitor have been reported to reduce the incidence of graft coronary arterial disease, although more data are necessary.[67] Overall, the effects of maintenance immunosuppression on graft arteriosclerosis remains an ill-defined area.

Cytomegalovirus (CMV)

CMV is an important cause of increased morbidity and mortality in allograft recipients. It may also play a role in chronic rejection by influencing vascular obliterative changes; studies of endomyocardial biopsy specimens from human heart allografts, for instance, showed that CMV linked immune activation and typical subendothelial inflammation may contribute to the later development of graft arteriosclerosis.[68] Experimental work has provided further support for the hypothesis that early post-transplant CMV infection accelerates immune-mediated chronic rejection lesions.[69] Acute vascular rejection in CMV infected heart transplant recipients has also been associated with a subendothelial inflammatory response in large epicardial arteries.[70] Cardiac arteriosclerosis appears to develop earlier and more frequently in patients with CMV infection,[71,72] although not all reports agree with this correlation.[73] The presence of circulating cytomegalic cells in CMV-infected patients also supports the hypothesis that the infection is associated with vascular endothelial cell damage.[74]

The mechanisms involved in the putative relationship between viral and perhaps other infections and rejection has been explained in several ways. A subgroup of kidney recipients with late acute rejection unresponsive to treatment, showed the presence of large numbers of memory-type CD8+ T cells in their peripheral blood associated with asymptomatic CMV infection; antiviral therapy improved graft function in the majority.[74,75] An immediate early gene of the virus able to code for a protein with sequence homology and immunologic cross-reactivity with the HLA-DR β chain may increase the alloimmune response of the host to donor antigens.[76] In addition, CMV encodes a glycoprotein homologous to the heavy chain of MHC class I antigen which can bind to its light chain.[77] Active infection has been strongly associated with expression of vascular adhesion molecule-1 (VCAM-1) on capillary endothelium; its effects on leukocyte adhesion and infiltration may contribute to long-term changes in the graft.[78] Cytokines associated with the inflammatory response of CMV infection may also increase endothelial cell damage and vasculopathic changes.

The Common Pathway

Detailed assessment of the chain of events leading to chronic graft dysfunction has been limited primarily by the availability of few reproducable and clinically relevant animal models.[79] It appears increasingly that local deposition of complement, inflammatory mediators and cytokines initiate vascular damage and graft activation. This sequence of events occurs irrespective of whether the injury is caused by immunological mechanisms such as acute rejection, or nonimmunological damage such as ischemia/reperfusion.[80] In ischemia, the endothelial phenotype may change and express new antigens that may enhance immune complex deposition, activation of the classical complement pathway, and loss of membrane complement regulatory proteins.[81,82] Antibody formation may be enhanced in a nonspecific fashion, as demonstrated in burn patients where increased amounts of antibodies are produced by mitogen-stimulated lymphocytes in conjunction with signs of complement activation.[83] In addition to endothelial cell damage and activation, platelets may accumulate on collagen exposed to the circulation; the activated clotting cascade may promote fibrin deposition.

Inflammatory mediators, including thromboxane, PDGF, leukotrienes, and platelet-activating factors, are released by injured graft endothelial cells; as a result, circulating leukocytes enter the graft interstitium after adhering to the activated endothelium, a process mediated by various adhesion molecules.[84] Although leukocyte infiltration increases early after injury, mononuclear cells, primarily lymphocytes and macrophages, may persist in both isografts and allografts with or without clinical evidence of graft dysfunction. These cells produce various cytokines and growth factors which may contribute to vessel wall or mesangial cell proliferation and ultimately, interstitial fibrosis.[7,79] It appears increasingly, based on several experimental studies, that chronic graft dysfunction, regardless of its etiology, shares features similar to those in other inflammatory processes that involve the cytokine-adhesion molecule cascade.

Macrophages seems particularly important in the process. Subendothelial plaques, consisting of macrophages in foamy transformation, have been noted in human renal allografts undergoing chronic rejection.[85] Similarly, macrophages of various phenotypes have been noted in biopsies of heart grafts experiencing graft arteriosclerosis in cholesterol fed and immunosuppressed rabbit recipients,[86] as well as in chronically rejecting organ allografts in rats.[87] Organ-specific differences in infiltrating macrophage phenotypes have been difficult to define; in addition, differences in between syngeneic and allogeneic grafts tend to be quantitative rather than qualitative. Macrophages or their associated products may contribute to tissue remodeling; these include IL-1, IL-6, TNF-α, MCP-1, b-FGF, TGF-β, and PDGF, the presence of which coincides closely with the development of irreversible vascular, glomerular, and interstitial lesions in both syngeneic and allogeneic grafts.[85,86] Adhesion molecules, primarily ICAM-1 and VCAM-1, are present on the vascular endothelium as well as on extravascular structures; LFA-1 and VLA-4, the leukocyte counter-receptors for these molecules, are upregulated on graft infiltrating host cells.[88,89] At least some of these molecules direct the positioning of host cell populations through the extracellular matrix of the graft and may modulate chronic changes.

Although not well understood, it is likely that individual graft components may respond differently to injury. Vascular obliterative lesions have been conceptualized as developing from repetitive episodes of endothelial cell injury, with intimal proliferation, hypertrophy and repair, all leading to gradual luminal narrowing.[90] Although the vascular lesions may arise in response to vessel wall damage incurred by alloimmune reactions, other mechanisms may cause similar lesions as illustrated by conditions like malignant hypertension, scleroderma, radiation nephritis, and hemolytic uremic syndrome.[91]

The glomerular lesions developing in chronically rejecting kidneys are variable and consist of focal and global sclerosis or a constellation of changes described under several synonyms, including transplant glomerulopathy.[92] Infiltrating macrophages and their products, particularly IL-6, may increase mesangial cell proliferation and cause increased mesangial matrix deposition.

Altered eicosanoid metabolism by glomerular capillary endothelium may be activated by hypertension, increasing the activities of cytokines and adhesion molecules and leading to progressive glomerulosclerosis.[93] Gradual functional deterioration secondary to glomerular fibrosis and vascular obliteration may also produce systemic hypertension and glomerulosclerosis, which may itself contribute to deterioration and fibrosis of the remaining functioning nephrons, establishing a self-sustaining vicious cycle.[94] Whether this type of "hyperfiltration" hypothesis, presumably critical in renal deterioration, has relevance where the heart is considered is not known, although may be implicated when considering the less optimal results of hearts from small donors placed in large recipients.[74]

Conclusions

Based in part on the interesting clinical observation that renal allografts from living-unrelated donors do as well, both acutely and over the long-term, as those from living-related sources, the hypothesis that antigen-independent events are important in the etiology and pathophysiology of chronic graft dysfunction (chronic rejection) has been examined, with particular reference to cardiac allografts. The multifaceted etiology of this phenomenon appears to include both immunological and nonspecific events, perhaps relating to each other in synergy. Therapeutic means to prevent or ameliorate the chronic process may have to broadened beyond those commonly employed.

References

1. Billingham ME. Histopathology of graft coronary disease. J Heart Lung Transplantation 1992; 11:S38-44.
2. Schoen FJ, Libby P. Cardiac transplant graft arteriosclerosis. Trends Cardiovasc Med 1991; 1:216-223.
3. Terasaki PI, Cecka JM, Gjertson DW et al. A ten year prediction for kidney transplant survival. In: Terasaki PI, Cecka JM, eds. Clinical Transplants 1992. UCLA Tissue Typing Laboratory, 1993; 33.
4. Thorogood J, Van Houwelingen HC, Van Rood J et al. Long-term results of kidney transplantation in Eurotransplant In: Paul LC, Solz K, eds. Organ Transplantation: Long-term Results. New York, Basel, Hong Kong: Marcel Dekker, Inc., 1992: 33.
5. Perfrey PS, Hollomby DJ, Gilmore NK et al. Glomerular sclerosis in a renal isograft and identical twin donor. Transplantation 1984; 38: 343-346.
6. Tilney NL. Renal transplantation between identical twins: A review. World J Surgery 1986; 10:381-388.
7. Tullius SG, Heeman UW, Hancock UW et al. Long-term kidney isografts develop functional and morphological changes which mimic those of chronic allografts rejection. Ann Surg 1994; 220:425-432.
8. Izutani H, Miyagawa S, Shirakura R et al. Evidence that graft coronary arteriosclerosis begins in the early phase after transplantation and progresses without chronic immunoreaction. Transplantation 1995; 60:1073-1079.
9. Tullius SG, Hancock WW, Heeman UW et al. Reversibility of chronic renal allograft rejection: critical effect of time after transplantation suggests both host immune dependent and independent phases of progressive injury. Transplantation 1994; 58:93-99.
10. Hauptman PJ, Davis SF, Miller L et al. The role of nonimmune risk factors in the development and progression of graft arteriosclerosis: Preliminary insights from a multicenter intravascular ultrasound study. J Heart Lung Transplant 1995; 14:S238-242.
11. Hosenpud JD, Novick RJ, Bennet LE et al. The registry of the International Society for Heart and Lung Transplantation: Thirteenth official report-1996. J Heart Lung Transplant 1996; 15:655-674.
12. Gao SZ, Schroder JS, Hunt S et al. Retransplantation for severe accelerated coronary artery disease in heart transplant recipients. Am J Cardiol 1988; 62:876-881.
13. Eich D, Thompson JA, Ko DJ et al. Hypercholesterolemia in long-term survivors of heart transplantation: An early marker of accelerated coronary artery disease. J Heart Lung Transplant 1991; 10:45-49.
14. Sharples LD, Caine N, Mullins P et al. Risk factor analysis for the major hazards following heart transplantation: Rejection, infection, and coronary occlusive disease. Transplantation 1991; 116:177-183.
15. Fricker FJ, Griffth BP, Hardesty RI et al. Experience with heart transplantation in children. Pediatrics 1987; 79:138-146.
16. Bieber CP, Hunt SA, Schwinn DA et al. Complications in long-term survivors of cardiac transplantation. Transplant Proc 1981; 13:207-211.
17. Odland MD, Kasiske BL. Kidneys from female donors are at increased risk for chronic

allograft rejection. Transplant Proc 1993; 25:912.

18. Brenner BM, Cohen RA, Milford EL. In renal transplantation, one size may not fit all. J Am Soc Nephrol 1992; 3:162-169.

19. Yuge J, Cecka JM. Sex and age effects in renal transplantation, In: Terasaki P, ed. Clinical Transplants 1991. Los Angeles, CA: UCLA Tissue Typing Laboratory, 1991, 257.

20. Arita K, Uozumi T, Oki S et al. The function of the hypothalamo-pituitary axis in brain dead patients. Acta Neurochir 1993; 123:64-75

21. Takada M, Nadeau KC, Chandraker A et al. Brain death selectively stimulates expression of renal inflammatory mediators and cytokine expression in means of inhibition. Surg Forum 1997 (in press).

22. Day JD, Rayburn BK, Gaudin PB et al. Cardiac allograft vasculopathy: The central pathogenetic role of ischemia-induced endothelial cell injury. J Heart Lung Transplantation 1995; 14:S143-149.

23. Johnson MR. Transplant coronary disease: Nonimmunological factors. J Heart Lung Transplantation 1992; 11:S124-132.

24. Libby P. Transplantation-associated arteriosclerosis: possible mechanisms. In: Tilney NL, Strom TB, and Paul LC, eds. Transplantation Biology: Cellular and Molecular Aspects. Philadelphia, PA: Lippincott-Raven Publishers, 1996: 577-586.

25. Ladoux A, Frelin C. Hypoxia is a strong inducer of vascular endothelial growth factor mRNA expression in the heart. Biochem Biophys Re Commun 1993; 195:1005-1010.

26. Levy AP, Levy NS. Wegner S et al. Transcriptional regulation of the rat vascular endothelial growth factor gene by hypoxia. J Biol Chem 1995; 270:13333-13340.

27. Kourembanas S, Hannan RL, Faller DV. Oxygen tension regulates the expression of the platelets-derived growth factor-B chain gene in human endothelial cells. J Clin Invest 1990; 86:670-674.

28. Lindney V, Reidy MA. Proliferation of smooth muscle cells after vascular injury is inhibited by an antibody against basic fibroblast growth factor. Proc Natl Acad Sci USA 1991; 88:3739-3743.

29. Dvorak HF, Brown LF, Detmar M et al. Vascular permeability factor/vascular endothelial growth factor, microvascular hypermeability, and angiogenesis. Am J Pathol 1995; 146:1029-1039.

30. Powell JS, Clozel JP, Muller RK et al. Inhibitors of angiotensin-converting enzyme prevent myointimal proliferation after vascular injury. Science 1989; 245:186-188.

31. Plissonier D, Amichot G, Duriez M et al. Effect of converting enzyme inhibition on

allograft-induced arterial wall injury and response. Hypertension 1991; 18(suppl 1): 1147-1154.

32. Furukawa Y, Matsumori A, Hirozane T et al. Angiotensin II receptor antagonist TCV-116 reduces graft coronary artery disease and preserves graft status in a murine model. A comparable study with captopril. Circulation 1996; 93:333-339.

33. Springer T. Adhesion receptors of the immune system Nature 1990; 346:425.

34. Shoskes DA, Parfrey NA, Halloran PF. Increased major histocompatibility complex antigen expression unilateral ischemic acute tubular necrosis in the mouse. Transplantation 1990; 49:201-207.

35. Goes N, Sims T, Urmson J et al. Disturbed MHC regulation in the IFN-γ knockout mouse. Evidence for three states of MHC expression with distinct roles for IFN-γ. J Immunol 1995; 155:4559-4566.

36. Linas SL, Whittenburg D, Parsons PE et al. Mild renal ischemia activates primed neutrophils to cause acute renal failure. Kidney Int 1992; 42:610-616.

37. Yamazaki T, Seko Y, Tamatani T et al. Expression of inter-cellular adhesion molecule-1 in rat hearts with ischemia/reperfusion and limitation of infarct size by treatment with antibodies. Am J Path 1993; 143:410.

38. Biolodeau M, Fitchett DH, Guerraty A et al. Complications in long-term survivors of cardiac transplantation. J Heart Transplant 1989; 8:454-459.

39. Stamler JS, Vaughan DE, Rudd MA et al. Frequency of hypercholesterolemia after cardiac transplantation. Am J Cardiol 1988; 62:1268-1272.

40. Becker DM, Chamberlain B, Swank R et al. Prevalence of hyperlipidemia in heart transplant recipients. Transplantation 1987; 44:323-324.

41. Becker DM, Chamberlain B, Swank R et al. Relationship between corticosteroid exposure and plasma lipid levels in heart transplant recipients. Am J Med 1988; 85:632-638.

42. Farmer JA, Ballantyne CM, Frazier OH et al. Lipoprotein (a) and apolipoprotein changes after cardiac transplantation. J Am Coll Cardiol 1991; 18:926-930.

43. Renlund DG, Bristow MR, Crandall BG et al. Hypercholesterolemia after heart transplantation: amelioration by corticosteroid-free maintenance immunosuppression. J Heart Transplant 1989; 8:214-219.

44. Öst L. Effects of cyclosporine on prednisolone metabolism. Lancet. 1984; 1:451.

45. Superko HR, Haskell WL, Di Ricco CD. Lipoprotein and hepatic lipase activity and high-density lipoprotein subclasses after cardiac transplantation. Am J Cardiol 1990; 66:1131-1134.

46. Keane WF, Kasiske BL, O'Donnell MP. Lipids and progressive glomerulosclerosis. Am J Nephrol 1988; 8:261-271.

47. Alonso DR, Storek PK, Monick R. Studies on the pathogenesis of atherosclerosis induced in rabbit cardiac allografts by the synergy of graft rejection and hypercholesterolemia. Am J Pathol 1977; 87:415-442.

48. Kuwahara M, Jacobson J, Kuwahara M et al. Coronary artery ultrastructural changes in cardiac transplant atherosclerosis in the rabbit. Transplantation 1991; 52:759-765.

49. Adams DH, Karnovsky MJ. Hypercholesterolemia does not exacerbate arterial intimal thickening in chronically rejecting rat cardiac allografts Transplant Proc 1989; 21: 437-439.

50. Mennander A, Tikkanen MJ, Räsänen-Sokolowski AK et al. Chronic rejection in rat allografts. VI. Effects of hypercholesterolemia in allograft atherosclerosis. J Heart Lung Transplant 1993; 12:123-131.

51. Tanaka H, Sukhova G, Libby P. Interaction of the allogeneic state and hypercholesterolemia in arterial lesion formation in experimental cardiac allografts. Arteriosclerosis Thrombosis 1994; 14:734-745.

52. Räsänen-Sokolowski AK, Tilly-Kiesi M, Ustinov J et al. Hyperlipidemia accelerates allograft arteriosclerosis (chronic rejection) in the rat. Arteriosclerosis Thrombosis 1994; 14:2032-2042.

53. Brazy PC, Pirsch JD, Belzer FO. Factors affecting renal allografts function in long-term recipients. Am J Kidney Dis 1992; 19: 558-566.

54. Dimeny E, Wahlberg J, Lithell H et al. Hyperlipidemia in renal transplantation-risk factor for long-term graft outcome. Eur J Clin Invest. 1995; 25:574-583.

55. Vollmer E, Bosse A, Bogeholz J et al. Apolipoproteins and immunohistological differentiation of cells in the arterial wall of kidneys in transplant arteriopathy. Morphological parallels with atherosclerosis. Pathol Res Pract 1991; 187:957-962.

56. Eich D, Thompson JA, Ko DJ et al. Hypercholesterolemia in long-term survivor of heart transplantation: An early marker of accelerated coronary arterial disease. J Heart Lung Transplant 1991; 10:45.

57. Kobashigawa JA, Katznelsosn S, Laks H et al. Effect of pravastatin on outcomes after cardiac transplantation. N Engl J Med 1995; 53:621-627.

58. Myers BD, Ross J, Newton L et al. Cyclosporine-associated chronic nephropathy. N Engl J Med 1984; 311:699-705.

59. Gao SZ, Schroeder JS, Alderman EL et al. Clinical and laboratory correlates of accelerated coronary artery disease in the cardiac transplant patient. Circulation 1987; 76: V56-61.

60. Cramer DV, Chapman FA, Wu G-D et al. Cardiac transplantation in the rat. II. Alteration of the severity of donor graft arteriosclerosis by modulation of the host immune response. Transplantation 1990; 50:554-558.

61. Paul LC, Davidoff A, Benediktsson H. Cardiac allograft atherosclerosis in the rat. The effect of histocompatibility factors, cyclosporine, and an angiotensin-converting enzyme inhibitor. Transplantation 1994; 57: 1767-1772.

62. Mennander A, Tiisala S, Paavonen T et al. Chronic rejection of the rat aortic allograft. II. Administration of cyclosporine induces accelerated allograft arteriosclerosis. Transplant Int 1991; 4:173-179.

63. Meiser BM, Billingham ME, Morris RE. Effects of cyclosporine, FK506, and rapamycin on graft-vessel disease. Lancet 1991; 338:1297-1298.

64. Gao SZ, Schroeder JS, Alderman EL et al. Prevalence of accelerated coronary artery disease in heart survivors. Comparison of cyclosporine and azathioprine regimens. Circulation 1989; 80:III100-105.

65. Olivari MT, Homans DC, Wilson RF et al. Coronary artery disease in cardiac transplant patients receiving triple-drug immunosuppressive therapy. Circulation 1989; 80: III111-115.

66. Uretsky BF, Murali S, Reddy PS et al. Development of coronary artery disease in cardiac transplant patients receiving immunosuppressive therapy with cyclosporine and prednisone. Circulation 1987; 76:827-834.

67. Morris RE. Rapamycins: antifungal, antitumor, antiproliferative, and immunosuppressive macrolides. Transplant Rev 1992; 6:39-87.

68. Koskinen PK, Krogerus LA, Nieminen MS et al. Cytomegalovirus infection-associated generalized immune activation in heart allograft recipients: a study of cellular events in peripheral blood and endomyocardial biopsy specimens. Transplant Int 1994; 7: 163-171.

69. Lemström KB, Bruning JH, Bruggeman CA et al. Cytomegalovirus infection enhances smooth muscle cell proliferation and intimal thickening of rat aortic allografts. J Clin Invest 1993; 92:549-558.

70. Norman SJ, Salomon DR, Leelachaikul P et al. Acute vascular rejection of the coronary arteries in human heart transplantation: pathology and correlations with immunosuppression and cytomegalovirus infection. J Heart Lung Transplant 1991; 10:674-687.

71. Grattan MT, Moreno-Cabral CE, Starnes VA et al. Cytomegalovirus infection is

associated with cardiac allograft rejection and atherosclerosis. JAMA 1989; 261:3561-3566.

72. McDonald K, Rector TS, Braulin EA et al. Association of coronary artery disease in cardiac transplant recipients with cytomegalovirus infection. Am J Cardiol 1989; 64: 359-362.

73. Weimar W, Balk AHMM, Metselaar HJ et al. On the relation between cytomegalovirus infection and rejection after heart transplantation. Transplantation 1991; 52:162-164.

74. Grefte A, Blom N, van der Giessen M et al. Ultrastructual analysis of circulating cytomegalic cells in patients with active cytomegalovirus infection: Evidence for virus production and endothelial origin. J Infect Dis 1993; 168:1110-1118.

75. Reinke P, Fietze E, Ode-Hakim S et al. Late acute renal allograft rejection and symptomless cytomegalovirus infection. Lancet 1994; 344:1737-1738.

76. Fujinami RS, Nelson JA, Walker L et al. Sequence homology and immunologic cross reactivity of human cytomegalovirus with HDL-DR beta chain: a means for graft rejection and immunosuppression. J Virol 1988; 62:100-105.

77. Beck S, Barrel BG. Human cytomegalovirus encodes a glycoprotein homologous to MHC class I antigens. Nature 1988; 331:269-272.

78. Koskinen PK. Induction of vascular cell adhesion molecule-1 is associated with cytomegalovirus antigenemia in human heart allograft. Transplantation 1993; 56:1103-1108.

79. Hancock WW, Whiteley DW, Tullius SG et al. Cytokines, adhesion molecules, and the pathogenesis of chronic rejection of rat renal allografts. Transplantation 1993; 56:643-650.

80. Tullius SG, Tilney NL. Both alloantigen-dependent and -independent factors influence chronic allograft rejection. Transplantation 1995; 59:313-318.

81. Väkevä A, Laurila P, Meri S. Regulation of complement membrane attack complex formation in myocardial infarction. Am J Pathol 1993; 143:65-75.

82. DeHeer E, Davidoff A, van der Wal A et al. Chronic renal allograft rejection: Transplantation-induced antibodies against basement membrane antigen. Lab Invest 1994; 8:313-318.

83. Moore FD Jr, Davis C, Rodrick M et al. Neutrophil activation in thermal injury as assessed by complement receptor upregulation. N Engl J Med 1986; 314: 948-953.

84. Wiles ME, Hechtman HB, Morel NML et al. Hypoxia reoxygenation-induced injury of cultured pulmonary microvessel endothelial cells. J Leukocyte Biol 1993; 53:490.

85. Porter KA. Renal transplantation. In: Heptinstall RH, ed. Pathology of the Kidney, 4th ed. Boston, MA: Little Brown and Co., 1992; 1799-1933.

86. Sasaguri S, Eishi Y, Tsukada T et al. Role of smooth muscle cells and macrophages in cardiac allograft arteriosclerosis in rabbits. J Heart Transplant 1990; 9:18-24.

87. Paul LC, Grothman GT, Benediktsson H et al. Macrophage subpopulations that infiltrate normal and rejecting heart and kidney grafts in the rat. Transplantation 1992; 53: 157-162.

88. Moolenaar W, Bruijn JA, Schrama E et al. T-cell receptors and ICAM-1 expression in renal allografts during rejection. Transplant Int 1991; 4:140-145.

89. Herskowitz A, Mayne AE, Willoughby SB et al. Patterns of myocardial cell adhesion molecule expression in human endomyocardial biopsies after cardiac transplantation. Induced ICAM-1 and VCAM-1 related to implantation and rejection. Am J Pathol 1994; 145:1082-1094.

90. Ross R. The pathogenesis of atherosclerosis: A perspective for 1990s. Nature 1993; 362:801-809.

91. Häyry P, Mennander A, Yilmaz J et al. Towards understanding the pathophysiology of chronic rejection. Clin Invest 1992; 70: 780-790.

92. Habib R, Broyer M. Clinical significance of allograft glomerulopathy. Kidney Int 1993; 44 (suppl 43):S95-98.

93. Shankland SJ, Ly H, Thai K et al. Increased glomerular capillary pressure alters glomerular cytokine expression. Circ Res 1994; 75:844-853.

94. Brenner BM, Milford EL. Nephron underdosing: A programmed cause of chronic renal allograft failure. Am J Kidney Dis 1993; 21:66-72.

CHAPTER 6

Tissue Remodeling and Extracellular Matrix Proteins

Marlene Rabinovitch

A variety of matrix proteins are implicated in the pathogenesis of chronic allograft vascular disease including fibronectin, tenascin, collagen, glycosaminoglycans and proteoglycans. Alterations in these matrix components can influence a range of cellular functions including proliferation, migration, differentiation, adhesion and apoptosis. Our laboratory has shown that central to the pathobiology of allograft vascular disease, there is an increase in synthesis and secretion of the extracellular matrix glycoprotein fibronectin in response to cytokines and liberation of proteolytic enzymes, specifically elastases. Fibronectin creates a two-way highway, both encouraging the transendothelial migration of inflammatory cells and promoting smooth muscle cell migration into the subendothelium. Elastases also release mitogenically active growth factors from proteoglycan stores and upregulate the glycoprotein tenascin, which plays a critical role in amplifying the proliferative response of the vascular smooth muscle cells to mitogens. The following is a discussion of the mechanisms which lead to changes in fibronectin and tenascin and studies which support their critical roles in SMC migration and proliferation, related to occlusive changes in coronary arteries following placement of a cardiac allograft.

Fibronectin and Smooth Muscle Cell Migration

Our previous studies showed that intimal cushions which are necessary for postnatal closure of the fetal ductus arteriosus[1] in response to oxygen are formed by migration of vascular smooth muscle cells into the subendothelium and that this process is dependent on increased production of the matrix glycoprotein fibronectin.[2-4] The expression of the receptor for hyaluronan mediated motility (RHAMM) at the leading edges of these cells also plays an important role in their migratory phenotype.[2] Blocking migration of these cells can be achieved by preventing their interaction with fibronectin or hyaluronan. The mechanism upregulating production of fibronectin was further addressed. Since ductus arteriosus smooth muscle cells have reduced elastin binding proteins on their surfaces resulting in poor assembly of elastin fibers and consequent generation of elastin peptides,[5,9] we investigated whether elastin peptides stimulate fibronectin production. In aortic smooth muscle cell cultures, elastin peptides do indeed upregulate fibronectin production as does chondroitin sulfate which removes elastin binding proteins thereby impairing elastin assembly and producing elastin peptides.[6]

We extended these studies to investigate the molecular mechanism involved in the

Transplant-Associated Coronary Artery Vasculopathy, edited by Marlene L. Rose.
©2001 Eurekah.com.

upregulation of fibronectin in these migratory ductus cells and found that there is heightened efficiency of translation of the fibronectin mRNA.[10] This appears to be due to an increase in the expression of an RNA binding protein which interacts with the A+U rich element in the 3'untranslated region of the fibronectin mRNA and recruits polysomes. The RNA binding protein was purified and found to be light chain 3 of microtubule associated proteins 1A and 1B. Transfection of aortic smooth muscle cells with LC3 resulted in increased fibronectin mRNA translation and protein synthesis. Moreover, transfection of decoy RNA to sequester these RNA binding proteins resulted in reduced fibronectin production in cultured cells. In the intact fetal lamb, administration of this decoy RNA to the ductus by gene transfer approach, reduced fibronectin synthesis, and resulted in marked inhibition of neointimal formation, and ductal patency owing to increased luminal patency in the full term lamb.[11]

In parallel studies ongoing in our laboratory, we investigated whether similar upregulation of fibronectin might underlie the coronary arteriopathy observed in association with heart transplantation. We therefore characterized the coronary artery changes that occur consequent to the placement of a heterotopic heart transplant in a piglet immunosuppressed with subtherapeutic doses of cyclosporine.[12] Increased production and deposition of fibronectin were indeed observed, accompanied by increased expression of the cytokines, interleukin 1 beta (IL-1β) and tumor necrosis factor alpha (TNFα) and increased adhesion of T cells along the endothelium and in the subendothelium as well as in the adventitia (Fig. 6.1). Based on these findings, we proposed and later showed that the cytokines produced by inflammatory cells might be inducing the upregulation of fibronectin in coronary artery smooth muscle cells and endothelial cells by paracrine and autocrine mechanisms.[13,14] We also documented increased synthesis of fibronectin in coronary artery smooth muscle cells harvested from the donor compared to

the host heart in these piglets and showed that this could be abrogated by treating the cultures with an IL-1 receptor antagonist[14] (Fig. 6.2). There was a decrease in both the mRNA level for fibronectin as well as in the protein measured by metabolic labeling with [35S]-methionine and gelatin sepharose extraction.

Further studies indicated that not only is there increased endogenous expression of cytokines in donor coronary artery smooth muscle and endothelial cells when compared to host cells, but that there is actually reciprocal coinduction of IL-1β and TNFα regulating the increase in fibronectin production.[15] That is, neutralizing antibodies to TNFα can repress IL-1β-mediated induction of FN and conversely IL-1 antibodies can abrogate the TNF-induced upregulation of fibronectin by these cells (Fig. 6.3A and 6.3B).

In subsequent studies, we used a rabbit heterotopic heart transplant model to show that increased expression of cytokines is important in the pathobiology of coronary artery disease.[16] Specifically, injection of the rabbits with the TNFα soluble receptor largely prevented the development of coronary artery neointimal lesions. The TNF soluble receptor did not, however, decrease the severity of myocardial rejection in this model. The rabbits were not immunosuppressed and by 8 days after the procedure 70% of coronary arteries, developed neointimal formation compared to 35% in the treated group (Fig. 6.4).

Further studies were then carried out to address the mechanism whereby cytokine induction specifically of fibronectin might be implicated in the pathogenesis of coronary artery disease after a heart transplant. Studies by other groups had shown that inflammatory cells migrate toward a FN gradient and we had confirmed that smooth muscle cell production of FN by cytokines will induce T cell transendothelial migration[17] and that this process can be blocked by CS1 peptides (CS1) encoding a domain unique to cell-associated fibronectin which interacts with the α4β1 integrin (Fig. 6.5). Thus, we reasoned that preventing native fibronectin interaction with smooth muscle or with inflammatory cells

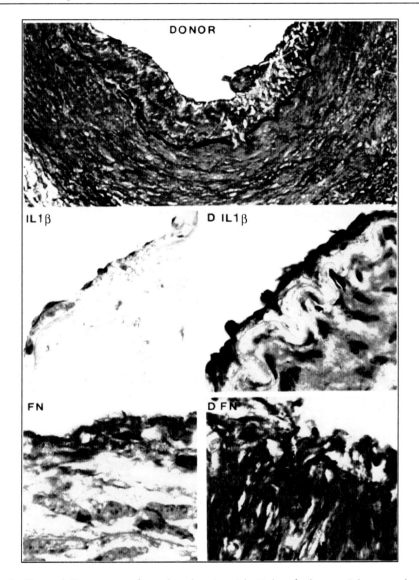

Fig. 6.1. Top panel: Coronary artery from a donor heart in a piglet 10 days after heterotopic heart transplant on subtherapeutic doses of cyclosporine shows a developing neointima on top of a fragmented elastic lamina. Middle panels show immunostaining for interleukin-1β (IL-1β). In representative host (left) and donor (right) coronary arteries. Lower panels show immunostaining for fibronectin (FN) in representative host (left) and donor (right) coronary arteries. Note increased intensity of immunostaining for IL-1β and FN in donor compared to host vessels.

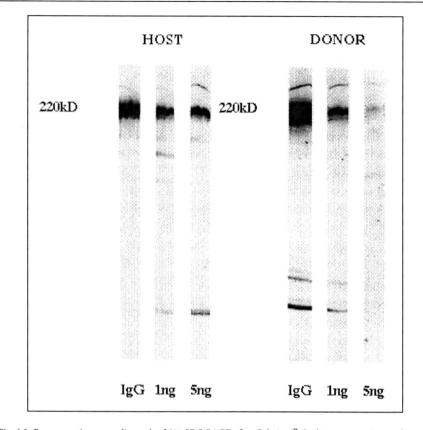

Fig. 6.2. Representative autoradiograph of 5% SDS-PAGE after Gelatin 4β-Sepharose extraction to determine fibronectin synthesis in host and donor coronary artery smooth muscle cells (harvested from a CsA-treated piglet at 10 d after transplant) in the presence and absence of IL-1ra. The apparent reduction in the host cell fibronectin synthesis at 5 ng/ml was not confirmed when the values were normalized by cell DNA content (data not shown). A progressive reduction in fibronectin synthesis is observed in the donor cells at 1 and 5 ng/ml of IL-1 ra. (Reproduced with permission from Clausell N and Rabinovitch M, J Clin Invest 1993; 92:1850-858.)

would prevent the trafficking of the inflammatory cells necessary for release of cytokines in the subendothelium and would also abrogate the smooth muscle cell migratory response. In addition, the CS1 peptide will also block T cell interactions with cell adhesion molecules such as vascular cell adhesion molecule (VCAM) which are also upregulated on the surfaces of coronary artery endothelial cells after transplant.[18] CS1 peptides were in fact shown to prevent the development of coronary artery disease in the overwhelming majority of vessels in the donor heart of the rabbit following heart transplant (Fig. 6.6A and B). There was arrested transendothelial T cell migration as judged by immunohistochemistry and repressed fibronectin expression.[19]

Further unpublished studies in our laboratory suggest that IL-1β increases fibronectin mRNA levels, whereas TNFα affects a posttranscriptional mechanism involved in FN production.[20] Since cytokines such as TNFα can induce production of nitric oxide, we investigated whether nitric oxide might be the link between TNFα stimulation and

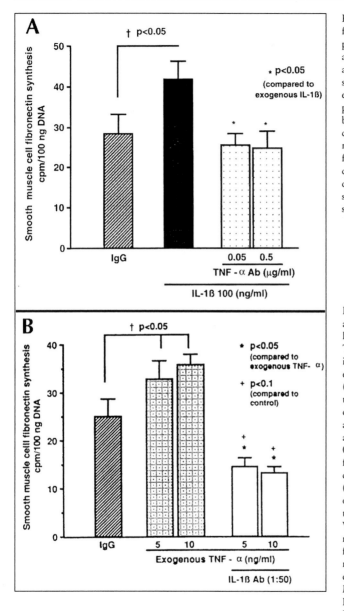

Fig. 6.3.A. IL-1β-stimulated fibronectin synthesis in the presence of neutralizing TNFα antibodies. Interleukin-1β at a concentration of 100 ng/ml simulated fibronectin, but this effect was abrogated in the presence of neutralizing antibodies (Ab) to TNFα in concentrations of 0.05 and 0.5 μg/ml (P < 0.05), lowering fibronectin production to control (IgG values). Values depicted in the graph represent mean values ± SE from six different experiments.

Fig. 6.3.B. Effect of IL-1β antibodies on TNFα stimulated synthesis of fibronectin. The exogenous TNFα increase in fibronectin levels compared to control (P < 0.05), at concentrations of 5 and 10 ng/ml, was downregulated when IL-1β antibodies (AB) were added at a dilution of 1:50 (P < 0.05, respectively). Notably, fibronectin production decreases to below control (IgG) levels in the presence of IL-1β antibodies trended toward significance (P < 0.1). Values depicted in the graph represent mean values ± SE from three different experiments. (3A & B Reproduced with permission from Molossi S, Clausell N and Rabinovitch M, J Cell Physiol 1995; 163:19-29.)

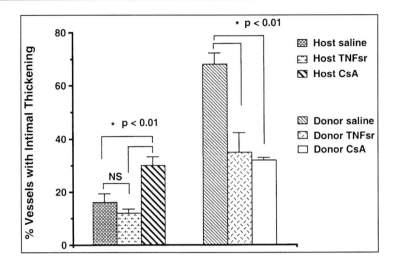

Fig. 6.4. Bar graph of effect of administration of tumor necrosis factor-soluble receptor (TNFsr) or cyclosporine A (CsA) on the number of vessels with intimal proliferation from host and donor hearts. A decrease in the number of vessels with intimal proliferation is observed in the donor hearts from the TNFsr and CsA groups compared with the saline group (P < 0.01). In the host hearts, there is no significant difference related to the use of TNFsr compared with the saline group with respect to the incidence of vessels with graft arteriopathy, but there is an increase in the number of vessels affected in the CsA group compared with the saline and TNFsr groups (P < 0.01). In the host hearts, a total number of 643 coronary arteries in the saline-treated group, 729 in the TNFsr-treated group, and 246 in the CsA-treated group were analyzed. In the donor hearts, a total number of 566 coronary arteries in the saline-treated group, 721 in the TNFsr-treated group, and 240 in the CsA-treated group were analyzed. (Reproduced with permission from Clausell N, Molossi S, Sett S and Rabinovitch M, Circulation 1994; 89:2768-2779.)

fibronectin production. In coronary artery smooth muscle cells, we showed that TNFα does indeed increase nitric oxide production by smooth muscle cells. By promoting nitric oxide production, fibronectin synthesis is enhanced and conversely by preventing nitric oxide production or availability, fibronectin production is not increased despite the presence of cytokines. The posttranscriptional mechanism identified is similar to that observed initially in fibronectin producing ductus arteriosus smooth muscle cells. That is, coronary artery smooth muscle cells in response to TNFα-mediated nitric oxide release, produce increased amounts of LC3, a microtubule-associated protein, which increases efficiency of fibronectin mRNA translation to protein by ribosome recruitment.

Further studies suggested that elastin peptides generated by elastase activity also contribute to the posttranscriptional regulation of FN in a manner synergistic with cytokines. IL-1β-mediated induction of FN can be abrogated by specific selective and nonselective serine elastase inhibitors such as elafin and alpha 1 proteinase inhibitor-respectively. The effect of elafin can be overridden by the addition of elastin peptides suggesting that elafin prevents cytokine-mediated induction of FN by preventing the degradation of elastin via a serine elastase and the production of elastin peptides necessary for the cytokine mediated response (our unpublished data).

There is also direct evidence for the activity of a serine elastase in the pathobiology of postcardiac transplant coronary artery

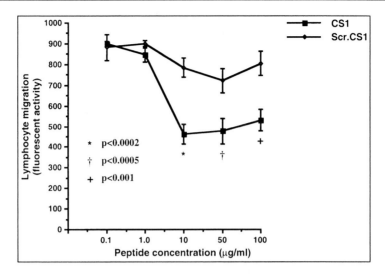

Fig. 6.5. Transendothelial migration of lymphocytes in the presence of increasing doses of CS1 peptide. The migratory behavior of nonactivated (resting) lymphocytes was effectively inhibited by 70% at an optimal dose of 10 μg/ml of CS1 peptide compared to control (scrambled [Scr.] CS1) peptide (*P < 0.0002). Further increase in CS1 concentrations, i.e., 50 and 100 μg/ml, also showed a significant inhibitory effect compared to control peptide (†P < 0.0005 and +P < 0.001, respectively), albeit not higher than the effect achieved with the concentration of 10 μg/ml. Values depicted in the graph represent mean values ±SE of four to seven experiments. (Reproduced with permission from Molossi, S, Elices M, Arrhenius T and Rabinovitch M. J Cell Physiol 1995; 164:620-633.)

disease related to proliferation of inflammatory as well as smooth muscle cells.[21] We have shown that elastases can release mitogenic factors from the extracellular matrix such as basic fibroblast growth factor (FGF-2).[22] In addition, elastases can activate matrix metalloproteinases which proteolyse collagen.[23] Proteolysed collagen induces beta 3 integrin-dependent transcriptional upregulation of the matrix glycoprotein tenascin in vascular smooth muscle cells (Fig. 6.7).[24,25] In turn, tenascin clusters $\alpha_v\beta_3$ integrins and rearranges the actin cytoskeleton into focal adhesion contacts which induce a tyrosine phosphorylation signal and cluster receptors for growth factors. Once the clustered growth factor receptors are ligated, there is induction of rapid phosphorylation culminating in the signal for cell growth. Thus tenascin enhances smooth muscle cell responsivity to growth factors by inducing

efficient receptor phosphorylation and its upregulation is mediated by elastases and by matrix metalloproteinases. Most recent work suggests that withdrawal of matrix metalloproteinases or elastases shut down TN synthesis and this not only prevents smooth muscle cell proliferation but actually induces apoptosis and regression of vascular disease.

Based on the evidence that there was increased serine elastase activity in coronary arteries from piglets following cardiac transplant,[21] we initiated studies in which the serine elastase inhibitor, elafin[26] was given to rabbits following heterotopic heart transplant. We showed that elafin infusion markedly reduced the number of coronary arteries with intimal lesions as well as the severity of these lesions in affected vessels (Fig. 6.8). In addition, the elafin prevented the induction of fibronectin as well as smooth muscle cell

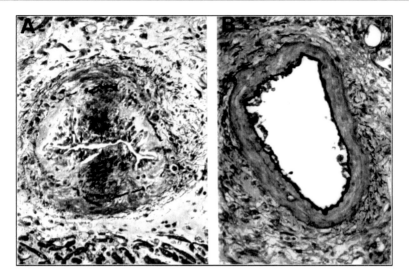

Fig. 6.6. A) Representative photomicrographs of Movat pentachrome staining of small coronary arteries in donor control (scrambled CS1) and donor CS1-treated groups. Control animals had extensive intimal thickening in allograft small coronary arteries (A, arrows pointing to severe luminal occlusion), which contrasts with a markedly attenuated intimal lesion observed in allograft small coronary arteries from CS1-treated animals (B). B) Effect of CS1 peptide treatment on the number of coronary arteries with intimal lesions in both host and donor hearts. The vast majority of vessels with intimal thickening were seen in small (diameter £ 100mm) and medium (diameter > 100 £ 500 μm) size coronary arteries. The number of affected vessels in the CS1-treated group was significantly reduced compared with the control (scrambled CS1) group (P < 0.001 for small size vessels and P < 0.05 for medium size vessels), where a total of 617 vessels and 827 vessels, respectively, were analyzed. In the host coronary arteries, no differences were seen in both groups for small and medium size vessels, where a total of 1,054 vessels in the CS1-treated group and a total of 999 vessels in the control group were analyzed. (Reproduced with permission from Molossi S, Elices M, Arrhenius R, Diaz R, Coulber C and Rabinovitch M. J Clin Invest 1995; 95:2601-2610.)

and T cell proliferation as judged by immunohistochemistry. Of particular interest was the fact that elafin also prevented the degradation of myocytes associated with myocardial rejection and necrosis. Thus, it appeared that increased elastase was playing a role in the pathobiology of cardiac dysfunction.

To further address whether chronic inflammation and the associated development of cardiac dysfunction may be related to increased elastase activity, we have subsequently used 1K (ZD 0892) (Zeneca) a selective serine elastase inhibitor in mice following experimental encephalomyocarditis. Infection with this virus causes acute myocardial inflammation and the subsequent development of myocardial fibrosis resulting in left ventricular dilatation and dysfunction. We were able to show that administration of the selective serine elastase inhibitor 1K in the drinking water of infected mice significantly reduced the area of inflammatory necrosis and associated microperfusion abnormalities and the consequent increased deposition of collagen in the myocardium and the evolution of

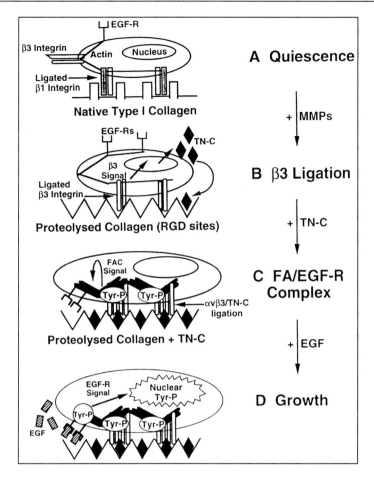

Fig. 6.7. Hypothetical model for the regulation and function of tenascin-C in vascular smooth muscle cells. A) Vascular smooth muscle cells attach and spread on native type I collagen using β_1 integrins. Under serum-free conditions, the cells withdraw from the cell cycle and become quiescent. B) Degradation of native type I collagen by matrix metalloproteinases (MMPs) leads to the exposure of cryptic RGD sites that preferentially bind β_3 subunit-containing integrins. In turn, occupancy and activation of β_3 integrins signals the production of TN-C. (C) Incorporation of multivalent TN-C protein into the underlying substrate leads to further aggregation and activation of β_3-containing integrins ($\alpha_v\beta_3$), and to the accumulation of tyrosine-phosphorylated (Tyr-P) signaling molecules and actin into a focal adhesion complex (FAC). Note that even in the absence of the EGF ligand, the TN-C-dependent reorganization of the cytoskeleton leads to clustering of actin-associated EGF-Rs. (D) Addition of EGF ligand to clustered EGF-Rs results in rapid and substantial tyrosine phosphorylation of the EGF-R and activation of downstream pathways culminating in the generation of nuclear signals leading to cell proliferation. (Reproduced with permission from Jones PL, Crack J and Rabinovitch M. J Cell Biol 1997; 139:279-293.)

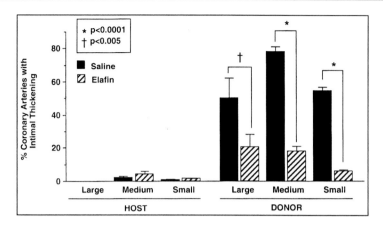

Fig. 6.8. Effect of recombinant human elafin treatment on the number of coronary arteries with intimal lesions in both host and donor hearts. Intimal thickening was observed in large (>500 mm diameter), medium (>100-500 mm diameter), and small (≤ 100 μm diameter) coronary arteries. The number of vessels in the elafin-treated group were significantly reduced compared to the control group ([†]P < 0.005 for large coronary artery and [*]P < 0.0001 for medium and small coronary artery, respectively). In host coronary arteries, no differences were observed in groups for large-, medium-, and small-size vessels. (Reproduced with permission from Cowan B, Baron O, Crack J, Coulber C, Wilson G and Rabinovitch M. J Clin Invest 1996. 97:2452-2468.)

chronic dysfunction as judged by measurements in the isolated perfused mouse heart.[27]

Conclusions

Thus, we have shown through experimental studies, that matrix proteins are of critical importance in the remodeling process associated with postcardiac transplant coronary artery disease. Cytokine-mediated induction of fibronectin by coronary artery endothelial and smooth muscle cells influence both smooth muscle cell and T cell migration and the subsequent formation of the neointima. Heightened elastase activity is a necessary cofactor in this reaction. In addition to the important synergism between elastin peptides and cytokines in mediating induction of fibronectin, increased elastase activity has been shown in other systems to liberate growth factors from the extracellular matrix that are mitogenic for smooth muscle cells and to upregulate the glycoprotein tenascin-C which amplifies the proliferative response to these molecules. The use of cytokine blockade or the prevention of fibronectin-cell interaction

with the CS1 peptide (alpha 4 beta 1 integrin inhibitor) were both highly effective strategies in reducing the incidence and severity of coronary artery lesions following experimental heart transplant. Inhibition of serine elastases had the added advantage of protecting the myocardium from rejection, and although not directly tested, could conceivably be protective against late myocardial dysfunction and fibrosis.

References

1. Grittenberger-de-Groot A, Van Ertbruggen L, Moulaert AJ et al. The ductus arteriosus in the preterm infant. Histologic and clinical observations. J Pediatr 1980; 96:88-93.
2. Boudreau N, Clausell N, Boyle J et al. Transforming growth factor-β regulates increased ductus arteriosus endothelial glycosaminoglycan synthesis and a posttranscriptional mechanism controls increased smooth muscle fibronectin, features associated with intimal proliferation. Lab Invest 1992; 67:350-359.
3. Boudreau N, Turley E, Rabinovitch M. Fibronectin, hyaluronan and a hyaluronan binding protein contribute to increased duc-

tus arteriosus smooth muscle cell migration. Dev Biol 1991; 143:235-247.

4. Boudreau N, Rabinovitch M. Developmentally regulated changes in extracellular matrix in endothelial and smooth muscle cells in the ductus arteriosus may be related to intimal proliferation. Lab Invest 1991; 64:187-199.

5. Hinek A, Mecham RP, Keeley F et al. Impaired elastin fiber assembly related to reduced 67-kD elastin-binding protein in fetal lamb ductus arteriosus and in cultured aortic smooth muscle cells treated with chondroitin sulfate. J Clin Invest 1991; 88:2083-2094.

6. Hinek A, Boyle J, Rabinovitch M. Vascular smooth muscle cell detachment from elastin and migration through elastic laminae is promoted by chondroitin sulfate-induced "shedding" of the 67-kDa cell surface elastin binding protein. Exp Cell Res 1992; 203:344-353.

7. Zhu L, Dagher E, Johnson DJ et al. A developmentally regulated program restricting insolubilization of elastin and formation of laminae in the fetal lamb ductus arteriosus. Lab Invest 1993; 68:321-331.

8. Hinek A, Rabinovitch M. 67-kD elastin-binding protein is a protective "companion" of extracellular insoluble elastin and intracellular tropoelastin. J Cell Biol 1994; 126:563-574.

9. Hinek A, Rabinovitch M. The ductus arteriosus migratory smooth muscle cell phenotype processes tropoelastin to a 52-kDa product associated with impaired assembly of elastic laminae. J Biol Chem 1993; 268:1405-1413.

10. Zhou B, Boudreau N, Coulber C et al. Microtubule-associated protein 1 light chain 3 is a fibronectin mRNA-binding protein linked to mRNA translation in lamb vascular smooth muscle cells. J Clin Invest 1997; 100:3070-3082.

11. Mason CAE, Bigras J-L, O'Blenes SB et al. Gene transfer in utero biologically engineers a patent ductus arteriosus in lambs by arresting fibronectin-dependent neointimal formation. Nat Med 1999 (February, in press)

12. Clausell N, Molossi S, Rabinovitch M. Increased interleukin-1β and fibronectin expression are early features of the development of the postcardiac transplant coronary arteriopathy in piglets. Am J Pathol 1993; 142:1772-1786.

13. Molossi S, Clausell N, Rabinovitch M. Coronary artery endothelial interleukin-1β mediates enhanced fibronectin production related to postcardiac transplant arteriopathy in piglets. Circulation 1993; 88:248-256.

14. Clausell N, Rabinovitch M. Upregulation of fibronectin synthesis by interleukin-1β in coronary artery smooth muscle cells is associated with the development of the postcardiac transplant arteriopathy in piglets. J Clin Invest 1992; 92:1850-1858.

15. Molossi S, Clausell N, Rabinovitch M. Reciprocal induction of tumor necrosis factor-α and interleukin-1β activity mediates fibronectin synthesis in coronary artery smooth muscle cells. J Cell Physiol 1995; 163:19-29.

16. Clausell N, Molossi S, Sett S et al. In vivo blockade of tumor necrosis factor-α in cholesterol-fed rabbits after cardiac transplant inhibits acute coronary artery neointimal formation. Circulation 1994; 89:2768-2779.

17. Molossi S, Elices M, Arrhenius T et al. Lymphocyte transendothelial migration toward smooth muscle cells in interleukin-1β stimulated cocultures is related to fibronectin interactions with $\alpha_4\beta_1$ and $\alpha_5\beta_1$ integrins. J Cell Physiol 1995; 164:620-633.

18. Molossi S, Clausell N, Sett S et al. ICAM-1 and VCAM-1 expression in accelerated cardiac allograft arteriopathy and myocardial rejection are influenced differently by cyclosporine A and tumour necrosis factor-a blockade. J Pathol 1995; 176:175-182.

19. Molossi S, Elices M, Arrhenius T et al. Blockade of very late antigen-4 integrin binding to fibronectin with connecting segment-1 peptide reduces accelerated coronary arteriopathy in rabbit cardiac allografts. J Clin Invest 1995; 95:2601-2610.

20. Mason C, Rabinovitch M. TNF-α induction of fibronectin synthesis is mediated by nitric oxide by a posttranscriptional mechanism involving increased expression and binding of LC3 to an ARE in the 3'UTR of FN mRNA. FASEB J (in press).

21. Oho S, Rabinovitch M. Postcardiac transplant arteriopathy in piglets is associated with fragmentation of elastin and increased activity of a serine elastase. Am J Pathol 1994; 145:202-210.

22. Thompson K, Rabinovitch M. Exogenous leukocyte and endogenous elastases can mediate mitogenic activity in pulmonary artery smooth muscle cells by release of extracellular-matrix bound basic fibroblast growth factor. J Cell Physiol 1996; 166:495-505.

23. Itoh Y, Nagase H. Preferential inactivation of tissue inhibitor of metalloproteinases-1 that is bound to the precursor of matrix metalloproteinase 9 (progelatinase B) by human neutrophil elastase. J Biol Chem 1995; 28:16518-16521.

24. Jones PL, Crack J, Rabinovitch M. Tenascin-C acts as a vascular smooth muscle cell survival factor and interacts with the $\alpha v\beta 3$

integrin to promote epidermal growth factor receptor phosphorylation and growth. J Cell Biol 1997; 139:279-293.

25. Jones PL, Jones FS, Zhou B et al. Denatured type I collagen induction of vascular smooth muscle cell tenascin-C gene expression is dependent upon a β3 integrin-mediated mitogen-activated protein kinase pathway and a 122 base pair promoter element. J Cell Sci (in press)

26. Cowan B, Baron O, Crack J et al. Elafin, a serine elastase inhibitor, attenuates post-cardiac transplant coronary arteriopathy and reduces myocardial necrosis in rabbits following heterotopic cardiac transplantation. J Clin Invest 1996; 97:2452-2468.

27. Lee J-K, Liu P, Dawood F et al. Elastase inhibitor reduces sequelae of experimental murine myocarditis. Circulation 1997; 96:I-738.

Abbreviations

IL-1β, interleukin 1 beta

TNFα, tumor necrosis factor alpha

VCAM, vascular cell adhesion molecule

FGF-2, basic fibroblast growth factor

CHAPTER 7

Insights Regarding the Pathogenesis of Transplant Arteriopathy from Experiments with Animals

Paul S. Russell, Catharine M. Chase and Robert B. Colvin

Much interest has focused in recent years upon the obstructive, atherosclerotic process that often appears in arterial vessels serving organs transplanted between allospecific or xenospecific individuals. The lesions concerned were first convincingly identified in kidneys transplanted to patients as early as the nineteen fifties.[1] Particular note was also made of impressive arteriosclerotic changes that appeared promptly in the coronary arteries of the first hearts transplanted in South Africa.[2] These coronary lesions raised special concern as the potential lethal consequences of lesions in this location were readily appreciated. Since then it has been recognized that these progressive and frequently severe lesions are a central feature of what has been termed "chronic rejection". As such, they clearly contribute to the ongoing functional impairment of any affected organ even though this may not take the form of sudden death as it can with coronary occlusion.

One interesting aspect of the phenomenon of chronic rejection is that although it typically includes certain widespread pathological features, such as the presence of focal collections of infiltrating inflammatory cells and varying degrees of fibrosis and atrophy, the vascular manifestations of the process often stand out as much more advanced and severe than anything else. Experimental evaluations of the process of transplant arteriopathy will be the subject of the present account.

The arteriosclerotic process that occurs in the vessels of transplanted organs has been called by quite a number of names, including "accelerated atherosclerosis", "chronic allograft arteriosclerosis", etc. We have generally referred to it as *transplant arteriopathy*, a term that is meant to embrace all of the various alterations that can be observed under different circumstances. Although suspicions from clinical evidence ran high from the beginning that these effects could be traced to some kind of continuing manifestation of the rejection reaction, it was not easy to be certain just how this inciting influence might relate to numerous other factors, such as the lipid environment in the recipient or the effects of ischemic damage to the involved organ in the process of transplantation, each of which could have a plausible role to play in bringing about the end result in certain recipients. Furthermore, the exact manner in which the immune vectors of rejection might themselves participate was not easily established from clinical observations. Differing opinions also appeared in the literature about such matters as the relationship between the severity of early rejection activity and the histocompatibility relationship between donor and recipient, to the later appearance of transplant arteriopathy.[3-5]

The complexity of the atherosclerotic process itself, and of the circumstances surrounding its appearance, have accordingly made it difficult to sort out from clinical observations alone the importance of various factors that may have roles in inciting or potentiating these vascular changes. Only in the controlled circumstances possible in planned experiments with experimental animals can definitive information be obtained that can lead to a more complete understanding of the process and thence to the development of enlightened programs of management.

A matter of some interest is the question of how much the pathogenesis of arteriopathic changes found in transplanted organs has in common with the ordinary form of atherosclerosis found in native vessels. This opens to possible consideration a wide ranging literature devoted to this subject. We will not attempt, however, to present evidence bearing upon this possibility in any systematic way, interesting as it is. Another process that clearly bears considerable similarity to transplant arteriopathy can be found in the kind of proliferative endovascular changes that occur after various forms of injury to native vessels, such as endovascular balloon stripping of the vascular lining, or in veins placed into the arterial tree to bypass points of obstruction. This related subject will be mentioned only where particularly relevant.

We will first discuss the main systems that have been used for experiments on transplant arteriopathy using animals along with some advantages and disadvantages of each. We will then summarize some of the combined findings that have been arrived at using these various experimental systems to illustrate the progress that has been made by their application. We will naturally draw upon our own experimental evidence for a number of examples as a matter of convenience.

Experimental Systems

Heterotopic Heart Transplants

Hearts, transplanted to a heterotopic location in such a way that the myocardium is well perfused but the beating chambers carry little blood, is a technique that has been extensively used in transplantation studies. The option of placing transplanted hearts in ectopic locations in this way was first advocated by Frank Mann.[6] Transplants are generally placed in the abdominal cavity in rodents but are often inserted into a suitable subcutaneous pocket in the neck in larger species. The donor aorta is joined to the recipient's aorta, or to another major artery such as the external carotid, and the donor pulmonary artery is attached to the vena cava or another major vein. This results in a situation in which normal coronary flow is maintained. Little blood finds its way into the chambers of the transplanted heart which do not support the circulation of the recipient, and this blood is mainly derived from venous return through the coronary veins. When needed, remedies for this shortcoming are now available (see below).

Rat

Heterotopic transplantation of the heart in rats has been used extensively in transplantation research. We owe the technique for transplantation in the rat to Ono and Lindsey.[7] The donor aorta is joined to the recipient aorta in the abdomen below the renal vessels and the pulmonary artery to the inferior vena cava with end-to-side connections. The survival of such transplanted hearts is usually assessed by a subjective judgment of the strength of its contractions by palpation through the abdominal wall of the recipient or by electrocardiographic tracings. The former method is generally quite adequate for the purposes of these experiments.

Transplants between rats in strain combinations sharing the same major histocompatibility complex (MHC) of genes generate a relatively mild rejection reaction so that they may survive and continue to contract with gradually diminishing vigor for a number of weeks without immunosuppression. This can offer an opportunity to observe vascular changes since advanced lesions are allowed to develop as hearts are not decisively rejected in the early weeks after transplantation. Most experiments without immunosuppression, however, tend to yield hearts that reach a fairly advanced stage of parenchymal rejection before full development of arteriopathy can take place, and in several studies the transplanted hearts have ceased to contract by the time their vessels are examined. Inbred strains of rats are available, of course, although their immunogenetic status is not as precisely defined for certain systems as for mice.

Mouse

An experimental system similar to that employed in the rat involving heterotopic transplantation of the heart was adapted to the mouse in 1970 by Corry et al.[8] Relevant structures in this species are much smaller than in the rat making the procedure considerably more demanding. Primary circulation to the transplant is provided from the aorta and vena cava in the abdominal cavity as in the rat employing microsurgical techniques and appropriate magnification. This method has been applied extensively in recent studies of coronary arteriopathy in which selected inbred strains of mice offer special opportunities for focusing upon certain features of the pathogenetic process of arteriopathy with results that will be described below.

The advantages of this technique for studies of transplant arteriopathy thus include the availability not only of highly defined inbred strains of animals but also of special strains that have undergone targeted genetic alterations. Strains differing from one another by well characterized histocompatibility factors, or by the presence or absence of selected substances that participate actively in certain immune responses, offer particularly attractive subjects for study. Such selected genetic deletions might include those specifying the production of adhesion substances, growth factors, cytokines, elements of the complement system or agents of importance to lipid metabolism such as apoE. Animals that have undergone specific alteration of selected traits by genetic manipulation has been accomplished and will continue to find a useful place in experiments in the future.

The disadvantages of the murine heart transplant system relate mainly to technical challenges because of the small size of the structures with which one must deal. This tends to restrict the numbers of subjects available for study, although the technique can be performed reliably by experienced hands. It is also true, as in the rat, that arteriopathic changes can only be assessed after the transplanted hearts have been excised so that the progress of changes in an individual vessel cannot be observed. The small size of the subject hearts also means that the preparation of appropriate sections to include vessels that can be examined adequately, especially coronary vessels at their point of departure from the donor aorta, demands special skill. Since hearts transplanted between mice or rats differing in respect of antigens specified by the MHC are normally rejected in 8-10 days, typical arteriopathic lesions do not form in these hearts before they are obliterated by acute rejection. Thus, as mentioned above for the rat, in order to study such lesions one has either to select donors and recipients that differ only by a few weak transplantation antigens, or, if one wants to investigate the effects of major antigens of different types, some form of recipient immunosuppression must be used. This requirement need not be considered a disadvantage as it is not dissimilar to the circumstances obtaining in patients undergoing chronic rejection in the presence of varying degrees of immunosuppression.

Rabbit

Rabbit hearts can be transplanted with vascular connections similar to those described above placing the donor heart either in the abdominal cavity or in a subcutaneous pocket fashioned in the neck with junction of donor vessels to the carotid artery and internal jugular vein.[9,10]

Larger animals, such as the pig, dog, and certain primates lend themselves to similar techniques and studies of transplant arteriopathy have been undertaken using hearts transplanted heterotopically to baboons,[11] pigs,[12] and dogs.[6] Indeed the dog was the first species to be employed for heart transplants of any kind by Carrel.[13] In these larger animals it has often been found useful to make a window between the atria to prevent pooling of blood on its return through the coronary veins, which may result in clotting within heart chambers. This expedient has also been used occasionally in smaller animals.[14] The disadvantage that such transplanted hearts do not continue to perform their normal work load, and consequently tend to gradually atrophy, is generally not of prime importance in experiments on transplant arteriopathy. Nevertheless, this relative drawback has been remedied by several alternative, ingenious modifications, including a recent one by Klima, et al who have contrived to retain working ventricles in heterotopically transplanted hearts in pigs and rats by redirecting the outflow of one pulmonary artery back into the transplanted heart (see Fig. 7.1).[15] The advantages of experimental systems in larger animals include the possibility of performing repeated biopsies on individual surviving hearts and the opportunity to follow the evolution of vascular changes (either progression or regression) in individual vessels of transplanted hearts by angiography or by the newer technique of endovascular ultrasound imaging.[16]

Orthotopic Heart Transplants

Dog

A substantial literature exists regarding the intrathoracic, orthotopic transplantation of hearts between dogs. The first reliable methods for accomplishing orthotopic heart replacement in the dog were reported by Lower and Shumway in 1960.[17] The added technical demands, which include the need for cardiopulmonary bypass, have made this technique less attractive for studies of chronic rejection and vasculopathy, but it must be acknowledged that it is this approach that mimics most closely the circumstances encountered in clinical transplantation. Thus, further experimentation with orthotopic heart transplants in dogs, and other larger animals, will doubtless be of value in the future.

Arterial Segment Transplants

Arterial segments inserted in series directly, with appropriate vascular anastomoses, into the arterial tree of recipients offer relatively simple technical alternatives to organ transplantation. An allogeneic (or xenogeneic) segment of vessel presented to a recipient in this way will be met by an immunologic attack from its recipient and will be subjected to a normal physiological environment in other respects including pressure and flow changes within it. Studies of transplant arteriopathy have been conducted with arterial segments orthotopically transplanted to the abdominal aorta in the rat,[18] mouse,[19] and rabbit.[20] Carotid arterial segments have also been inserted in an orthotopic fashion (i.e., into a carotid arterial defect in the recipient).[21,22] The lesions that develop in these arterial segments are generally similar in appearance to those found in the vessels of transplanted organs. The fact that the tissue for study is confined to a single vessel of standard size can make the quantitative evaluation of any subsequent alterations considerably easier than it is when an entire organ has to be considered (see below). Disadvantages of this approach include the fact that vessels transplanted in this way are deprived of their vasa vasorum so that their intrinsic architecture is somewhat abnormal. Furthermore, tissues immediately surrounding the transplanted arteries, that may have an important influence upon intravascular events in transplanted organs, are not included in the tissue to be studied. Vessels of different sizes are also not available for examination and the parenchyma of an accompanying transplanted organ is likewise not included.

Other Organ Transplants

As the changes of transplant arteriopathy are regularly found in any organ transferred to a genetically dissimilar recipient with primary vascular union other organs than the heart are entirely suitable for studies in this field. The dual vascular supply of the lung and the liver, while offering interesting opportunities for special attention, may make vascular studies somewhat more complicated in these organs, and the management of operative procedures required to transfer these organs, as well as the support of their recipients, make them more difficult subjects than is the case for the alternatives described above.

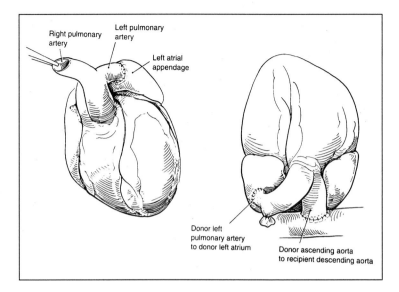

Fig. 7.1. Donor heart preparation (left) and transplant in place in recipient (right). The left pulmonary artery is first anastomosed to the left atrial appendage. Subsequently, the donor aorta is anastomosed to the recipient aorta and the right pulmonary artery is ligated. Reprinted with permission from: Klima U, Guerrero JL, Levine RL, Vlahakes GJ. Transplantation 1997; 64: 215-222. © 1997 Williams and Wilkins Co.

Transplanted kidneys in the rat[23] and the mouse[24] manifest typical proliferative endovascular alterations and accordingly may be good subjects for certain studies of the vascular manifestations of chronic rejection as well as other late alterations peculiar to this organ. Transplanted kidneys appear to be more subject to certain changes, especially when the total renal mass is reduced from the normal level provided by two kidneys. Tilney's findings that a variety of late changes, not unlike "chronic rejection", can be found in kidneys transplanted even to isologous recipients are of interest.[23] This observation offers appealing opportunities for further study, but such studies, at present, remain of uncertain applicability in interpreting the vascular changes found in other settings.

Quantification of Vascular Changes

A reliable system for the quantitative evaluation of the extent or severity of the changes under study is essential in experiments of the kind we are considering. Transplant arteriopathy involves changes in the arterial walls of both larger and smaller vessels reaching into more distal portions of the distribution of the arterial system within an organ. Most studies of chronic rejection have focused only on the complex events involved in vasculopathy and have not considered concomitant changes throughout the remainder of the transplanted organ. Even with this simplifying step one is left with a set of alterations that are difficult to quantify. As mentioned above, if one is observing only a single transplanted arterial segment, the problem of quantifying encroachment upon its lumen is relatively simple. The same is not the case for an entire organ. It is well recognized that arteriopathic changes are usually fairly diffuse in distribution, taking the coronary system which has been most extensively studied as an example. They can, however, be segmental, and, although they usually tend to be circumferential around the entire lining of an affected vessel, they can involve a given vessel asymmetrically as well as segmentally. This situation offers a challenge when one seeks a quantitative summation of the degree of involvement of the entire arterial tree of a given organ. If angiography is possible in the experiment planned, a more complete picture of the vessels at risk for vasculopathy may be available, but even with this information in hand its conversion to a single measure of involvement requires an arbitrary step. Most experiments have depended upon microscopic sections of a transplanted vessel or organ in which a variable number of arterial vessels can be evaluated.

Some form of standardized sampling of the tissue to be examined is first required. Next, individual vessels can be examined and classified according to a predesignated system for judging the severity of the disease in a cross section of a given vessel. One such system, that has been widely adapted, was devised to classify the severity of coronary lesions in transplanted rat hearts on the basis of the percentage of the vessel lumen that is encroached upon by the arteriopathic process.[25] We have adapted this approach and have defined four stages ranging from an uninvolved vessel to one that manifests high grade lumenal encroachment (Table 7.1). The assignment of a vessel to a certain class includes a subjective element so that some studies using this method have included blinded observations by more than one observer. An alternative, more objective method, that usually involves the optical projection of images of vessels will yield figures that can be subjected to morphometric analysis. Thus, numerical estimates of the magnitude of neovascular deposits within the internal elastic lamina of affected vessels have been devised.[26,27] It must be recognized, however, that even though an objective method of estimating changes in individual vessels can be adopted it can only be applied to those vessels that are selected for examination so that sampling variations must still be taken into account. A careful analysis of this problem, as presented by hearts transplanted between mice, has recently been published, and this provides reassuring information about the reliability of examining ves-

sels in selected sections through excised hearts in terms of their reliability as representative samples.[27] Most experiments demand that data resulting from this first stage of evaluation of individual transplanted organs be combined with those from examination of vessels from other similar test organs and the summated results related to controls. Statistical comparisons must then be made where appropriate. One way of combining information from a group of similarly treated subjects is simply to combine the results from all of the vessels examined in all of the animals of a single group. This result can then be compared with a similar result from controls. We have employed this approach and have used the statistical method for comparisons of this kind described by Mehta.[28]

Some Findings from Animal Experiments

Description of Standard Arterial Lesions

The lesions encountered in arterial vessels of various transplanted organs in different species of animals are quite similar to one another. In larger animals the major vessels, e.g., aorta or renal artery, are generally less involved than smaller vessels, and when they are the involvement comes later. The process usually extends into quite distal branches such as the intramyocardial branches of the coronary system. In smaller animals, especially the mouse, the donor aortic segment of heart transplants can be extensively involved, and there is often advanced involvement of the coronary vessels at their bifurcation from the aorta in every phase of the process. Alterations can be found throughout the coronary systems of affected organs, however, and smaller, more distal vessels may become affected early in the process. Arterial bifurcations are often affected preferentially, and an artery may be involved segmentally and asymmetrically over its distribution with affected portions giving way to segments of normal vessel, as mentioned above. Although it is generally felt that the venous systems of

Table 7.1. Stages of vascular involvement[a]

Stage 0:	vessel unaffected
Stage 1:	accumulation of inflammatory cells along intimal surfaces but with less than 10% occlusion of the lumen
Stage 2:	more advanced changes including definite intimal proliferation and thickening but with less than 50% occlusion of the lumen
Stage 3:	high grade occlusion of the vessel with more than 50% occlusion of its lumen. Affected vessels with stage 3 involvement often manifested only a small lumen, or none at all

[a] This scoring system was applied to all of the medium and large arterial vessels in each specimen. An involvement index was then determined for each specimen, which represented the average stage of involvement of all such vessels encountered. The involvement indices of all hearts in a group were combined to determine the mean average vascular involvement.
Adapted from: Russell PS, Chase CM, Winn HJ et al. J Immunol 1994; 152:5135-5141. © 1994 The American Association of Immunologists

transplanted organs do not manifest similar changes there is some evidence that occasional, and usually mild, manifestations of intimal inflammation can be found in some veins as well.[27] The similarity of typical, advanced arterial lesions in several species, including human beings, is illustrated in Figure 7.2 and a general assessment of the comparative characteristics of experimental systems for the study of vascular changes in transplanted structures is set out in Table 7.2.[29]

The time of onset of the earliest changes detectable along the endothelial surfaces of arterial vessels depends in part upon the degree of histoincompatibility between donor and recipient and upon the amount of immunosuppression employed to treat the recipient to prevent acute, generalized organ rejection. Usually at the end of about a week, or earlier, inflammatory cells of recipient origin adhere to endothelial cells. Recipient derived inflammatory cells include T lymphocytes of both the CD4 and CD8 subclasses, as well as macrophages in large numbers and natural killer cells (Fig. 7.3A). This endothelialitis may progress steadily leading to considerable encroachment upon the vascular lumen. Thereafter, progressive dissolution of the normal architecture of the media becomes evident, and this is accompanied by proliferation of smooth muscle cells as they migrate toward the vascular lumen

where they become a major component of the thickening and obstructing neointima[30,31] (Fig. 7.3B). This neointima also includes copious deposits of fibrous elements and ground substance probably contributed by the active smooth muscle cells that are so prominently involved (Fig. 7.3C). The process tends to advance gradually over the weeks after transplantation. As this occurs the makeup of the neointimal substance includes a smaller proportion of recipient derived inflammatory cells and a greater one of smooth muscle cells from the media and of ground substance to yield typical "fibrous" lesions.

Immunogenetic Requirements for Transplant Arteriopathy

Experiments utilizing several animal species have now made it clear that transplant arteriopathy is virtually absent from transplants performed between genetically identical pairs of individuals (syngeneic or isogeneic transplants) for an extended period of time. It has been suggested by Tilney and colleagues that some evidence of myointimal proliferative changes can be found in the coronary vessels of some rat hearts transplanted to isogeneic recipients after 200 days of observation, and even more, after 300 days.[32] The mechanisms behind these interesting effects remain to be established.

Heart transplants between members of strains of mice that differ only in respect of a

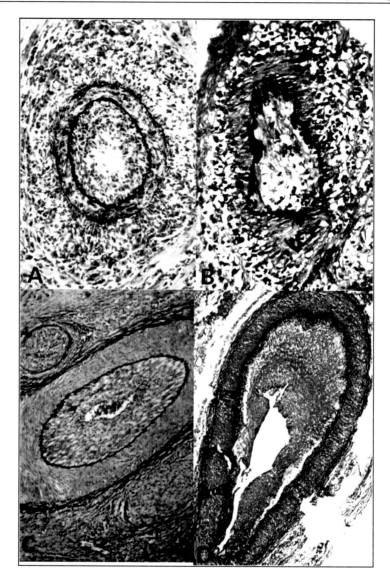

Fig. 7.2. Coronary arteries from mouse (A), human (B,D) and pig (C) cardiac allotransplants. The chronic arteriopathy in the mouse and pig is similar to that in the human in vessels of corresponding size. Both large and small coronary arteries show intimal fibrosis and luminal narrowing. The perivascular infiltrate is prominent in the small arteries and extends into the media (A,B). (A,B,C,D) Elastic tissue stains. A) B10.A to B10.BR, day 56 x 80; B) human cardiac allotransplant at 5 years, x 100; C) pig allotransplant at 22 days, x 40; and D) human cardiac allotransplant at 2 years, x 20. Figure 2A reprinted with permission from: Russell PS, Chase CM, Winn HJ, Colvin RB. Am J Pathol 1994; 144:260-274. © 1994 American Society for Investigative Pathology; Figure 2C reprinted with permission from: Madsen JC, Sachs DH, Fallon JT, Weissman NJ. J Thoracic Cardiovascular Surg 1996; 111:1230-1239. © 1996 Mosby –Year Book Inc..

Table 7.2. Experimental allograft arteriopathy in several species

	Mouse	Rat	Rabbit	Pig
Organs Transplanted				
Heart	+	+	+	+
Kidney	+	+		+
Arterial/Aortic Segment	+	+	+	+
Vascular Similarity to Human				
Class II Expression	−	±	+	+
Lipid Deposition	+[a]	−	±	++
ICAM-1	+	+		
VCAM-1	+		±	
Size	−	−	−	−
Defined Immunogenetics	+++	+	−	+
Monoclonal Ab/DNA Probes	+++	++	+	+
Low Cost	++	++	+	−
Transgenic Subjects	+++	−	−	−

[a] ApoE deficient strain
Adapted and reprinted with permission from: Colvin RB, Chase CM, Winn HJ et al. In: Orosz CG, Sedmark DD, Ferguson RM, eds. Transplant Vascular Sclerosis. Austin, TX: RG Landes, 1994; 7-34. ©1995 RG Landes Co.

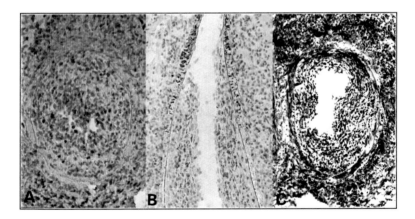

Fig. 7.3. Coronary arteries in mouse cardiac transplants at day 56 (B10.A to B10.BR). A) CD3[+] cells diffusely infiltrate the expanded neointima. B) α-smooth muscle actin[+] cells are present in the expanded intima. Loss of smooth muscle actin from the media has also occurred. C) Collagen deposition (blue stain) is prominent in the neointima. Immunoperoxidase stains for A) anti-CD3 x 50 and B) anti-α-smooth muscle actin x 80; C) Masson trichrome, x 50. Figure 7.3A reprinted with permission from: Russell PS, Chase CM, Winn HJ, Colvin RB. Am J Pathol 1994; 144:260-274. ©1994 American Society for Investigative Pathology.

single antigenic specificity determined by a gene of class I of the MHC have been shown to produce entirely typical lesions of transplant arteriopathy.[33] Some of the results from these experiments are set out in Table 7.3. The same result occurred when the two test strains differed only in respect of a class II gene or only when multiple non-MHC determined incompatibilities were present.[33] Results with similar experiments using members of rat strains selected to differ only by a class I determined specificity, or by non-MHC determined differences, have also been reported with results consistent with the above,[25,34,35] although recent evidence from experiments with partially inbred pigs has suggested a much more prominent role for class I than class II determined antigens in this species.[12]

Immunological Mechanisms for the Production of Transplant Arteriopathy

Evidence cited in the previous paragraph, i.e., that transplants will not develop arteriopathic changes unless they are performed between individuals that are histoincompatible with one another, supports the conclusion that these incompatibilities provoke an immune reaction that is then responsible, at least for setting into motion, the process that culminates in the inflammatory vascular changes we are considering. We have explored several aspects of the question of immune complicity in transplant arteriopathy. In the standard mouse strain combination with which we have become most familiar (B10.A to B10.BR, a class I antigen disparate combination) a full conventional immunological response occurs when living tissues are transplanted, calling into play both the entire range of cell-mediated mechanisms as well as humoral antibodies of both the IgG and IgM classes. Interestingly enough, however, it had been found by colleagues in our laboratory (H.J. Winn and Thomas Fuller) that tissues transplanted between these strains in the reverse direction (B10.BR to B10.A), although undergoing rejection at about the same rate and with the same

apparent vigor, did not call forth any detectable humoral component of the immune response. Coronary arteriopathy was found in the vessels of transplanted hearts in both of these combinations. On close appraisal of the severity of these lesions, however, it was found that arteriopathic lesions emerging in the presence of humoral immunity were significantly more severe than those formed in its absence (see Table 7.4). Furthermore, the supplementation of the immune response with antiserum injections to those recipients incapable of raising such immunity on their own (the B10.BR to B10.A combination) resulted in a very significant increase in the severity of vascular lesions in a dose-dependent manner.[36]

An additional experiment confirmed our conclusion that humoral antibodies could play an important role in generating arteriopathic lesions. This made use of a strain of mouse almost entirely devoid of immune reactivity, a mutant strain with "severe combined immunodeficiency", the C.B-17 *scid* strain. As expected, allogeneic heart transplants to members of this strain will be accepted with virtually no evidence of rejection. Such transplants survived for many weeks maintaining vigorous contractions, and examination of sections of myocardium from these hearts, and of their coronary vessels, revealed only trifling evidence of inflammatory changes in the myocardium and occasional collections of migratory cells adherent to coronary endothelium. By contrast, if an antiserum, prepared in fully reactive recipients of the same genotype as the *scid* recipients against donor specific antigens, is administered to transplant bearing subjects impressive arteriopathic changes can be found when hearts are removed for examination at about four weeks (see Figs. 7.4A and 7.4B). Taken together, these findings provide evidence that humoral antibodies cannot only be responsible for participating in and exacerbating the severity of arteriopathic changes but that they can actually be the primary instigators of arteriopathic changes on their own.

More recently, additional experiments have been conducted with B cell deficient mice.

Table 7.3. Evolution of vascular involvement

Strain Combination* (incompatibility)	4 Weeks			8 Weeks		
	No. of Tx	Involv. index**	Most Affect. Vessel	No. of Tx	Involv. Index	Most Affect. Vessel
B10.A to B10.BR (class I)	17	1.32	3.0	8	2.31	3.0
bm12 to BL/6 (class II)	4	0.12	1.0	3	2.10	3.0
129 to BL/6 (non-MHC)	4	0.11	1.0	2	1.20	2.0
B10.A to B10.A (isograft)	5	0	0.0	2	0	0.0

* CD4/CD8 immunosuppression given to recipients at a dose of 0.1 ml of each on days -6, -3, and -1 prior to transplantation.
** See Table 7.1
Adapted and reprinted with permission from: Russell PS, Chase CM, Winn HJ et al. Am J Pathol 1994; 144:260-274. ©1994 American Society for Investigative Pathology.

These mice were of the C57BL/6-*Igh-6*-*tm1Cgn* strain (commonly referred to as μMT mice). They have been rendered devoid of the transmembranous portion of the immunoglobulin heavy chain by homologous recombination technology and are accordingly devoid of B lymphocytes and of all immunoglobulins. They were employed because the other recipients we had previously used, although putatively without the specific antibodies of concern, could not be declared to be entirely without any humoral immunity, and low levels of such antibodies, or even higher levels intermittently, might have escaped detection. Examination of the coronary vessels of transplants to these recipients turned out to be particularly revealing.[37] They showed quite obvious endothelialitis with adherence of T lymphocytes, macrophages and killer cells to the arterial linings. These were especially prominent at the takeoff of the coronary vessels from the aorta and in the larger coronary vessels. When compared with the more familiar arteriopathic lesions characteristically found in fully reactive recipients, however, a significant difference in the character of the inflammatory process became clear. Whereas lesions produced in the absence of antibody involved the mobilization of a population of migratory cells quite similar to those seen in standard lesions, and some of these cells penetrated to the media where evidence of destruction of smooth muscle cells could be found using stains specific for α-actin, there was very little evidence of migration of smooth muscle cells toward the vascular lumen to participate in neointimal formation (see Figs. 7.5A and 7.5B). Also striking was the near complete lack of newly formed fibrous ground substance in affected vessel walls that so typically characterizes the usual "fibrotic" lesion of transplant arteriopathy. Accordingly, we have concluded that the humoral component of the immune reaction not only can incite arteriopathic lesions unaided by any other immunologically

Table 7.4. Atherosclerotic involvement of heart transplants

	Group	Treatment*	No. of Transplants	Mean Average Vascular Involv.**	Ranking of Individual Specimens by Average Vascular Involvement				% of Hearts with Vasc. Lesions	No. with Ab titer/ No. Tested (Titer Range)	Comparison of Groups p Values
					0	0.1-1.0	1.1-2.0	2.1-3.0			
A	B10.A to B10.BR	none	26	2.1	3	4	7	12	88%	14/15 (1:2 to 1:16)	A-B p < 0.00001
B	B10.BR to B10.A	none	16	0.25	12	3	1	0	25%	0/5	B-C p < 0.11
C	B10.BR to B10.A	0.2 ml anti-BR Ab	17	0.5	8	6	3	0	41%		B-D
D	B10.BR to B10.A	0.4 ml anti-BR Ab	5	2.2	0	1	2	2	100%		p < 0.0003
E	B10.BR to B6AF1	none	8	0.68	3	3	2	0	62%	0/5	B-E p < 0.09

* All recipients received immunosuppression with anti-CD4 and anti-CD8 antibodies; ** See Table 7.1

Adapted and reprinted with permission from: Russell PS, Chase CM, Winn HJ et al. J Immunol 1994; 152:5135-5141. ©1994 The American Association of Immunologists.

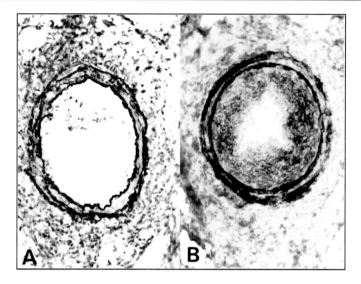

Fig. 7. 4. Cardiac allotransplants in immunodeficient mouse recipients (C.B-17 SCID, *scid*). A) The allogeneic coronary artery shows little or no intimal thickening in an untreated *scid* control 29 days after transplantation. B) In contrast, the allogeneic arteries in a *scid* mouse that received anti-donor class I antibody developed marked intimal fibrosis 40 days after transplantation. Both elastic tissue stain, x 80. Figure 7.4B reprinted with permission from: Russell PS, Chase CM, Winn HJ, Colvin RB. J Immunol 1994; 152:5135-5141. ©1994 The American Association of Immunologists.

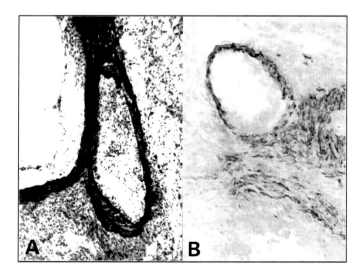

Fig. 7.5. Cardiac allotransplants to B cell deficient mice. A) An allogeneic coronary artery has a florid mononuclear internal infiltrate, but little or no fibrosis. B) The thickened intima has no α-smooth muscle actin+ cells. A) Elastic tissue stain, day 44, x 50; B) immunoperoxidase stain, anti-γ-smooth muscle actin, day 44, x 120. Reprinted with permission from: Russell PS, Chase CM, Colvin RB. Transplant 1997; 64:000-000, 1997. ©1997 Williams and Wilkins Co.

specific component, and can potentiate their severity in the presence of other factors, but it can also change the character of the arteriopathic changes that occur, transforming the lesion from a cellular endothelialitis with only mild accompanying destructive effects to a mature, fibrous alteration with copious deposition of ground substance and extensive destruction of arterial wall architecture in association with proliferation and migration of medial smooth muscle cells. The reported observation that IgG can be found localized on endothelial and smooth muscle cells of concordant arterial xenografts (from hamsters to rats), and that IgM is preferentially localized in the same way in discordant xenografts (from guinea pigs to rats), supports the possibility that similar mechanisms may be at work in the arteriopathy that is observed in xenospecific combinations.[38]

Much remains to be learned about the mechanism of action of antibodies and their cofactors in bringing about these changes and the subject remains under active investigation. Although, as mentioned above, this chapter cannot take fully into account the extensive literature in the general field of atherosclerotic disease as it may relate to our present more restricted subject, it should be acknowledged that the possibility that both may depend upon immune mechanisms offers an intriguing connection. Repeated assertions can be found over a number of years that the inflammatory process associated with atherosclerosis suggests the involvement of specific immune events in its production. For example, it seems clear that atherosclerotic plaques taken from appropriate patients contain not only macrophages and smooth muscle cells but activated T lymphocytes as well.[39] This interesting subject has also been reviewed.[40]

Involvement of Cytokines, Adhesion Molecules and Growth Factors

There is now a lot of information regarding the presence, and the roles, of adhesion factors in the localization of migratory cells in various types of inflammatory lesions, and it is now well appreciated that this class of substances plays a central part in their development. Additional substances in this class are being discovered and additional functions are being attached to them singly and in concert with other adhesion substances and with certain cytokines under different circumstances. The manifestations we consider here are a special case in this general category.

Adhesion Molecules

A full classification and list of properties of the growing number of cell surface factors that act to promote adherence of inflammatory cells to target structures will not be attempted here, but there is much evidence to support their central role in localizing inflammatory cells (see ref. 39, for example). One such system is exemplified by the lymphocyte function associated antigen (LFA-1, or CD11, CD18), a glycoprotein that is a heterodimeric, divalent cation-dependent member of the family of adhesion molecules termed integrins. This substance is prevalent on a variety of migratory inflammatory cells. LFA-1 finds its ligand on sessile structures, such as the linings of vessels, in the form of the intercellular adhesion molecule-1 (ICAM-1)[41] or of variants of ICAM-1 now designated ICAM-2 and ICAM-3. Expression of these substances can vary widely and can be shown to be under the control of certain cytokines, such as interferon-γ (INFγ)[42] which, incidentally, is also capable of strongly upregulating the expression of histocompatibility antigens on a variety of tissues in vivo, as exemplified in the mouse.[43] The relationships between these factors are far from simple, however, as they can have overlapping patterns of activation with a variety of cytokines.[44]

ICAM-1 and LFA-1 can be found quite prominently displayed in association with the development of arteriopathic lesions in transplanted organs.[33] By means of appropriate staining with monoclonal antibodies (mAbs) directed to these substances, and with the use of immunoperoxidase techniques, strikingly increased expression of the former can be demonstrated in the intima of coronary arteries in transplanted mouse hearts as the process of arteriopathy is progressing (Fig. 7.6A). Likewise LFA-1 is readily demonstrable on the surfaces of a variety of inflammatory cells which contribute in such a significant way to the neointima, particularly in the early phases, as well as on such cells as they permeate through the entire vessel wall and distribute themselves around affected vessels in a halo configuration (Fig. 7.6B). Because of the finding of strikingly upregulated expression of ICAM-1 and LFA-1 we elected to test the effects on the evolution of coronary transplant arteriopathy of blocking their function by repeatedly administering mAbs to them (see below). It was, perhaps, surprising that blocking of this single pathway was as effective as it proved to be since there is not only redundancy within this system itself, as ICAM-2 and ICAM-3 can act as alternative adhesion targets, but even more because of the existence of other inducible adhesion pathways, such as VCAM-1 that binds to the integrin, VLA-4.[45,46] It is altogether likely that these additional adhesion systems can have significant roles in the generation of arteriopathic changes. Other systems, too, have been suggested to have similar functions, and it is likely that more evidence will be forthcoming to help define their roles in detail in the production of transplant arteriopathy.

Cytokines and Growth Factors

The phenomenon of upregulation of expression of a wide variety of substances known to be involved in inflammatory processes has been prominent in studies of the evolution of transplant arteriopathy. As might be expected, there is ample evidence that such well known contributors to inflammatory processes as INFγ and tumor necrosis factor(s) (TNF) and various interleukins are readily demonstrable in association with evolving arteriopathic lesions.[21] A variety of other substances that can be suspected of contributing to some aspect of the process can also be found. For example, a cytokine-inducible form of nitric oxide synthase has been found in heterotopically transplanted rat hearts the

presence of which has been reported to bear a relationship to chronic rejection and arteriopathy.[47] Other examples include increased expression of acidic fibroblast growth factor in similar transplants in relation to transplant arteriopathy[48] and insulin-like growth factor-I.[49] Vasoconstrictor and mitogenic substances, such as endothelin, have also been associated with chronic rejection of rat cardiac grafts, and especially with endothelial thickening and interstitial fibrosis.[50,51] The lesions in question offer fertile opportunities to search for new and previously unidentified molecules that may appear as the process advances, and newer techniques of molecular biology offer means whereby small amounts of such substances may be identified and expanded for characterization.[52] The subject of upregulation or increased expression of immunologically relevant substances cannot be left without mentioning the markedly increased expression of histocompatibility antigens that is a regular accompaniment of rejection activity. This has been known for some time to be the case in transplanted kidneys in mice, as mentioned above.[53] It has been demonstrated that one strong stimulus for this remarkable effect can be found in levels of INFγ as systemic administration in vivo of recombinant INFγ results in marked increased expression of histocompatibility factors.[43] The precise importance of modulation of histocompatibility antigen expression to the progress of chronic rejection and transplant arteriopathy has not been independently evaluated. It is entirely plausible, however, that it could make an important contribution. Since the early manifestations of transplant arteriopathy consist especially of binding of recipient inflammatory cells to the endothelial lining of vessels in transplanted organs, it is of interest that, even in the presence of transplant rejection, only class I, and not class II, determined antigens, are expressed on endothelial cells in the mouse. As the vessels in heart transplants between mice differing only in respect of class II-determined antigens have been shown to develop typical, although less severe, transplant arteriopathy, incompatibilities expressed in medial smooth muscle cells, and elsewhere in the donor vessel wall, appear to be sufficient to provoke an inflammatory reaction that results in arteriopathy.

Modulating or Secondary Influences

Shortly after the early observations of arteriopathic changes in organs transplanted to patients clinicians naturally considered what factors in their management might be altered in efforts to prevent or ameliorate arteriopathy. Systemic arterial hypertension, the lipid environment provided to the donor organ, and certain other factors, such as the impact of viral infections, especially cytomegalovirus,[54] were suspected of complicity. Experiments with animal systems have now begun to clarify impressions of this kind. Enhanced smooth muscle proliferation has been observed in aortic allografts in rats in the presence of rat cytomegalovirus infection.[55] This effect could result from the virus causing an increased expression of certain adhesion substances and interleukins, and evidence has been recorded that these effects can be found in culture systems using appropriate human cells and that they result from a direct effect of virus upon the affected cells.[56]

The importance of hyperlipidemia to the evolution of the process has received considerable attention as it has been suspected of involvement from clinical observations for some time.[57] A possible confluence of allergic or immunological factors along with the effects of hyperlipidemia has also been suspected on the basis of findings from experiments with rabbits, a favorite species for investigations in the field of hyperlipidemia.[9,58] Rats and rabbits receiving transplanted hearts in the presence of very high cholesterol diets reportedly develop arteriopathic lesions bearing increased amounts of lipid material, but the lesions were felt not to be more severe than in controls in the sense that they did not encroach any more upon the lumena of affected vessels than lesions in normal recipients.[59] A somewhat similar conclusion was reached from experiments with transplanted aortic segments in rats,[60] but a careful analysis of the effects of hypercholesterolemia

alone, compared with those when hypertriglyceridemia was added, suggested that more severe proliferative vascular wall changes occurred when both were elevated together.[61] In the mouse, animals devoid of the lipid clearing substance apoE[62] by "knockout" technology can be used as recipients. Such mice maintain blood cholesterol levels in the 400-500 mg/dl range on normal diets. Native hearts of these mice show cholesterol deposits in their aortic roots (Fig. 7.7A), a finding never observed in normal mice. Hearts transplanted to such recipients develop arteriopathic lesions as expected, but they contain large amounts of lipid even in the smaller coronary branches, although lipid is deposited most heavily in major coronary vessels and in the aorta (Fig. 7.7B).[63]

Reversibility of Transplant Arteriopathy by Retrotransplantation

The susceptibility of evolving arteriopathic lesions to being reversed with restoration of an affected vessel to an immunologically neutral environment is of considerable importance in considering strategies for management. Lesions involving advanced destructive changes in the arterial media with loss of smooth muscle cells in association with the deposition of collagen and ground substance would not appear to be reversible simply by interrupting the causes for their initiation in the first place. Thus, one could suspect from studying the appearance and makeup of advanced lesions, that inhibiting the continuing effects of the immune response to donor specific antigens would be unlikely to reverse them. Several experiments have been reported in which evidence bearing upon these points has been sought. They have consisted of removing transplanted organs or vascular segments at various times and transplanting them back to recipients identical with their original donors. Studies of this kind using aortic segments in the rat suggested that vascular lesions are not appreciably reversible after as short a time as 10 days residence in a primary recipient,[64] or, in one report with rat heart transplants, as soon as 3-5 days.[65] Other

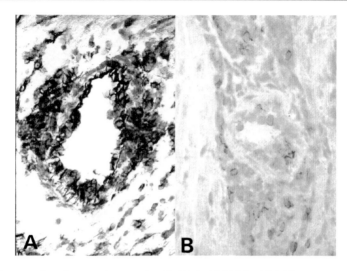

Fig. 7.6. Mouse cardiac allotransplant arteries. A) Marked ICAM-1 (CD54) expression is evident in the arterial endothelium and the adventitial cells. B) LFA-1 (CD11a) a receptor for ICAM-1, is expressed by the mononuclear cells infiltrating the endothelium and adventitia. Immunoperoxidase stains. A) 129 to C57BL/6, day 56, x250; B) B10.A to B10.BR, day 28, x250. Reprinted with permission from: Russell PS, Chase CM, Winn HJ, Colvin RB. Am J Pathol 1994; 144:260-274. ©1994 American Society for Investigative Pathology.

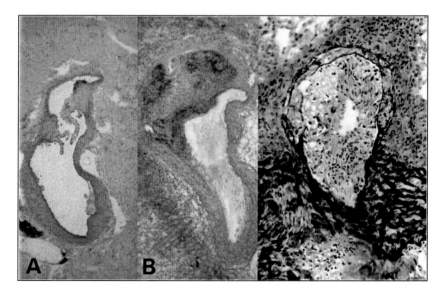

Fig. 7.7. Cardiac transplants to apolipoprotein E deficient mice. An oil Red O stain reveals a marked accumulation of fat in the thickened aorta of a transplanted heart (B), which contrasts with that in the native aorta (A) in the same animal. C) The allogeneic coronary artery near its origin has prominent intimal foam cells along the internal elastic lamina. A,B) x20; C) elastic tissue stain x100. Reprinted with permission from: Russell PS, Chase CM, Colvin RB. Am J Path 1996; 149 (1): 91-99. ©1996 American Society for Investigative Pathology.

experiments with rat heart transplants have yielded somewhat different findings in respect of the degree of reversibility of changes and the time after initial transplantation during which they respond to reverse transplantation. Thus, in one report, retransplantation to the donor strain resulted in only some reduction in "mononuclear infiltration"[66] whereas in another, in which a more compatible strain combination was used, considerable reversal of arteriopathic changes was found after retransplantation following as long as 40 days of residence in the primary recipient. Under certain circumstances, however, including an appropriate strain combination of rat, combined with the use of cyclosporine immunosuppression, reversibility of the effects of rejection of transplanted kidneys was found to be possible for as long as 12 weeks.[67] Thereafter, when changes became more advanced, reversibility could not be demonstrated, and it was concluded that further progress of the process had eventually become immunologically independent.[68] These experiments might have been considered imperfect in that they did not take into account the fact that the lesions to be reversed were composed of cells of both donor and recipient origin. Thus, recipient derived cells would have elicited an immune response when the organ was transferred to a member of its original donor strain. This consideration has, however, been taken into account in a few experiments in which transplants of rat aortic segments were evaluated when transferred secondarily into members of an F1 strain between the original donor and recipient strains.[64] In this case the results appear to have been much the same as in the previous experience, that is reversibility was only observed after a short time of residence in the primary recipient.

Pathogenesis of Transplant Arteriopathy

The impression one gets from reviewing all of the current information about the initiation and later evolution of transplant arteriopathy from animal experiments can be summarized as follows. The process is initiated as a direct consequence of specific immune reactivity by the recipient against foreign histocompatibility factors of any of the major classes in the donor tissue. The possible implication of tissue specific antigens, i.e., antigens confined only to cells in affected vessels, such as endothelial or smooth muscle cells, has not been formally ruled out, although there is sufficient evidence available now to declare that such antigens need not be present. A full understanding of the mechanism(s) by which the immune response initiates the arteriopathic process in vessel walls is not yet in hand. It is clear, however, that both humoral and cellular components of the immune response can initiate forms of vasculopathy independently of one another, although there is evidence that humoral antibody will promote the formation of "fibrotic" lesions with deposition of ground substance and collagen and centripetal migration of smooth muscle cells to a much greater degree than will the cell mediated component of the immune response.

As an intimate and close consequence of specific immune reactivity a complex inflammatory response is set into motion involving the generation of adhesion molecules that promote the localization of several categories of recipient derived inflammatory cells. A variety of cytokines is also released by a process of upregulation of the activity of the cells by which they are produced. These have several prominent effects ranging from the increased expression of histocompatibility factors to a wave of mitotic division of endothelial and smooth muscle cells with migration of the latter into a neointima that leads to progressive obstruction of blood flow through affected vessels.

It is well known that very similar pathological changes can be produced in vessels by quite different inciting factors, including simple mechanical damage to the endothelial surfaces of vessels by such means as trauma from an endovascular balloon. Although transplant arteriopathy depends upon immunological insults upon the vascular lining for its inception it is not unlikely that other factors may play a part in some circumstances.

The segmental nature of transplant arteriopathy raises the question of why a process mediated systemically should sometimes show striking variations in severity in different segments of the same vessel. Mechanical factors may have a role in explaining this, including the effects of turbulence of flow with its resultant mechanical consequences at branching points of vessels, a factor that seems to be of considerable importance in ordinary atherosclerosis.

Some Approaches to the Control of Transplant Vasculopathy

According to the formulation put forward above, attempts to manage or modify transplant arteriopathy could be designed to intrude upon the process at any one of several points. Suppression or elimination of the immune response to the histocompatibility antigens in the transplanted organ should result in the most effective and thorough elimination of arteriopathy as the inception of the process would be blocked with a minimal amount of vascular damage. The experiments in which the effects of stopping the specific immune response to a transplant, mentioned above, suggest that control of the immune response, to be effective, must be accomplished early and be quite complete. Two general approaches have been considered in this regard. The first has been an exploration of various means simply of suppressing the immune response nonspecifically, or of modulating the effects of the immune response as they subsequently become effective. The second approach depends upon various ways of producing specific immune unresponsiveness to antigens of the transplant donor.

A suggestion that the first approach could be at least somewhat effective had been reported from clinical experience.[69] Trials in experimental animals have included the use of cyclosporine treatment that has been reported to reduce arteriopathic lesions in transplanted mouse hearts.[70] Interestingly, it was reported quite early by Laden that antilymphocyte serum treatment of rabbits and rats that had received allogeneic heart transplants would allow prolonged survival of the transplanted hearts but would not materially reduce arteriopathic changes within them. This was attributed to the fact that such sera, although able to suppress cell-mediated responses profoundly, are less effective against humoral immunity, and it was the latter that was concluded in this study to be of major import to the development of vasculopathy.[71] Other potent immunosuppressive agents can have similar effects in that they may be sufficient to prevent acute, early rejection quite well while still permitting the development of arteriopathy. Examples of this are brief courses of treatment with mAbs to CD4 and CD8 specificities[33] and some chemical agents, such as gallium nitrate.[72] Rapamycin, given at sufficient doses, has been found to have special potency in inhibiting arteriopathy in transplanted rat hearts. This agent was described as having not only immunosuppressive activity but also as being an active inhibitor of the production of several cytokines and growth factors. This broad range of efficacy was felt to be responsible for its capacity to inhibit the development not only of transplant arteriopathy but also the arteriopathic changes that occur following mechanical injury.[73] The mechanism for this effect of rapamycin has been investigated and attributed to decreased expression of the gro/melanoma-growth stimulatory gene as well as macrophage inflammatory protein-2 genes. These effects result in inhibition of INFγ and granzyme B production in test transplants.[74] Another, more precisely directed, approach has been the use of antisense phosphorothioate oligonucleotides to inhibit the synthesis of intracellular macromolecules thought to have an influence on cellular proliferation in vessel walls. This has been tested with some encouraging results in respect of the "proliferating cell nuclear antigen" (PCNA) by Simons et al in a system involving endovascular injury of the carotid artery in the rat.[75]

Inhibition of the release or the effectiveness of certain cytokines and growth factors that are produced in the course of the immune response may have definite suppressive effects on the evolution of transplant arteriopathy.

For example, treatment with an antibody to platelet derived growth factor (PDGF) has been reported to suppress the accumulation of smooth muscle cells in lesions produced in the carotid arteries of rats by balloon trauma.[76] A related observation is that, when combined with cyclosporine therapy, dipyridamole, an "antiplatelet agent" that prevents high levels of 3'-5'-cyclic AMP, will reduce arteriopathic lesions in transplanted rat hearts.[25] Other elements of the clotting mechanism have also been suspected of participating in the arteriopathic process, and support for this has come from experiments in which heparin was used at various doses to inhibit neointimal hyperplasia after endovascular trauma in rats.[77]

Treatment with a monoclonal antibody to INFγ can suppress significantly the development of arteriopathic lesions in transplanted mouse hearts.[78] Agents known to have general modulating effects upon inflammatory reactions through incompletely defined mechanisms have also been shown to reduce the severity of arteriopathic lesions in transplanted organs. These include a diet deficient in essential fatty acids.[30] Treatment with the somatostatin analog angiopeptin has also been reported to have a somewhat similar effect,[10] which has also been reported following treatment with a vitamin D analog, MC1288.[79] Interestingly, arteriopathic lesions in aortic segments transplanted to male rabbits have been found to be considerably reduced in the presence of estradiol treatment with reduction of smooth muscle proliferation and preservation of normal endothelial cells.[80]

Another possibility, suggested perhaps by the clinical observation that hypertension could be associated with chronic rejection of transplanted hearts, has been the use of various antihypertensive agents. Thus, administration of captopril, an angiotensin converting enzyme inhibitor, reduces the intimal hyperplasia that can occur in transplanted venous segments.[81] A similar conclusion was reached in experiments in which renal allografts in rats were studied in the presence of various combinations of reserpine, hydralazine, hydrochlorothiazide, cilazapril, or L-158,809, an angiotensin II receptor blocker.[82]

An active field of inquiry has to do with questions surrounding the importance of Th1 versus Th2 pathways of activation in chronic rejection and arteriopathy. Many of these studies make use of cytokine determinations in samples of various kinds from suitable animal subjects in efforts to characterize the type of response in progress. Th1 type responses have been associated with rejection whereas Th2 responses with less rejection and less vasculopathy.[21] The evolution of responsiveness through these pathways may be altered by treatment with monoclonal antibodies to CD4 determinants[83] or by other means.[84] It has also been suggested that donor specific blood transfusion will tend to shift the immune response toward one of the Th2 type with reduction in the severity of subsequent vasculopathy.[85,86]

Another class of possibilities for affecting the outcome after the arteriopathic process has been initiated by specific immune factors are options for influencing the evolution of the inflammatory reaction. The complexity and apparent redundancy of the pathways involved in cell mobilization and in cytokine release and the subsequent influence of released cytokines would appear to make it unlikely that intrusion at any one point could have observable effects. This has not been the case for all such attempts, however. Continuing treatment of murine recipients of cardiac transplants with mAbs to ICAM-1 and LFA-1 together has been found to suppress greatly the severity of arteriopathic changes in the transplanted hearts. Either antibody alone was insufficient to do this.[87] In this connection it is of interest that hearts from donors entirely deficient in ICAM-1 do not appear to survive longer on transplantation than controls.[88]

In considering the second category of ways of managing transplant arteriopathy, namely the induction of specific unresponsiveness to donor antigens, a broad range of possibilities is available for consideration as "operational unresponsiveness" can, of course, be brought about by a number of mechanisms. Thus, in recent years immune unresponsiveness has been obtainable in adult individuals of several species by a number of quite different forms of treatment, and these have been shown to depend upon several mechanisms. The form of unresponsiveness that was described first, and which probably had the most impact upon the conceptual framework employed in the field, was termed "actively acquired immunological tolerance" instituted at the late fetal or neonatal stage by the injection of living adult allogeneic lymphoid cells. This state was declared to be a "central failure" of the immune response to a specific antigen or set of antigens.[89] Quite complete ablation of the immune response to allogeneic antigens can also be achieved, at least so far as survival of test transplants and most forms of in vitro immune responsiveness, by alternative means. In general, it has seemed that unresponsiveness is easier to accomplish in smaller species of experimental animals than larger ones, but it must be remembered that the initial observations of tolerance were made in cattle (albeit of tolerance generated in utero). This fact should lend encouragement to those who lament the difficulties accompanying more recent attempts to induce unresponsiveness in larger animals. The efficacy of these various forms of unresponsiveness in preventing transplant arteriopathy has not been systematically investigated, although some encouraging information is available. Most of the treatments that have been tested to date probably depend upon the phenomenon, or phenomena, of "peripheral" unresponsiveness, a condition that may depend upon suppressor cells or other states that might, in principal, be reversible. A possible exception to this may be the use of direct inoculation of antigenic material from the intended donor into the thymus of the recipient, an option that has been reported to have some efficacy against arteriopathy in experiments using rats.[90] A similar report has been made in regard to the state of unresponsiveness that can follow irradiation and donor-specific bone marrow infusion,[91] a result that may well depend upon mechanisms similar to those at work in association with donor-specific blood transfusion,

mentioned above. Another interesting option is the form of hyporesponsiveness that results following recipient treatment with antigens or antigenic peptides of donor origin, often administered orally.[92,93] This alternative has recently been lent considerable interest by encouraging results, not only in respect of transplant survival but also in suppressing the development of arteriopathy in transplanted rat hearts following the repeated injection of a synthetic peptide corresponding to nonpolymorphic regions of an MHC class I sequence along with cyclosporine treatment.[94] Another report has come recently from Madsen and colleagues, who have described circumstances in which acceptance of a renal transplant between pigs, differing mainly in respect of MHC class I disparities, can be induced with short term, high dose cyclosporine treatment. Following this, if a later cardiac transplant from the same donor source is performed in the presence of the surviving renal transplant, the transplanted heart is spared acute rejection and also later arteriopathic changes.[95] A quite similar result has also been reported recently in which the tolerance-inducing preliminary organ transplant were livers transplanted to allogeneic mice. Such transplants are known to induce a form of specific unresponsiveness in their recipients as do kidney transplants in many recipients in this species.[24] Thus, a later segmental aortic transplant from the donor strain proved to be protected from developing transplant arteriopathy.[96] Other experiments along these lines will surely be reported before long and will be awaited with interest.

Summary

In this chapter we have endeavored to illustrate why animal experiments are essential to the elucidation of the mechanisms, and accordingly to the development of enlightened programs for the management of transplant arteriopathy. Evidence is advanced to support the notion that the initiation of the process that leads to this form of arteriopathy is heavily dependent upon the specific immune response of the transplant recipient to foreign histocompatibility antigens present in the donor. After the train of events that leads to advanced lesions is started, nonspecific events typical of an evolving inflammatory process are called forth, and these tend to make the entire, ongoing process independent of its initiating stimulus. It appears that a full quenching or avoidance of the specific immune reaction, that can include both cell mediated and humoral components, can obviate the appearance of arteriopathy, but that the later events, that include complex interactions of adhesion molecules and cytokines can also be inhibited by various means with favorable results. Furthermore, the character of the inflammatory process can also be influenced by secondary features, such as altering the lipid environment in which the vasculopathy unfolds.

Information regarding this interesting and important manifestation of transplant rejection is emerging rapidly, and newer options for its control can be expected to appear as knowledge increases.

References

1. Porter KA, Thomson WB, Owen K et al. Obliterative vascular changes in four human kidney homotransplants. Br Med J 1963; 2:639-645.
2. Thompson JG. Provisional report on the autopsy of L.W. S Afr Med J 1969; 41:1277.
3. Uretsky BF, Murali S, Reddy PS et al. Development of coronary artery disease in cardiac transplant recipients receiving immunosuppressive therapy with cyclosporine and prednisone. Circulation 1987; 76:827-834.
4. Gao SZ, Schroeder JS, Hunt S et al. Retransplantation for severe accelerated coronary artery disease in heart transplant recipients. Am J Cardiol 1988; 62:876-881.
5. Bieber CP, Stinson EB, Shumway NE et al. Cardiac transplantation in man. VII. Cardiac allograft pathology. Circulation 1970; 61: 753-772.
6. Mann FC, Priestley JT, Markowitz J et al. Transplantation of the intact mammalian heart. Arch Surg 1933; 26:219.
7. Ono K, Lindsey ES. Improved technique of heart transplantation in rats. J Thorac Cardiovasc Surg 1969; 57:225-229.
8. Corry RJ, Winn HJ, Russell PS. Primarily vascularized allografts of hearts in mice. The role of H-2D, H-2K and non-H-2 antigen in rejections. Transplantation 1973; 16: 343-350.
9. Alonso DR, Starek PK, Minick CR. Studies on the pathogenesis of atheroarteriosclerosis induced in rabbit cardiac allografts by the synergy of graft rejection and hypercholesterolemia. Am J Path 1977; 87 (2):415-442.
10. Foegh ML. Accelerated cardiac transplant atherosclerosis/chronic rejection in rabbits: Inhibition by angiopeptin. Transpl Proc 1993; 25 (2):2095-2097.
11. Michler RE, McManus RP, Smith CR et al. Technique for primate heterotopic cardiac xenotransplantation. J Med Primatol 1985; 14 (6):357-362.
12. Madsen JC, Sachs DH, Fallon JT et al. Cardiac allograft vasculopathy in partially inbred miniature swine. J Thorac Cardiovasc Surg 1996; 111:1230-1239.
13. Carrel A, Guthrie CC. The transplantation of veins and organs. Amer Med 1905; 10:1101.
14. Steinbrüchel DA, Nielsen B, Salomon S et al. A new model for heterotopic heart transplantation in rodents: Graft atrial septectomy. Transpl Proc 1994; 26:198-199.
15. Klima U, Guerrero JL, Levine RL et al. A new, biventricular working heterotopic heart transplant model. Transplantation 1997; 64:215-222.
16. St. Goar FG, Pinto FJ, Alderman EL et al. Intracoronary ultrasound in cardiac transplant recipients. In vivo evidence of "angiographically silent" intimal thickening [see comments]. Circulation 1992; 85 (3): 979-987.
17. Lower RR, Shumway NE. Studies on the orthotopic transplantation of the canine heart. Surg Forum 1960; 11:18-19.
18. Mennander A, Tiisala S, Halttunen J et al. Chronic rejection in rat aortic allografts. An experimental model for transplant arteriosclerosis. Arterioscler Thromb 1991; 11 (3): 671-680.
19. Shi C, Russell ME, Bianchi C et al. Murine model of accelerated transplant arteriosclerosis. Circ Res 1994; 75 (2):199-207.
20. Jacobson, Cheng L, Lyly L et al. Effect of estradiol on accelerated atherosclerosis in rabbit heterotopic aortic allografts. J Heart Lung Transpl 1992; 11:1188-1193.
21. Hancock WW, Shi C, Picard MH et al. Lew-to-F344 carotid artery allografts: Analysis of a rat model of posttransplant vascular injury involving cell-mediated and humoral responses. Transplantation 1995; 60:1565-1572.
22. Wehr S, Rudin M, Joergensen J et al. Allo- and autotransplantation of carotid artery—a new model of chronic graft vessel disease. Transplantation 1997; 64:20-27.
23. Tilney NL, Whitley WD, Diamond JR et al. Chronic rejection—an undefined conundrum. Transplantation 1991; 52:389.
24. Russell PS, Chase CM, Colvin RB et al. Kidney transplants in mice. An analysis of

the immune status of mice bearing long-term, H-2 incompatible transplants. J Exp Med 1978; 147:1449-1468.

25. Lurie KG, Billingham ME, Jamieson SW et al. Pathogenesis and prevention of graft arteriosclerosis in an experimental heart transplant model. Transplantation 1981; 31: 41-47.

26. Aherne WA, Dunhill MS. Morphometry. 1982.

27. Armstrong AT, Strauch AR, Starling RC et al. Morphometric analysis of neointimal formation in murine cardiac allografts. Transplantation 1997; 63:941-947.

28. Mehta CR, Patel NR, Tsiatis AA. Exact significance testing to establish treatment equivalence with ordered categorical data. Biometrics 1984; 40:819-825.

29. Colvin RB, Chase CM, Winn HJ et al. Chronic allograft arteriopathy: Insights from experimental models. In: Orosz CG, Sedmak DD, Ferguson RM, eds. Transplant Vascular Sclerosis. Austin TX: RG Landes, 1995: 7-34.

30. Adams DH, Wyner LR, Steinbeck MJ et al. Inhibition of graft arteriosclerosis by modulation of the inflammatory response. Transpl Proc 1993; 25 (2):2092-2094.

31. Geraghty JG, Stoltenberg RL, Sollinger HW et al. Vascular smooth muscle cells and neointimal hyperplasia in chronic transplant rejection. Transplantation 1996; 62:502-509.

32. Schmid C, Heeman U, Tilney NL. Factors contributing to the development of chronic rejection in heterotopic rat heart transplantation. Transplantation 1997; 64:222-228.

33. Russell PS, Chase CM, Winn HJ et al. Coronary atherosclerosis in transplanted mouse hearts. I. Time course, immunogenetic and immunopathological considerations. Am J Path 1994; 144:260-274.

34. Cramer DV, Qian S, Harnaha J et al. Cardiac transplantation in the rat. I. The effect of histocompatibility differences on graft arteriosclerosis. Transplantation 1989; 47: 414-419.

35. Adams DH, Tilney NL, Collins JJ, Jr. et al. Experimental graft arteriosclerosis. I. The Lewis-to-F-344 allograft model. Transplantation 1992; 53 (5):1115-1119.

36. Russell PS, Chase CM, Winn HJ et al. Coronary atherosclerosis in transplanted mouse hearts. II. Importance of humoral immunity. J Immunol 1994; 152:5135-5141.

37. Russell PS, Chase CM, Colvin RB. Alloantibody and T cell mediated immunity in the pathogenesis of transplant atherosclerosis: Lack of progression to sclerotic lesions in B cell deficient mice. Transplantation 1997; 64.

38. Allaire E, Mandet C, Bruneval P et al. Cell and extracellular matrix rejection in arterial concordant and discordant xenografts in the rat. Transplantation 1996; 62:794-803.

39. Stemme S, Holm J, Hansson GK. T lymphocytes in human atherosclerotic plaques are memory cells expressing CD45RO and the integrin VLA-1. Arterioscl Thromb 1992; 12:206-211.

40. Hansson GK, Jonasson L, Seifert PS et al. Immune mechanism in atherosclerosis. Arteriosclerosis 1989; 9:567-578.

41. Dustin ML, Springer TA. Lymphocyte function-associated antigen-1 (LFA-1) interaction with intercellular adhesion molecule-1 (ICAM-1) is one of at least three mechanisms for lymphocyte adhesion to cultured endothelial cells. J Cell Biol 1988; 107: 321-331.

42. Dustin ML, Rothlein R, Bhan AK et al. Induction by IL1 and interferon g: Tissue distribution, biochemistry, and function of a natural adherence molecule (ICAM-1). J Immunol 1986; 137 (1):245-254.

43. Skoskiewicz MJ, Colvin RB, Schneeberger EE et al. Widespread and selective induction of major histocompatibility complex-determined antigens in vivo by γ interferon. J Exp Med 1985; 162:1645-1664.

44. Pober JS, Gimbrone MA, Jr., Lapierre LA et al. Overlapping patterns of activation of human endothelial cells by interleukin 1, tumor necrosis factor, and immune interferon. J Immunol 1986; 137 (6):1893-1896.

45. Pelletier RP, Morgan CJ, Sedmak DD et al. Analysis of inflammatory endothelial changes, including VCAM-1 expression, in murine cardiac grafts. Transplantation 1993; 55 (2):315-320.

46. Morgan CJ, Pelletier RP, Harnandez CJ et al. Alloantigen-dependent endothelial phenotype and lumphokine mRNA expression in rejecting murine cardiac allografts. Transplantation 1993; 55:919-923.

47. Russell ME, Wallace AF, Wyner LR et al. Upregulation and modulation of inducible nitric oxide synthase in rat cardiac allografts with chronic rejection and transplant atherosclerosis. Circulation 1995; 92:457-464.

48. Zhao X-M, Blanton RH, Becker YT et al. Increased expression of acidic fibroblast growth factor (aFGF) and FGF receptor-1 (FGFR-1) in rat cardiac allografts versus isografts and normal hearts. Circulation 1994; 90(suppl I):I-361.

49. Motomura N, Lou H, Maurice P et al. Acceleration of arteriosclerosis of the rat aorta allograft by insulin growth factor-I. Transplantation 1997; 63 (7):932-936.

50. Forbes RD, Cernacek P, Zheng S et al. Increased endothelin expression in a rat cardiac allograft model of chronic vascular rejection. Transplantation 1996; 61 (5):791-7.

51. Watschinger B, Sayegh MH, Hancock WW et al. Upregulation of endothelin-1 mRNA and peptide expression in rat cardiac allografts with rejection and arteriosclerosis. Am J Pathol 1995; 146:1065.

52. Utans U, Quist WC, McManus BM et al. Allograft inflammatory factory-1. A cytokine-responsive macrophage molecule expressed in transplanted human hearts. Transplantation 1996; 61 (9):1387-1392.

53. Benson E, Colvin RB, Russell PS. Induction of IA antigens in murine renal transplants. J Immunol 1985; 134:7-9.

54. Grattan MT, Moreno-Cabral CE, Starnes VA et al. Cytomegalovirus infection is associated with cardiac allograft rejection and atherosclerosis. JAMA 1989; 261:3561-3566.

55. Lemström KB, Bruning JH, Bruggeman CA et al. Cytomegalovirus infection enhances smooth muscle cell proliferation and intimal thickening of rat aortic allografts. J Clin Invest 1993; 92:549-558.

56. Craigen JL, Grundy JE. Cytomegalovirus induced up-regulation of LFA-3(CD58) and ICAM-1 (CD54) is a direct viral effect that is not prevented by gangcyclovir or foscarnet treatment. Transplantation 1996; 62:1102-1108.

57. Hess ML, Hastillo A, Mohanakumar T et al. Accelerated atherosclerosis in cardiac transplantation: Role of cytotoxic B-cell antibodies and hyperlipidemia. Circulation 1983; 68(suppl II):II-94-101.

58. Minick CR, Murphy GE. Experimental induction of atheroarteriosclerosis by the synergy of allergic injury to arteries and lipid-rich diet. Am J Pathol 1973; 73:265-300.

59. Adams DH, Karnovsky MJ. Hypercholesterolemia does not exacerbate arterial intimal thickening in chronically rejecting rat cardiac allografts. Transpl Proc 1989; 21 (1 Pt 1):437-439.

60. Mennander A, Tikkanen MJ, Räisänen-Sokolowski A. Chronic rejection in rat aortic allografts. IV. Effect of hypercholesterolemia in allograft arteriosclerosis. J Heart Lung Transpl 1993; 12 (1 Pt 1):123-132.

61. Räisänen-Sokolowski A, Tilly-Kiesi M, Ustinov J et al. Hyperlipidemia accelerates allograft arteriosclerosis (chronic rejection) in the rat. Arterioscl Thromb 1994; 14 (12):2032-2042.

62. Breslow JL. Transgenic mouse models of lipoprotein metabolism and atherosclerosis. Proc Natl Acad Sci USA 1993; 90:8314-8318.

63. Russell PS, Chase CM, Colvin RB. Accelerated atheromatous lesions in mouse hearts transplanted to apolipoprotein-E-deficient recipients. Am J Path 1996; 149 (1):91-99.

64. Mennander A, Häyry P. Reversibility of allograft arteriosclerosis after retransplantation to donor strain. Transplantation 1996; 62:526-529.

65. Izutani H, Miyagawa S, Shirakura R et al. Evidence that graft coronary arteriosclerosis begins in the early phase after transplantation and progresses without chronic immunoreaction. Histopathological analysis using a retransplantation model. Transplantation 1995; 60 (10):1073-1079.

66. Schmid C, Heemann U, Tilney NL. Retransplantation reverses mononuclear infiltration but not myointimal proliferation in a rat model of chronic cardiac allograft rejection. Transplantation 1996; 61 (12):1695-1699.

67. Tullius SG, Hancock WW, Heemann U et al. Reversibility of chronic renal allograft rejection. Critical effect of time after transplantation suggests both host immune dependent and independent phases of progressive injury. Transplantation 1994; 58 (1):93-99.

68. Forbes RD, Zheng SX, Gomersall M et al. Irreversible chronic vascular rejection occurs only after development of advanced allograft vasculopathy: A comparative study of a rat cardiac allograft model using a retransplantation protocol. Transplantation 1997; 63 (5):743-749.

69. Ballester M, Obrador D, Carrio I et al. Reversal of rejection-induced coronary vasculitis detected early after heart transplantation with increased immunosuppression. J Heart Transpl 1989; 8:413-417.

70. Hirozane T, Matsumori A, Furukawa Y et al. Experimental graft coronary artery disease in a murine heterotopic transplant model. Circulation 1995; 91:386-392.

71. Laden AMK. The effects of treatment on arterial lesions of rat and rabbit cardiac allografts. Transplantation 1972; 13:281-290.

72. Orosz CG, Wakely E, Bergese SD et al. Prevention of murine cardiac allograft rejection with gallium nitrate. Transplantation 1996; 61:783-791.

73. Gregory CR, Huie P, Billingham ME et al. Rapamycin inhibits arterial intimal thickening caused by both alloimmune and mechanical injury. Transplantation 1993; 55:1409-1418.

74. Wieder KJ, Hancock WW, Schmidbauer G et al. Rapamycin treatment depresses intragraft expression of KC/MIP-2, granzyme B, and INF-g in rat recipients of cardiac allografts. J Immunol 1993; 151:1158-1166.

75. Simons M, Eleman ER, Rosenberg RD. Antisense proliferating cell nuclear antigen oligonucleotides inhibit hyperplasia in a rat carotid artery injury model. J Clin Invest 1994; 93:2351-2356.

76. Ferns GAA, Raines EW, Sprugel KH et al. Inhibition of neointimal smooth muscle accumulation after angioplasty by an antibody to PDGF. Science 1991; 253:1129-1132.

77. Rogers C, Karnovsky MJ, Edelman ER. Inhibition of experimental neointimal hyperplasia and thrombosis depends on the type of vascular injury and the site of drug administration. Circulation 1993; 88 (3): 1215-1221.

78. Russell PS, Chase CM, Winn HJ et al. Coronary atherosclerosis in transplanted mouse hearts. III. Effects of recipient treatment with a monoclonal antibody to interferon-γ. Transplantation 1994; 57: 1367-1371.

79. Räisänen-Sokolowski AK, Pakkala IS, Samila SP et al. A vitamin D analog, MC1288, inhibits adventitial inflammation and suppresses intimal lesions in rat aortic allografts. Transplantation 1997; 63:936-941.

80. Cheng LP, Kuwahara M, Jacobsson J et al. Inhibition of myointimal hyperplasia and macrophage infiltration by estradiol in aorta allografts. Transplantation 1991; 52:967-972.

81. O'Donohoe MK, Schwartz LB, Radic ZS et al. Chronic ACE inhibition reduceds intimal hyperplasia in experimental vein grafts. Ann Surg 1991; 20:727-732.

82. Benediktsson H, Chea R, Davidoff A et al. Antihypertensive drug treatment in chronic renal allograft rejection in the rat. Transplantation 1996; 62:1634-1642.

83. Mottram PL, Han W-R, Purcell LJ et al. Increased expression of IL-4 and IL-10 and decreased expression of IL-2 and interferon-γ in long-surviving mouse heart allografts after brief CD4-monoclonal antibody therapy. Transplantation 1995; 59:559-565.

84. Bach FH, Ferran C, Hechenleitner P et al. Accommodation of vascularized xenografts: Expression of "protective genes" by donor endothelial cells in a host Th2 environment. Nature Medicine 1997; 3:196-204.

85. Wood PJ, Roberts IS, Yang CP et al. Prevention of chronic rejection by donor-specific blood transfusion in a new model of chronic cardiac allograft rejection. Transplantation 1996; 61 (10):1440-1443.

86. Carlquist JF, Edelman LS, White W et al. Cytokines and rejection of mouse cardiac allografts. Transplantation 1996; 62:1160-1166.

87. Russell PS, Chase CM, Colvin RB. Coronary atherosclerosis in transplanted mouse hearts. IV. Effects of treatment with monoclonal antibodies to intercellular adhesion molecule-1 and leukocyte function-associated antigen-1. Transplantation 1995; 60 (7): 724-729.

88. Schowengerdt KO, Zhu JY, Stepkowski SM et al. Cardiac allograft survival in mice deficient in intercellular adhesion molecule-1. Circulation 1995; 92:82-87.

89. Billingham RE, Brent L, Medawar PB. Actively acquired tolerance of foreign cells. Nature 1953; 172:603-606.

90. Shin YT, Adams DH, Wyner LR et al. Intrathymic tolerance in the Lewis-to-F344 chronic cardiac allograft rejection model. Transplantation 1995; 59 (12):1647-1653.

91. Orloff MS, DeMara EM, Coppage ML et al. Prevention of chronic rejection and graft arteriosclerosis by tolerance induction. Transplantation 1995; 59 (2):282-288.

92. Sayegh MH, Khoury SJ, Hancock WW et al. Induction of immunity and oral tolerance with polymorphic class II major histocompatibility complex allopeptides in the rat. Proc Natl Acad Sci 1992; 89:7762-7766.

93. Weiner HL. Oral tolerance: Immune mechanisms and treatment of autoimmune disease. Immunol Today 1997; 18 (7):335-342.

94. Murphy B, Kim KS, Buelow R et al. Synthetic MHC class I peptide prolongs cardiac survival and attenuates transplant arteriosclerosis in the Lewis-Fischer 344 model of chronic allograft rejection. Transplantation 1997; 64:14-19.

95. Madsen JC, Yamada K, Allan JS et al. Prevention of cardiac allograft vasculopathy across class I MHC disparities by the induction of transplantation tolerance. submitted for publication.

96. Subbotin V, Sun H, Aitouche A et al. Abrogation of chronic rejection in a murine model of aortic allotransplantation by prior induction of donor-specific tolerance. Transplantation 1997; 64:690-695.

CHAPTER 8

Cytomegalovirus as a Contributing Factor in Transplant Vascular Sclerosis

W. James Waldman, Deborah A. Knight and Adriana Zeevi

Overview

Transplant vascular sclerosis (TVS) is a multifactorial, delayed complication of solid organ transplantation. An accelerated form of arteriosclerosis, TVS develops over a period of months to years following transplantation, frequently compromising the circulation within solid organ allografts. With the advent of dramatically improved immunosuppressive therapy which can now effectively control acute rejection, TVS has emerged as a major limitation to long-term graft survival especially in heart transplant recipients.[1] Many potential contributing etiologic factors have been identified, some of which show clear association with the development of TVS, and others whose specific etiologic relationships remain somewhat elusive. Among the latter, cytomegalovirus (CMV) infection holds a prominent position. This can perhaps be explained by our present incomplete understanding of the nature of the immune reaction to CMV in the allogeneic setting. Furthermore therapeutic intervention with a growing battery of antiviral agents, which in most cases effectively control acute disease, may result in the appearance of delayed sequelae whose relationship with the initial causative agent is difficult to resolve. This review begins with a brief description of CMV. Evidence accumulated from clinical studies and animal experiments which argue for or against a contributing role for CMV in the development of TVS is then summarized. Documentation supporting the central role of the graft vascular endothelium in the modulation of graft/host equilibrium is presented, followed by a description of potential mechanisms by which CMV might perturb equilibrium at the graft endothelial interface.

Biology of Human Cytomegalovirus

Human cytomegalovirus (CMV) is a betaherpesvirus with worldwide distribution. Although prevalence varies with geographical location and socioeconomic status,[2-4] it has been estimated that 90% of the population has been exposed to this virus by age 60. Generally, CMV poses little threat to those individuals with a mature, competent immune system, with most infections resulting in mild or subclinical disease. However, for those individuals with compromised immunity, such as AIDS patients, cancer patients undergoing chemotherapy, or organ transplant recipients, CMV infection can lead to severe life-threatening complications such as pneumonitis, gastrointestinal mucosal ulceration, retinitis, and hepatitis, as well as destructive inflammatory lesions in a number of other locations.[5] In addition, CMV has been implicated as an exacerbating factor in allograft rejection, in particular, chronic vascular rejection or transplant vascular sclerosis (TVS).[6-8] As this list of potential complications implies, CMV infects a number of different cell types in vivo, including epithelial cells of the lung, salivary gland, and gastrointestinal tract, vascular endothelium, leukocytes such as monocytes and neutrophils, fibroblasts, and others.[9]

Since the initial isolation of CMV in the 1950s,[10-12] many details of the virus life cycle have been elucidated; yet much remains to be resolved. The mature virion measures ~200 nm in diameter and consists of a 240 kb double-stranded DNA genome (the largest of the herpesviruses) packaged in an icosahedral protein capsid surrounded by tegument and a lipid bilayer envelope. Embedded within the envelope are virally-encoded glycoproteins as well as host cellular proteins (Fig. 8.1).[13]

Relative to other herpesviruses, CMV replication progresses slowly (~72 hours for completion), a factor which may be partially responsible for its limited pathogenicity in the immunocompetent host.[13] Although a number of candidate host cell surface molecules have been proposed as receptors for CMV,[14-18] definitive identification of such has remained elusive. Upon entry into the host cell however, it has been shown that tegument proteins such as lower matrix protein pp65, upper matrix protein pp71, and basic tegument protein pp150, rapidly translocate along with viral DNA to the cell nucleus (Fig. 8.2A).[13] As is common among members of the herpesvirus family, CMV gene expression is temporally coordinated (Fig. 8.2B). Expression of the immediate early (IE) genes can be detected within 2-4 hours postinoculation and do not require de novo protein synthesis for expression, implying that virion-associated proteins (pp71) and/or host cellular proteins are responsible for activating IE gene expression. Indeed, the IE gene promoter has a very strong enhancer element that is responsive to both viral and cellular transcription factors.[13] IE gene products are largely regula-

tory, transactivating early and late gene expression. CMV early genes encode enzymes and DNA binding proteins involved in viral DNA synthesis, including viral DNA polymerase. Late genes, whose expression is largely dependent upon viral DNA synthesis, code primarily for structural components of the virion.[13]

Virion assembly in the permissive host cell begins in the nucleus where viral DNA is packaged within protein nucleocapsids. Capsids are then transported to the cytoplasm where they acquire tegument and ultimately, an external glycolipid envelope.[13] Mechanisms of virion trafficking and assembly are currently under intense investigation and much remains to be resolved.

As is characteristic of all herpesviruses, CMV establishes long-term latency in the infected host. In general terms latency is classically defined by demonstration of viral presence (i.e., viral DNA) in the absence of virus replication. However compared to what is known of latency of other herpesviruses (HSV, EBV, etc.), the latent state of CMV remains poorly defined. Neither the physical state of the latent viral genome nor the specific nature of its activity in latently infected cells has been clearly elucidated. Furthermore, although several candidate cell types, such as monocytes, endothelial cells, myeloid progenitors, and others, have been implicated as hosts for latent CMV,[9,19-24] many uncertainties persist and additional viral reservoirs likely remain to be discovered. In some cases CMV "latency" might, in fact, be more accurately defined as low-level persistence.

Regardless of the specific parameters defining the latent state, it is well-established that quiescent virus can reactivate to a replicative state under conditions of immune compromise. Following such reactivation or following primary infection, nascent CMV remains cell-associated,[25] disseminating primarily by direct cell-to-cell transmission. Compelling evidence supports a hematogenous component as a major contributing mechanism in systemic dissemination. Infectious virus can be recovered from circulating monocytes and neutrophils during

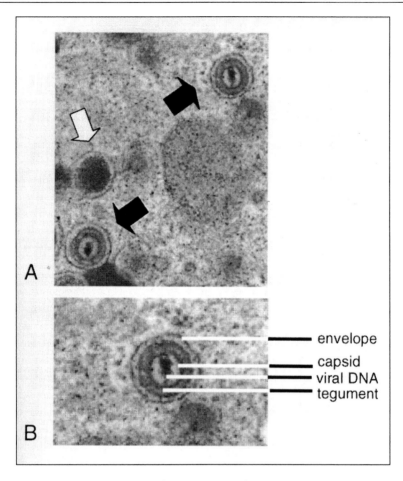

Fig. 8.1. Transmission electron micrographs of CMV within an infected human endothelial cell. A) peripheral cytoplasmic region of infected cell with enveloped virions measuring approximately 200 nm (black arrows) and viral dense body (white arrow). B) higher magnification of a single-enveloped virion showing viral envelope, tegument, nucleocapsid, and DNA core.

active disease, and leukocyte depletion of blood prior to transfusion has been shown to significantly decrease the transmission of CMV to seronegative newborns and transplant recipients.[26-29] Additionally, a contributing role for the endothelium in the process of hematogenous dissemination has been implied by studies which have demonstrated bidirectional transmission of infectious CMV between endothelial cells and adherent monocytes or neutrophils in vitro.[30-31] For this reason, and because of the immunologic significance of the graft endothelium as the ultimate interface between the engrafted organ and the host, interactions occurring at this interface will be explored further in forthcoming sections of this review.

CMV as a Contributing Factor in TVS: Clinical Observations and Animal Studies

Although consensus is not universal, evidence continues to accumulate implicating CMV as an exacerbating agent in the development of TVS. A viral contribution to

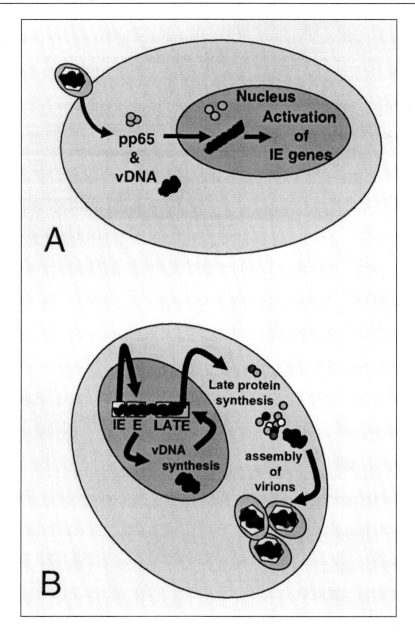

Fig. 8.2. Schematic diagram depicting major events in the lytic cycle of CMV. A) Upon viral entry into the host cell, viral DNA and several tegument proteins such as pp65 rapidly translocate to the cell nucleus. B) Immediate early gene products transactivate early gene expression, providing the enzymes and accessory proteins required for viral DNA synthesis. Finally, and largely dependent upon viral DNA synthesis, expression of the late set of viral genes, which code primarily for structural proteins, provide the necessary components for virion assembly.

allograft rejection was suggested as early as 1970.[32] However efforts directed toward documenting and understanding this phenomenon have intensified dramatically over the past decade. In 1986 von Willebrand et al[33] reported a much higher incidence of renal allograft rejection among patients with documented CMV disease (86%) than among disease-free recipients (17%). Over the next several years, as long-term graft survival continued to improve, a number of investigators documented an intriguing association between CMV disease and accelerated TVS. Among the first of these, Grattan et al[6] followed 301 consecutive cardiac transplant recipients and, in 1989, reported increases in rate of graft rejection and severity of TVS, as well as higher incidence of graft loss and patient death among the 91 patients with documented CMV disease as compared with the disease-free group. Nearly concurrently and shortly thereafter several other investigators, notably McDonald et al,[34] Loebe et al,[35] Everett et al[36] and Koskinen et al,[8] independently substantiated these findings, reporting similar phenomena among smaller populations of heart transplant recipients. More recently, results of several large-scale clinical studies have lent further support to the association between CMV infection and TVS. Mattila et al,[37] reporting on a 10 year, 190 patient study in Finland, and Rocca et al,[38] describing results of a 10 year, 156 patient study in Switzerland, both found CMV to be a significant risk factor in the development of graft coronary disease.

It should be noted that CMV infection has been implicated as a contributing factor in the development of chronic rejection in other organ transplants as well. The actuarial prevalence of obliterative bronchiolitis (a manifestation of chronic lung graft rejection) at 2 years posttransplantation has been reported to be much greater in patients with biopsy-proved CMV pneumonitis (74%) than in patients who experience no CMV disease (22%).[39-41] Finally, CMV hepatitis in the transplanted liver has been associated with vanishing bile duct syndrome, also a manifestation of chronic rejection.[42]

Additional evidence in support of a viral contribution to graft vascular disease has been provided by a number of animal experiments. The isolation and characterization of rat cytomegalovirus (RCMV, Maastricht strain) reported in 1982 by Bruggeman et al[43] facilitated the development of several rat transplantation models which have since been extensively exploited. The first such model, a cornerstone upon which multiple subsequent studies have been based, was reported by Lemström et al in 1993.[44] In this series of experiments segments of descending thoracic aorta were transplanted from DA (AG-B4, RT1ᵃ) rats into WF (AG-B2, RT1ᵛ) rats. Animals were inoculated with RCMV at day 1 or day 60 postoperatively and assessed at intervals (up to 12 months) for degree of intimal hyperplasia. Results of these studies demonstrated significant enhancement of graft vascular disease in transplanted aortic segments in acutely infected rats as compared to the uninoculated control animals.[44] These findings have subsequently been confirmed and extended by Li et al[45,46] employing different donor/recipient combinations. In the years since its development, this model has been modified to demonstrate similar associations between RCMV and TVS in heterotopic cardiac allografts,[47] as well as RCMV-enhanced chronic renal allograft rejection.[48-50]

As mentioned above, consensus on the role of CMV infection as a contributing factor in graft rejection is not universal. Weimar et al[51] studied the clinical course of 87 cardiac transplants and found no association between acute or chronic rejection and patient pretransplant CMV serostatus or posttransplant seroconversion. Stovin et al,[52] in a more limited study of 43 heart transplant recipients (22 of whom seroconverted), found no evidence of CMV-associated inflammation in endomyocardial biopsies taken up to the time of CMV seroconversion (approximately 57 days posttransplant), and no difference in rejection frequency between these patients and those who remained CMV-seronegative. In a retrospective analysis of 210 cardiac transplant recipients, Radovancevic et al[53] found correlations between graft coronary artery disease and previous acute rejection, postoperative arterial hypertension, and smoking, but not with CMV infection. These findings are difficult to resolve with those cited above; however they appear vastly overshadowed by the continuously accumulating evidence in support of a perturbing role for the virus. Several other investigators have reported a low frequency of CMV DNA within diseased arteries of cardiac or renal allografts.[54,55] However, based upon mechanisms proposed in forthcoming sections of this review, the absence of CMV in advanced lesions does not necessarily preclude its contribution to initiation and early development of TVS.

Graft Endothelial Activation and Host Inflammatory Responses

The vascular endothelium within the solid organ transplant assumes unique significance as the anatomical and functional interface between engrafted donor tissue and the host immune system. Once considered a complacent barrier whose assumed principal function was the inhibition of abnormal intravascular thrombosis, the endothelium has emerged as a dynamically interactive participant in immunomodulation. Endothelial cells (EC) have been demonstrated to: 1) regulate leukocyte migration,[56] 2) express HLA molecules and present antigen,[57,58] 3) elaborate and respond to immunomodulating cytokines,[59] and 4) inducibly express immunoreactive cellular adhesion molecules.[59] These properties imply a central role for the EC in the initiation, perpetuation, and/or attenuation of graft-associated alloimmune interactions. Under normal circumstances vascular EC remain in a quiescent state, expressing constitutive levels of HLA class I but little or no HLA class II. In addition quiescent cells express modest levels of intercellular adhesion molecule-1 (ICAM-1), no E-selectin, and (with the exception of limited subsets) no vascular cell adhesion molecule-1 (VCAM-1). However under inflammatory conditions, as a consequence of cytokine interactions EC are induced to express class II, VCAM-1, and

E-selectin, as well as enhanced levels of class I and ICAM-1.[59] Indeed these phenotypic responses are characteristic of graft endothelium during episodes of vascular rejection and in developing TVS lesions.[60-63]

Hruban et al[60] and Salomon et al[61] have demonstrated such aberrant expression of endothelial class II in affected coronary arteries within human cardiac allografts. These studies also showed an associated marked subendothelial infiltration of HLA DR+ macrophages and T cells of both the CD4+ and CD8+ subsets. Furthermore Briscoe et al[62] and Fuggle et al[63] have demonstrated enhanced endothelial adhesion molecule expression, particularly ICAM-1 and VCAM-1, in vascular lesions within human cardiac and renal transplants respectively. Again, such phenotypic changes in the endothelia were associated with inflammatory infiltrates consisting primarily of T cells and macrophages. While these findings provide compelling evidence for an association between graft endothelial activation and host inflammatory responses, they pose important questions regarding cause and effect. In an effort to determine the kinetic relationship of these events, Tanaka et al,[64] employing a rabbit cardiac allograft model, showed that graft endothelial adhesion molecule expression precedes rejection-associated mononuclear cell infiltration.

Collectively, these findings imply that forces which perturb equilibrium at the graft endothelial interface by enhancing foreign HLA and adhesion molecule expression can profoundly affect the outcome of the transplant. A substantial volume of data has accumulated implicating CMV as one such perturbing force. Von Willebrand et al[33] observed high levels of HLA class II expression on both EC and tubular epithelium recovered from needle aspiration biopsy of renal transplants immediately following the onset of CMV disease. Concomitantly or shortly thereafter, inflammatory infiltration was detected in nearly all grafts. Similar CMV-associated inflammatory enhancement has been observed by others in human cardiac transplants,[65,66] and in transplanted rat aortic segments,[44,46,65,67] hearts,[47] and kidneys.[50]

In all cases, these inflammatory changes were associated with rejection episodes and/or the development of TVS. Other reported perturbations include graft endothelial denudation,[45] endothelial proliferation,[45,47,66] and aberrant endothelial adhesion molecule expression.[68,69] In a series of 105 post-transplant human endomyocardial biopsies, Koskinen[68] observed induction of endothelial VCAM-1 and E-selectin associated with the onset of CMV antigenemia. Furthermore, Yilmaz et al,[69] employing the rat model, demonstrated an association between CMV-enhanced chronic renal allograft rejection and vascular endothelial and tubular epithelial ICAM-1 expression.

CMV as a Catalyst of TVS Development: Potential Mechanisms

Although the data implicating CMV as an exacerbating factor in the development of graft vascular disease are intriguing and provocative, mechanisms underlying such interactions remain to be fully resolved. It is well documented that EC are a common target for CMV infection in vivo regardless of the organ involved[70-72] (Fig. 8.3). Furthermore, EC can serve as fully permissive hosts for CMV in vitro provided the natural endothelial cytopathogenicity of the virus is preserved by propagation in EC.[73,74] We initially hypothesized that CMV might directly induce HLA class II on infected EC and thus enhance graft endothelial alloimmunogenicity. Although studies by Ustinov et al[75] suggest that this may occur in rat endothelia, our extensive work with human venous, arterial, and microvascular EC,[76-78] as well as studies by others,[79-81] demonstrate that CMV infection does not induce endothelial HLA DR, DP, or DQ in the human. Furthermore, although interferon-γ (IFNγ) is a potent inducing agent for endothelial HLA class II, infected cells are totally refractory to such induction as a consequence of CMV-mediated interruption of the IFNγ signal transduction pathway.[77,78,82,83]

Because of the potential significance of the graft vascular endothelium, considerable effort has been directed toward characterization of the impact of CMV upon the immunobiologic properties of EC. It is now well established that CMV inhibits HLA class I expression in infected fibroblasts,[84,85] and we have shown that the same is true of human EC.[77,86,87] In a manner analogous to class II, infected cells also lose class I responsiveness to IFNγ and tumor necrosis factor-α (TNFα)[77,86,87] With regard to adhesion molecule expression, it has been demonstrated that although the virus strongly enhances expression of ICAM-1 on infected endothelial cells and fibroblasts,[81,86-91] neither VCAM-1 nor E-selectin are expressed on infected EC, nor can these be induced by TNFα (as they can in uninfected cells).[89,91] Collectively these findings argue against a direct role for CMV in the enhancement of host responses to foreign HLA/adhesion molecules.

Despite their lack of key immunoreactive surface molecules, CMV-infected EC powerfully stimulate allogeneic or autologous, CMV-seropositive donor-derived T cells to produce IL2 and to proliferate.[86,92] Based upon these findings and the suggestion of von Willebrand et al[33] that endothelial HLA class II induction might be a consequence of IFNγ production by CMV-responsive T cells, we developed several coculture systems to model such phenomena in vitro. We initially cocultured CMV-seropositive donor-derived T cells directly with allogeneic uninfected EC or with endothelial monolayers harboring low-level CMV (< 0.5% infected cells).[82] While HLA DR induction on up to 70% of uninfected bystander EC was consistently observed in CMV-infected cultures, no such induction was noted in the absence of virus.[82] In a follow-up series of experiments, the system was modified by coculturing T cells with infected allogeneic EC in transwell inserts above uninfected endothelial monolayers.[93] Immunofluorescence flow cytometry and immunohistochemical staining of EC incubated beneath CMV-activated T cells revealed induction of HLA DR and VCAM-1, and enhancement of HLA class I and ICAM-1, effects never exhibited by EC

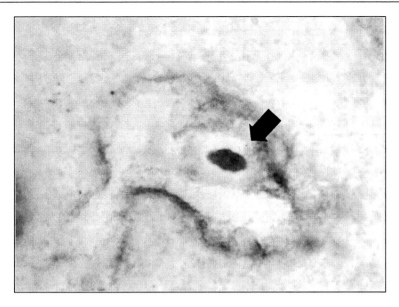

Fig. 8.3. Photomicrograph of a section of human colonic mucosa showing a small vessel containing a CMV-infected vascular endothelial cell, as demonstrated by simultaneous immunohistochemical staining for von Willebrand factor (granular cytoplasmic stain) and in situ hybridization with a biotin-labeled CMV-specific DNA probe (dense nuclear stain). (Roberts et al[72] and by permission from the American Medical Association.)

incubated beneath T cells alone, CMV-infected EC alone, or T cells cocultured with uninfected EC. Furthermore, the inclusion of blocking antibodies in these cocultures identified the inducing agents as IFNγ and TNFα.[93]

Since chronic graft endothelial injury has been implicated as a contributing factor in the development and progression of TVS, additional experiments were designed to assess the ability of CMV-infected EC to stimulate the generation of cytotoxic T cells, and to identify potential targets.[87] Following stimulation with CMV-infected or uninfected EC, limiting dilutions of CMV-seropositive (or seronegative) donor-derived T cells were tested for cytolytic activity against CMV-infected or uninfected EC by ^{51}Cr-release assay. Data generated by these experiments demonstrated that CMV-infected EC are resistant to T cell-mediated lysis regardless of the nature of T cell prestimulation. Interestingly, however, CMV-infected EC stimulate the development of cytolytic T cells capable of destroying corresponding uninfected EC.[87]

Based upon these results and documentation by Koskinen et al[94] of systemic immune activation in CMV-infected cardiac allograft recipients, we propose that CMV might act as a catalyst, promoting and intensifying immunopathologic interactions at the graft endothelial interface (Fig. 8.4); specifically, that CMV-infected EC within the graft vasculature (or possibly elsewhere) can initiate a host immune activation cascade, and that the cytokine storm which ensues can act upon uninfected bystander graft endothelia, enhancing alloimmunogenicity.[93] Since it has been shown that pretreatment of uninfected EC with IFNγ greatly augments their ability to stimulate allogeneic T cell IL2 production and proliferation,[95] we further propose that cytokine-activated graft EC could then stimulate recruitment of additional alloreactive T cell populations. Cytokines elaborated by these allo-activated T cells could then promote continuous expansion of the region of endothelial activation which progresses independent of the virus. Finally, the cytolytic interactions described above[87] suggest that CMV-infected EC have the potential to stimulate chronic host T-cell-mediated injury of uninfected bystander graft endothelia.

Since our ongoing studies have consistently shown CMV-infected EC to be highly efficient stimulators of T-cell activation,[82,92] we postulate that initiation of these processes would not require extensive graft endothelial infection. Indeed the virus may be completely cleared from the region by the time the lesion has progressed to the clinically detectable stage. Thus the infrequency of CMV DNA found by Gulizia et al[54] in sclerosed cardiac allograft vessels does not necessarily rule out a viral contribution to the initiation and progression of the disease process.

It should be noted that Craigen and Grundy[96] have demonstrated that neither ganciclovir nor phosphonoformate (foscarnet) prevent CMV-mediated ICAM-1 enhancement on infected cells, and we have likewise shown that these antiviral agents do not attenuate allogeneic T cell activation responses (unpublished observations). Furthermore, we have observed identical EC(CMV)/T cell interactions (albeit at reduced intensity) in the presence of concentrations of cyclosporine A or tacrolimus sufficient to completely attenuate the mixed lymphocyte reaction.[97] Thus it seems likely that the scenario proposed above (Fig. 8.4) could develop over time within graft vasculature even under immunosuppressive conditions and during or following a course of antiviral chemotherapy.

An alternative, but not mutually exclusive mechanism by which CMV might enhance inflammatory interactions at the graft endothelial interface relates to the growing family of intracellular molecular chaperones known as heat shock proteins (hsp). Santomenna and Colberg-Poley[98] have shown hsp-70 to be induced by CMV in infected fibroblasts, and we have observed this phenomenon in infected EC as well (unpublished observations). Based upon earlier observations of hsp-specific immunity in tissues affected by various autoimmune diseases,[99] Moliterno et al[100,101] searched for hsp-reactive lymphocytes in endomyocardial

biopsy tissues obtained from heart transplant recipients, and in cellular infiltrates within rat heterotopic cardiac allografts. In both cases such cells were successfully recovered from rejecting allografts, as determined by proliferation responses to a series of recombinant hsp preparations. Importantly, little or no hsp-reactivity was found in syngeneic rat grafts or in animals in which rejection was prevented by immunosuppressive therapy. Collectively these findings suggest the possibility that CMV-infected cells, by virtue of their coexpression of viral antigens and hsp, may stimulate the generation of activated T cells which express cross-reactivity against uninfected hsp-expressing cells within the graft.

Regardless of mechanisms of initiation, the ultimate focus of lesion development is the vascular intima. Thus, remaining to be elucidated are potential downstream consequences of endothelial perturbation which could account for the documented hyperplastic responses. Data which have accumulated collectively imply a process of protracted repair and tissue remodeling, presumably in response to chronic endothelial injury.[102] As summarized above, CMV-infected EC, although themselves resistant to cytolysis, are capable of stimulating the generation of T cells exhibiting lytic activity against uninfected endothelia.[87] These findings suggest that CMV-infected EC can persist within the graft, perpetuating inflammatory havoc and promoting chronic endothelial injury. Furthermore, since both increased endothelial adhesion molecule expression and leukocyte activation correlate with enhanced transendothelial migration,[103] our model predicts CMV-mediated intensification of graft vascular inflammatory infiltration. Since mononuclear leukocytes and EC, as well as smooth muscles cells (SMC) themselves represent important sources of SMC growth factors and chemoattractants (as described below), these processes could facilitate the delivery of such activating mediators to the intimal region.

PDGF, implicated as a major mediator in the genesis of TVS, is both chemotactic and mitogenic for SMC and is produced by

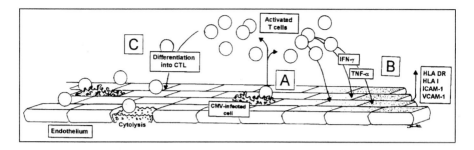

Fig. 8.4. Proposed model illustrating CMV-triggered events occurring at the graft/host endothelial interface that result in enhancement of graft endothelial alloimmunogenicity and lytic injury of graft endothelium. A) T-cell activation by CMV-infected graft endothelial cells. B) Enhancement of HLA and adhesion molecule expression upon proximal uninfected endothelium in response to cytokines elaborated by CMV-responsive T cells. C) Virally-enhanced lytic injury of uninfected bystander graft endothelia. (Waldman et al[87] and by permission from Transplantation/Lippincott, Williams & Wilkins.)

EC (particularly upon injury), infiltrating monocytes and, in vascular lesions, by SMC themselves.[102,104,105] FGF-2 (bFGF), another potent SMC mitogen, is produced by infiltrating T cells.[106,107] While a third mitogen, FGF-1 (aFGF), has been localized to sclerotic lesions, its precise source remains unclear.[108] TGFβ, a pleiotropic mediator which stimulates the production of extracellular matrix proteins and exhibits chemotactic properties for macrophages and fibroblasts, has been localized to multiple sources including EC, monocytes, platelets, T cells, and SMC.[109] In addition, Michelson et al[110] have shown that CMV infection of fibroblasts directly induces TGFβ transcription and secretion. Whether the same is true of infected EC remains to be determined.

In summary, multiple potential mechanisms exist by which CMV-related perturbation of equilibrium at the graft endothelial interface might adversely affect underlying tissues, all of which seem feasible, and all of which merit further investigation. In closing it should be noted that an alternative school of thought exists regarding the specific role of CMV in neointimal proliferation. Several investigators have observed viral infection of vascular SMC.[111,112] These findings, together with recent reports of CMV-induced resistance to apoptosis,[113,114] have generated the hypothesis that the accumulation of mass in the hyperplastic intima may directly result from failure of infected SMC to undergo programmed cell death. However, in contrast to the model proposed above, this scenario would seem to require the persistence of a substantial fraction of CMV-infected SMC in the advanced lesion, and this has not been apparent in studies performed to date. Still, considering the uncertainties that yet remain, this line of investigation deserves further attention.

Epilogue

Substantial progress in antiviral chemotherapy has provided the tools to largely control acute phase disease in many cases. However it must be remembered that these antiviral agents are not directly virocidal, but rather only virostatic, and, particularly in the case of herpesvirus infection, never completely eliminate the virus from the host. One consequence of the resultant decrease in morbidity and mortality is the potential for the development of delayed sequelae not obviously attributable to the initial causative agent. Similarly, the development of increasingly effective immunosuppressive agents which successfully control acute allograft rejection have promoted the emergence of new complications which appear to result from low level protracted immune interactions, but which are likewise difficult to fully resolve.

In spite of substantial controversy regarding the etiology of chronic rejection, none would disagree that causation is multifactorial. The increasing use of ganciclovir and other agents in the treatment and/or prophylaxis of CMV disease in transplant recipients, by controlling overt disease without completely eliminating viral presence from all tissues, further complicates resolution of a contributing role for this virus. Thus in determining the etiology of chronic disease in the current environment, it seems the time has come to accept the possible breakdown of Koch's postulates in this context. Although consensus is not universal, the bulk of evidence accumulated to date supports a contributing role for CMV in the genesis of TVS. Regarding mechanisms, we favor a scenario in which the virus acts as a catalyst for immunopathologic interactions within the graft vasculature, tipping the balance against maintenance of graft/host equilibrium. However, potential solutions for this complex problem remain elusive.

Our constantly accelerating rate of discovery and progress in medical intervention, while of net clinical benefit, will almost certainly continue to create new diseases. Thus we, as a community of scientists and clinicians, must be prepared with heightened intellectual prowess, ever-evolving creativity, active imagination, and open minds in facing the medical challenges of the coming millennium.

Acknowledgment

The authors' work cited in this review was supported in part by grant HL56482 (WJW) from the National Heart, Lung, and Blood Institute of the National Institutes of Health.

References

1. Johnson DE, Gao SZ, Schroeder JS et al. The spectrum of coronary artery pathologic findings in human cardiac allografts. J Heart Transplant 1989; 8:349-359.
2. Krech U. Complement-fixing antibodies against cytomegalovirus in different parts of the world. Bulletin of the World Health Organization. 1969; 49:103-106.
3. Preiksaitis JK, Larke RPB, Froese GJ. Comparative seroepidemiology of cytomegalovirus infection in the Canadian arctic and an urban center. J Med Virol 1988; 24:299-307.
4. White NH, Yow MD, Demmler GJ et al. Prevalence of cytomegalovirus antibody in subjects between the ages of 6 and 22 years. J Inf Dis 1989; 159:1013-1017.
5. Alford CA, Britt WJ. Cytomegalovirus. In: Roizman B, Whitley RJ, Lopez C, eds. The Human Herpesviruses. New York: Raven Press, Ltd., 1993: 227-255.
6. Grattan MT, Moreno-Cabral CE, Starnes VA et al. Cytomegalovirus infection is associated with cardiac allograft rejection and atherosclerosis. JAMA 1989; 261:3561-3566.
7. Normann SJ, Salomon DR, Leelachaikul P et al. Acute vascular rejection of the coronary arteries in human heart transplantation: pathology and correlation with immunosuppression and cytomegalovirus infection. J Heart Lung Transplant 1991; 10:674-687.
8. Koskinen PK, Nieminen MS, Krogerus LA et al. Cytomegalovirus infection accelerates cardiac allograft vasculopathy: Correlation between angiographic and endomyocardial biopsy findings in heart transplant patients. Transpl Int 1993; 6:341-347.
9. Sinzger C, Jahn G. Human cytomegalovirus cell tropism and pathogenesis. Intervirology 1996; 39:302-319.
10. Smith MG. Propagation in tissue cultures of a cytopathogenic virus from human salivary gland virus (SGV) disease. Proc Soc Exp Biol Med 1956; 92:424-430.
11. Weller TH, Macauley JC, Craig JM et al. Isolation of intranuclear inclusion producing agents from infants with illnesses resembling cytomegalic inclusion disease. Proc Soc Exp Biol Med 1957; 94:4-12.
12. Rowe WP, Hartley JW, Waterman S et al. Cytopathogenic agent resembling human salivary gland virus recovered from tissue cultures of human adenoids. Proc Soc Exp Biol Med 1956; 92:418-424.
13. Mocarski, ES, Jr. Cytomegalovirus Biology and Replication. In: Roizman B, Whitley RJ, Lopez C, eds. The Human Herpesviruses. New York: Raven Press, Ltd., 1993: 173-226.
14. Söderberg C, Giugni TD, Zaia JA et al. CD13 (human aminopeptidase N) mediates human cytomegalovirus infection. J Virol 1993; 67: 6576-6585.
15. Aldish JD, Lahijani RS, St. Jeor SC. Identification of a putative cell receptor for human cytomegalovirus. Virology 1990; 176: 337-345.
16. Nowlin DM, Cooper NR, Compton T. Expression of a human cytomegalovirus receptor correlates with infectibility of cells. J Virol 1991; 65:3114-3121.

17. Kari B, Gehrz R. A human cytomegalovirus glycoprotein complex designated gC-II is a major heparin-binding component of the envelope. J Virol 1992; 66:1761-1764.

18. Neyts J, Snoeck R, Schols D et al. Sulfated polymers inhibit the interaction of human cytomegalovirus with cell surface heparin sulfate. Virology 1992; 189:48-58.

19. Taylor-Weideman J, Sissons JGP, Borysiewicz LK et al. Monocytes are a major site of persistence of human cytomegalovirus in peripheral blood mononuclear cells. J Gen Virol 1991; 72:2059-2064.

20. Kondo K, Xu J, Mocarski ES. Human cytomegalovirus latent gene expression in granulocyte-macrophage progenitors in culture and in seropositive individuals. Proc Natl Acad Sci USA 1996; 93:11137-11142.

21. Hahn G, Jores R, Mocarski ES. Cytomegalovirus remains latent in a common precursor of dendritic and myeloid cells. Proc Natl Acad Sci USA 1998; 95:3937-3942.

22. Fish KN, Stenglein SG, Ibanez C et al. Cytomegalovirus persistence in macrophages and endothelial cells. Scand J Infect Dis Suppl 1995; 99:34-40.

23. Toorkey CB, Carrigan DR. Immunohistochemical detection of an immediate early antigen of human cytomegalovirus in normal tissues. J Infect Dis 1989; 160:741-751.

24. Meyerson D, Hackman RC, Nelson JA et al. Widespread presence of histologically occult cytomegalovirus. Hum Pathol 1984; 15:430-439.

25. Spector SA, Merrill R, Wolf D et al. Detection of human cytomegalovirus in plasma of AIDS patients during acute visceral disease by DNA amplification. J Clin Microbiol 1992; 30:2359-2365.

26. van Prooijen HC, Visser JJ, van Oostendorp WR et al. Prevention of primary transfusion-associated cytomegalovirus infection in bone marrow transplant recipients by the removal of white cells from blood components with high-affinity filters. Br J Haematol 1994; 87:144-147.

27. Gilbert GL, Hayes K, Hudson IL et al. Prevention of transfusion-acquired cytomegalovirus infection in infants by blood filtration to remove leukocytes. The Lancet June 1989:1228-1231.

28. Winston DJ, Ho WG, Howell CL et al. Cytomegalovirus infections associated with leukocyte transfusion. Ann Int Med 1980; 93:671-675.

29. Wilhelm JA, Matter L, Schopfer K. The risk of transmitting cytomegalovirus to patients receiving blood transfusions. J Infect Dis 1986; 154:169-171.

30. Waldman WJ, Knight DA, Huang EH et al. Bidirectional transmission of infectious cytomegalovirus between monocytes and vascular endothelial cells: an in vitro model. J Infect Dis 1995; 171:263-272.

31. Grundy JE, Lawson KM, MacCormac LP et al. Cytomegalovirus-infected endothelial cells recruit neutrophils by the secretion of C-X-C chemokines and transmit virus by direct neutrophil-endothelial cell contact and during neutrophil transendothelial migration. J Infect Dis 1998; 177:1465-1474.

32. Simmons RL, Weil R, Tallent MB. Do mild infections trigger the rejection of renal allografts? Transplant Proc 1970; 2:419-421.

33. von Willebrand E, Pettersson E, Ahonen J et al. CMV infection, class II antigen expression, and human kidney allograft rejection. Transplantation 1986; 42:364-367.

34. McDonald K, Rector TS, Braunlin EA et al. Association of coronary artery disease in cardiac transplant recipients with cytomegalovirus infection. Am J Cardiol 1989; 64:59-362.

35. Loebe M, Schüler S, Zais O et al. Role of cytomegalovirus infection in the development of coronary artery disease in the transplanted heart. J Heart Transplant 1990; 9:707-711.

36. Everett JP, Hershberger RE, Norman DJ et al. Prolonged cytomegalovirus infection with viremia is associated with development of cardiac allograft vasculopathy. J Heart Lung Transplant 1992; 11:S133-S137.

36. Matilla S, Heikkilä L, Sipponen J et al. Heart transplantation in Finland 1985-1995. Annales Chirurgiae et Gynaecologiae 1997; 86:113-120.

37. Brunner-La Rocca HP, Schneider J, Künzli A et al. Cardiac allograft rejection late after transplantation is a risk factor for graft coronary artery disease. Transplantation 1998; 65:538-543.

38. Keenan RJ, Lega ME, Dummer JS et al. Cytomegalovirus serologic status and postoperative infection correlated with risk of developing chronic rejection after pulmonary transplantation. Transplantation 1991; 51:433-438.

38. Duncan A, Paradis IL, Yousem SA et al. Sequelae of cytomegalovirus pulmonary infections in lung allograft recipients. Am Rev Respir Dis 1992; 146:1419-1425.

39. Paradis I, Yousem S, Griffith B. Airway obstruction and bronchiolitis obliterans after lung transplantation. Clin Chest Med 1993; 14:751-763.

40. O'Grady JG, Alexander GJ, Sutherland S et al. Cytomegalovirus infection and donor/recipient HLA antigens: interdependent cofactors in pathogenesis of vanishing bile-duct syndrome after liver transplantation. Lancet 1988; 2:302-305.

41. Bruggeman CA, Meijer H, Dormans PHJ et al. Isolation of a cytomegalovirus-like agent from wild rats. Arch Virol 1982; 73:231-241.

42. Lemström KB, Bruning JH, Bruggeman CA et al. Cytomegalovirus infection enhances smooth muscle cell proliferation and intimal thickening of rat aortic allografts. J Clin Invest 1993; 92:549-558.

43. Li F, Grauls G, Yin M et al. Initial endothelial injury and cytomegalovirus infection accelerate the development of allograft arteriosclerosis. Transplant Proc 1995; 27:3552-3554.

44. Li F, Yin M, Van Dam JG et al. Cytomegalovirus infection enhances the neointima formation in rat aortic allografts. Effects of major histocompatibility complex class I and class II antigen differences. Transplantation 1998; 65:1298-1304.

45. Lemström K, Koskinen P, Krogerus L et al. Cytomegalovirus antigen expression, endothelial cell proliferation, and intimal thickening in rat cardiac allografts after cytomegalovirus infection. Circulation 1995; 92:2594-2604.

46. Yilmaz S, Koskinen PK, Kallio E et al. Cytomegalovirus infection-enhanced chronic kidney allograft refection is linked with intercellular adhesion molecule-1 expression. Kidney Int 1996; 50:526-537.

47. Koskinen PK, Yilmaz S, Kallio E et al. Rat cytomegalovirus infection and chronic kidney allograft rejection. Transpl Int 1996; 9:S3-S4.

48. Lautenschlager I, Soots A, Krogerus L et al. CMV increases inflammation and accelerates chronic rejection in rat kidney allografts. Transplant Proc 1997; 29:802-803.

49. Weimar W, Balk AHMM, Metselaar HJ et al. On the relation between cytomegalovirus infection and rejection after heart transplantation. Transplantation 1991; 52:162-164.

50. Stovin PGI, Wreghitt TG, English TAH et al. Lack of association between cytomegalovirus infection of heart and rejection-like inflammation. J Clin Pathol 1989; 42:81-83.

51. Radovancevic B, Poindexter S, Birovljev S et al. Risk factors for development of accelerated coronary artery disease in cardiac transplant recipients. Eur J Cardio-Thorac Surg 1990; 4:309-313.

52. Gulizia JM, Kandolf R, Kendall TJ et al. Infrequency of cytomegalovirus genome in coronary arteriopathy of human heart allografts. Am J Pathol 1995; 147:461-475.

53. Nadasdy T, Smith J, Laszik Z et al. Absence of association between cytomegalovirus infection and obliterative transplant arteriopathy in renal allograft rejection. Mod Pathol 1994; 7:289-294.

54. Jutila MA, Berg EL, Kishimoto TK et al. Inflammation-induced endothelial cell adhesion to lymphocytes, neutrophils, and monocytes. Role of homing receptors and other adhesion molecules. Transplantation 1989; 48:727-731.

55. Hirschberg H, Bergh OJ, Thorsby E. Antigen-presenting properties of human vascular endothelial cells. J Exp Med 1980; 152:249s-255s.

56. Hughes CCW, Savage COS, Pober JS. The endothelial cell as a regulator of T cell function. Immunol Rev 1990; 117:85-102.

57. Pober JS, Cotran RS. Cytokines and endothelial cell biology. Physiol Rev 1990; 70:427-451.

58. Hruban RH, Beschorner WE, Baumgartner WA et al. Accelerated arteriosclerosis in heart transplant recipients is associated with T-lymphocyte-mediated endothelialitis. Am J Pathol 1990; 137:871-882.

59. Salomon RN, Hughes CCW, Schoen FJ et al. Human coronary transplantation-associated arteriosclerosis. Evidence for a chronic immune reaction to activated graft endothelial cells. Am J Pathol 1991; 138:791-798.

60. Briscoe DM, Schoen FJ, Rice GE et al. Induced expression of endothelial-leukocyte adhesion molecules in human cardiac allografts. Transplantation 1991; 51:537-539.

61. Fuggle SV, Sanderson JB, Gray DW et al. Variation in expression of endothelial adhesion molecules in pretransplant and transplanted kidneys: Correlation with intragraft events. Transplantation 1993; 55:117-123.

62. Tanaka H, Sukhova GK, Swanson SJ et al. Endothelial and smooth muscle cells express leukocyte adhesion molecules heterogeneously during acute rejection of rabbit cardiac allografts. Am J Pathol 1994; 144:938-951.

63. Koskinen P, Lemström K, Bruggeman C et al. Acute cytomegalovirus infection induces a subendothelial inflammation (endothelialitis) in the allograft vascular wall. A possible linkage with enhanced allograft arteriosclerosis. Am J Pathol 1994; 144: 1-50.

64. Koskinen PK, Krogerus LA, Nieminen MS et al. Quantitation of cytomegalovirus infection-associated histologic findings in endomyocardial biopsies of heart allografts. J Heart Lung Transplant 1993; 12:343-354.

65. Li FL, Grauls G, Yin M et al. Correlation between the intensity of cytomegalovirus infection and the amount of perivasculitis in aortic allografts. Transpl Int 1996; 9:S340-S344.

66. Koskinen PK. The association of the induction of vascular cell adhesion molecule-1 with cytomegalovirus antigenemia in human heart allografts. Transplantation 1993; 56:1103-1108.

67. Yilmaz S, Koskinen PK, Kallio E et al. Cytomegalovirus infection-enhanced chronic kidney allograft rejection is linked with intercellular adhesion molecule-1 expression. Kidney Int 1996; 50:526-537.

68. Craighead JE. Pulmonary cytomegalovirus infection in the adult. Am J Pathol 1971; 63:487-504.

69. Myerson D, Hackman RC, Nelson JA et al. Widespread presence of histologically occult cytomegalovirus. Hum Pathol 1984; 15: 430-439.

70. Roberts WH, Sneddon JM, Waldman WJ et al. Cytomegalovirus infection of gastrointestinal endothelium demonstrated by simultaneous nucleic acid hybridization and immunohistochemistry. Arch Pathol Lab Med 1989; 113:461-464.

71. Waldman WJ, Sneddon JM, Stephens RE et al. Enhanced endothelial cytopathogenicity induced by a cytomegalovirus strain propagated in endothelial cells. J Med Virol 1989; 28:223-230.

72. Waldman WJ, Roberts WH, Davis DH et al. Preservation of natural endothelial cytopathogenicity of cytomegalovirus by propagation in endothelial cells. Arch Virol 1991; 117:143-164.

73. Ustinov JA, Loginov RJ, Bruggeman CA et al. Cytomegalovirus induces class II expression in rat heart endothelial cells. J Heart Lung Transplantation 1993; 12:644-651.

74. Sedmak DD, Roberts WH, Stephens RE et al. Inability of cytomegalovirus infection of cultured endothelial cells to induce HLA class II antigen expression. Transplantation 1990; 49:458-462.

75. Sedmak DD, Guglielmo AM, Knight DA et al. Cytomegalovirus inhibits Major Histocompatibility class II expression on infected endothelial cells. Am J Pathol 1994; 144: 683-692.

76. Knight DA, Waldman WJ, Sedmak DD. Human cytomegalovirus does not induce human leukocyte antigen class II expression on arterial endothelial cells. Transplantation 1997; 63:1366-1369.

77. van Dorp WT, Jonges E, Bruggeman CA et al. Direct induction of MHC class I, but not class II, expression on endothelial cells by cytomegalovirus infection. Transplantation 1989; 48:469-472.

78. Hosenpud JD, Chou S, Wagner CR. Cytomegalovirus-induced regulation of major histocompatibility complex class I antigen expression in human aortic smooth muscle cells. Transplantation 1991; 52:896-903.

79. Scholz M, Hamann A, Blaheta RA et al. Cytomegalovirus- and interferon-related effects on human endothelial cells: Cytomegalovirus infection reduces upregulation of HLA class II antigen expression after treatment with interferon-γ. Hum Immunol 1992; 35:230-238.

80. Waldman WJ, Knight DA, Adams PW et al. In vitro induction of endothelial HLA class II antigen expression by CMV-activated CD4⁺ T cells. Transplantation 1993; 56: 1504-1512.

81. Miller DM, Rahill BM, Boss JM et al. Human cytomegalovirus inhibits major histocompatibility complex class II expression by disruption of the Jak/Stat pathway. J Exp Med 1998; 187:675-683.

82. Barnes PD, Grundy JE. Down-regulation of the class I HLA heterodimer and b_2-microglobulin on the surface of cells infected with cytomegalovirus. J Gen Virol 1992; 73: 2395-2403.

83. Beersma MFC, Bijlmakers MJE, Ploegh HL. Human cytomegalovirus down-regulates HLA class I expression by reducing the stability of class I H chains. J Immunol 1993; 151:4455-4464.

84. Waldman WJ, Knight DA, Huang EH. An in vitro model of T cell activation by autologous cytomegalovirus (CMV)-infected human adult endothelial cells: contribution of CMV-enhanced endothelial ICAM-1. J Immunol 1998; 160:3143-3151.

85. Waldman WJ, Knight DA, Adams PW. Cytolytic activity against allogeneic human endothelia: Resistance of cytomegalovirus-infected cells and virally activated lysis of uninfected cells. Transplantation 1998; 66:67-77.

86. Grundy JE, Downes KL. Up-regulation of LFA-3 and ICAM-1 on the surface of fibroblasts infected with cytomegalovirus. Immunology 1993; 78:405-412.

87. Sedmak DD, Knight DA, Vook NA et al. Divergent patterns of ELAM-1, ICAM-1, and VCAM-1 expression on cytomegalovirus-infected endothelial cells. Transplantation 1994; 58:1379-1385.

88. Burns LJ, Pooley JC, Walsh DJ et al. Intercellular adhesion molecule-1 expression in endothelial cells is activated by cytomegalovirus immediate early proteins. Transplantation 1999; 67:137-144.

89. Knight DA, Waldman WJ, Sedmak DD. Cytomegalovirus-mediated modulation of adhesion molecule expression by human arterial and microvascular endothelial cells. Transplantation: in press.

90. Waldman WJ, Adams PW, Orosz CG et al. T lymphocyte activation by cytomegalovirus-infected, allogeneic cultured human endothelial cells. Transplantation 1992; 54: 887-896.

91. Waldman WJ, Knight DA. Cytokine-mediated induction of endothelial adhesion

molecule and histocompatibility leukocyte antigen expression by cytomegalovirus-activated T cells. Am J Pathol 1996; 148: 105-119.

92. Koskinen PK, Krogerus LA, Nieminen MS et al. Cytomegalovirus infection-associated generalized immune activation in heart allograft recipients: A study of cellular events in peripheral blood and endomyocardial biopsy specimens. Transpl Int 1994; 7: 163-171.

93. Adams PW, Lee HS, Waldman WJ et al. Alloantigenicity of human endothelial cells: 1. Frequency and phenotype of human helper T lymphocytes that can react to allogeneic endothelial cells. J Immunol 1992; 148:3753-3760.

94. Craigen JL, Grundy JE. Cytomegalovirus induced up-regulation of LFA-3 (CD58) and ICAM-1 (CD54) is a direct viral effect that is not prevented by ganciclovir or foscarnet treatment. Transplantation 1996; 62:1102-1108.

95. Breth MR, Sedmak DD, Waldman WJ. Cytokine-mediated induction of endothelial activation by CMV-responsive T cells: persistence in the presence of cyclosporine (CsA). FASEB J 1996; 10:A1149.

96. Santomenna LD, Colberg-Poley AM. Induction of cellular hsp70 expression by human cytomegalovirus. J Virol 1990; 64: 2033-2040.

97. Kaufmann SH. Heat shock proteins and the immune response. Immunol Today 1990; 11:129-136.

98. Moliterno R, Woan M, Bentlejewski C et al. Heat shock protein-induced T-lymphocyte propagation from endomyocardial biopsies in heart transplantation. J Heart Lung Transplant 1995; 14:329-337.

99. Moliterno R, Valdivia L, Pan F et al. Heat shock protein reactivity of lymphocytes isolated from heterotopic rat cardiac allografts. Transplantation 1995; 59:598-604.

100. Tullius SG, Tilney NL. Both alloantigen-dependent and -independent factors influence chronic allograft rejection. Transplantation 1995; 59:313-318.

101. Oppenheimer-Marks N, Davis LS, Lipsky PE. Human T lymphocyte adhesion to endothelial cells and transendothelial migration. Alteration of receptor use relates to the activation status of both the T cell and the endothelial cell. J Immunol 1990; 145: 140-148.

102. Ross R, Masuda J, Raines EW et al. Localization of PDGF-B protein in macrophages in all phases of atherogenesis. Science 1990; 248:1009-1011.

103. Alpers CE, Hudkins KL, O'Brien KD. Mediator molecules in chronic renal vascular rejection, in Orosz CG, Sedmak DD, Ferguson RM (ed): Transplant Vascular Sclerosis., RG Landes Company, 1995: 61-69.

104. Alpers CE, Schelling ME, Hudkins KL. Localization of basic fibroblast growth factor (bFGF) and its receptor (FGFR1, flg) in fetal, mature, and transplanted human kidneys. Lab Invest 1994; 71:156A.

105. Blotnick S, Peoples GE, Freeman MR et al. T lymphocytes synthesize and export heparin-binding epidermal growth factor-like growth factor and basic fibroblast growth factor, mitogens for vascular cells and fibroblasts: Differential production and release by CD4+ and CD8+ T cells. Proc Natl Acad Sci USA 1994; 91:2890-2894.

106. Kerby JD, Verran DJ, Luo KL et al. Immunolocalization of FGF-1 and receptors in human renal allograft vasculopathy associated with chronic rejection. Transplantation 1996; 62:467-475.

107. Raines EW, Ross R. Smooth muscle cells and the pathogenesis of the lesions of atherosclerosis. Br Heart J 1993; 61:530-537.

108. Michelson S, Alcami J, Kim SJ et al. Human cytomegalovirus infection induces transcription and secretion of transforming growth factor β-1. J Virol 1994; 68: 5730-5737.

109. Wu TC, Hruban RH, Ambinder RF et al. Demonstration of cytomegalovirus nucleic acids in the coronary arteries of transplanted hearts. Am J Pathol 1992; 140:739-747.

110. Persoons MC, Daemen MJ, Bruning JH et al. Active cytomegalovirus infection of arterial smooth muscle cells in immunocompromised rats. A clue to herpesvirus-associated atherogenesis? Circ Res 1994; 75: 214-220.

111. Kovacs A, Weber ML, Burns LJ et al. Cytoplasmic sequestration of p53 in cytomegalovirus-infected human endothelial cells. Am J Pathol 1996; 149:1531-1539.

112. Zhu H, Shen Y, Shenk T. Human cytomegalovirus IE1 and IE2 proteins block apoptosis. J Virol 1995; 69:7960-7970.

CHAPTER 9

Lipids and Lipid Lowering Therapy

Jon A. Kobashigawa

In the past decade, improved management strategies in heart transplantation have dramatically increased allograft survival and reduced early posttransplant morbidity. However, these advances have been accompanied by the emergence of a new group of complications. One of these complications, hyperlipidemia, is an important concern in heart transplant recipients.

Because of the well-established correlation between lipid levels and atherosclerosis in nontransplant populations,[1] it is logical to expect that heart transplant recipients would also be placed at increased risk for cardiovascular events. In addition, there is emerging evidence linking elevated lipid levels to allograft vasculopathy, an unusually accelerated form of atherosclerotic vascular disease.[2] In fact, this complication has emerged as one of the primary causes of morbidity and mortality in long-term transplant survivors, surpassing even infection. This chapter will first discuss the mechanisms and clinical implications of hyperlipidemia in the heart transplant population and will conclude with discussion of treatment options.

Potential Causes of Hyperlipidemia

Potential causes of hyperlipidemia in transplant recipients include diet, genetic predisposition, and immunosuppressive medications. Many patients are at or below their ideal body weight before transplantation but become obese after successful procedures.

Obesity in heart transplantation has been closely associated with the development of hyperlipidemia.[3] Some patients with preoperative diagnosis of atherosclerotic cardiovascular disease have familial hyperlipidemia. This genetic predisposition contributes to the post-transplant hyperlipidemic state.[4]

Immunosuppressive agents such as corticosteroids and cyclosporine are implicated in the development of hyperlipidemia and possible mechanisms are described in Figure 9.1. It has been suggested that cyclosporine may inhibit the enzyme 26-hydroxylase which is important in the bile acid synthetic pathway.[6] Cyclosporine would thereby decrease the synthesis of bile acids from cholesterol and subsequently the transport of cholesterol to the intestines. Cyclosporine is also reported to bind to the low density lipoprotein (LDL) receptor, which results in increased serum levels of LDL cholesterol.[6] It is also thought that cyclosporine increases hepatic lipase activity and decreases lipoprotein lipase activity, resulting in impaired clearance of very low density lipoprotein (VLDL) and LDL. Corticosteroids are reported to enhance the activity of acetyl-CoA carboxylase and free fatty acid synthetase, increase hepatic synthesis of VLDL, down-regulate LDL receptor activity, increase the activity of HMG-CoA reductase, and inhibit lipoprotein lipase.[7-9] This results in increased VLDL, total cholesterol and triglyceride levels, and decreased high density lipoprotein (HDL) levels.

Transplant-Associated Coronary Artery Vasculopathy, edited by Marlene L. Rose.
©2001 Eurekah.com

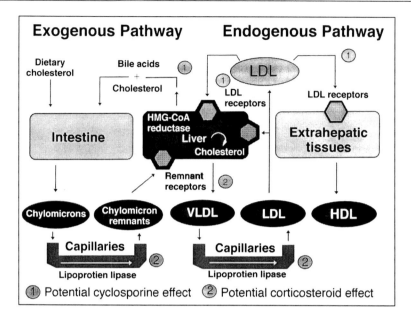

Fig. 9.1. Potential cyclosporine effects are designated as "1". Cyclosporine is suggested to inhibit the enzyme 26-hydroxylase which is important in the bile acid synthetic pathway. This would decrease the synthesis of bile acids from cholesterol and subsequently the transport of cholesterol to the intestines. Cyclosporine is also reported to bind to the low density lipoprotein (LDL) receptor, which results in increased serum levels of LDL-cholesterol. Potential corticosteroid effects are designated as "2" and are reported to enhance the activity of acetyl-CoA carboxylase and free fatty acid synthetase, and inhibit lipoprotein lipase. This results in increased levels of very low density lipoprotein (VLDL), total cholesterol, and triglyceride levels. (High density lipoprotein = HDL) Adapted from Kobashigawa et al.[5]

Clinical Course

Lipid abnormalities are reported in 60% to 80% of heart transplant patients receiving the standard triple-drug regimen consisting of cyclosporine, azathioprine, and prednisone.[10] Several groups have examined the patterns of lipid abnormalities following cardiac transplantation. Studies in this population show that increases in total cholesterol, low density lipoprotein cholesterol, apolipoprotein B, and triglyceride levels develop at 3-18 months.[11,12] Some studies suggest that these lipid levels slowly fall as the time after transplant lengthens.[3] The reports are more variable regarding levels of high density lipoprotein cholesterol.[13,14] Interestingly, lipoprotein (a) levels have been found to decrease by almost 40% after cardiac transplantation.[14]

Ballantyne et al[11] reported that mean total cholesterol values in 100 cardiac transplant recipients increased from pretransplant levels of 168 ± 7 to 234 ± 7 mg/dl at 3 months after transplant. During this same period, LDL-cholesterol rose from 111 ± 6 to 148 ± 6 mg/dl, HDL-cholesterol rose from 34 ± 1 to 47 ± 1, and triglyceride levels rose from 107 ± 6 to 195 ± 10 mg/dl. There were no further significant rises after the 3-month evaluation, but LDL-cholesterol and triglyceride levels remained elevated in 64% and 41% of patients, respectively, 6 months after dietary therapy was instituted.

The development of transplant coronary artery disease (CAD) in cardiac allografts is one of the major causes of graft failure in long-term survivors of cardiac transplantation and a primary contributor to overall patient morbidity and mortality. The incidence of this disease ranges from 1-18% at 1 year to 20-50% at 3 years.[15,16] Numerous immune and nonimmune risk factors are associated with the development of transplant CAD. Immune risk factors[10,17] as a cause for transplant CAD is evidenced by increased levels of cytotoxic B-cell antibodies, increased anti-HLA antibodies, a correlation between disease development and acute cellular rejection and humoral (antibody-mediated) rejection, cytomegalovirus infection, sensitization to monoclonal antibody OKT3, and detection of early and persistently elevated interleukin-2 receptor levels. Nonimmune risk factors[13] include hyperlipidemia, recipient age and gender, obesity, pretransplant diagnosis, and donor ischemic time. Among nonimmune risk factors for transplant CAD, the most consistently described relationship has been with cholesterol.

In an autopsy study, McManus et al carried out morphometric, immunohistochemical, ultrastructural, and biochemical studies in 23 explanted allografts and donor age-matched native coronary artery controls.[18] Mean total cholesterol, esterified cholesterol, free cholesterol content in the transplant arteriopathic coronaries were greater than tenfold higher than in comparable native coronary segments. Extent of lipids in the arterial walls was highly correlated with digitized percent luminal narrowing. The authors conclude that lipid accumulation is an important early and persistent phenomenon in the development of transplant CAD.

The relationship of elevated triglyceride levels to transplant CAD risk has not been fully defined.[13] Winters et al showed that higher versus lower triglycerides (328 vs 145 mg/dl) were associated with a marked difference in luminal narrowing through inspection of failed allografts.[19] Valantine reported a correlation of elevated triglycerides and low HDL-cholesterol to increasing intimal thickness from a multicenter intracoronary ultrasound study.[20]

Treatment of Hyperlipidemia

Clinical assessment for hyperlipidemia should be initiated soon after transplantation. It is controversial whether therapy for hyperlipidemia in heart transplant patients should follow the guidelines recommended for the general population and detailed in the report of the second Adult Treatment Panel of the National Cholesterol Education Program (NCEP).[1] In heart transplant patients, potential strategies include dietary therapy, reduced doses of immunosuppressive agents, and lipid-lowering agents.

Diet

In the heart transplant population, dietary modification is the safest form of treatment for elevated LDL-cholesterol. Patients should be asked to comply with the American Heart Association Step I or Step II Diet as recommended by the NCEP.[1] However, it has been shown that despite dietary intervention, many patients have persistently high lipid levels. Ballantyne et al measured mean plasma lipid values in 100 patients at 1, 3, 6, and 12 months after heart transplantation.[11] All patients were given instructions on the American Heart Association Step I Diet before hospital discharge. Lipid values did not change significantly during the 3 months in which patients were asked to comply with the Step I Diet, and many patients had persistent elevations of LDL-cholesterol and triglyceride levels. It is likely that despite dietary intervention and optimal medical management for hypertension and immunosuppression, many patients may have high lipid levels and require pharmacologic intervention.

Immunosuppressive Therapy

As previously discussed, both cyclosporine and prednisone have been independently linked with increased risk for hyperlipidemia; therefore, one strategy to reduce hyperlipidemia is to modify the dosage of one or both of these drugs. Several heart transplant programs have reported decreases in cholesterol levels after withdrawal of corticosteroids from the maintenance immunosuppressive regimen.[21-24] Effects on cholesterol reduction at 1 year after stopping corticosteroids range between 6% and 26%. However, not all patients may benefit from this strategy; steroid withdrawal in heart transplant patients was successfully carried out in 56-89% of patients.[21-23]

Drug Therapy for Cholesterol Lowering

The NCEP guidelines regarding lipid-lowering drug therapy are based for the most part on three major drug classes: the 3-hydroxy-3-methylglutaryl coenzyme A (HMG-CoA) reductase inhibitors, bile acid sequestrants, and nicotinic acid. The fibric acid derivatives are indicated for patients with very high triglyceride levels, and probucol is suggested only for those patients who have not tolerated or responded to the other cholesterol-lowering drugs. Potential interactions with immunosuppressive regimens should be considered when using lipid-lowering drugs in transplant recipients and are described in Table 9.1.

HMG-CoA Reductase Inhibitors

These drugs inhibit HMG-CoA reductase, a key rate-limiting enzyme in the pathway for cholesterol biosynthesis; at therapeutic doses, they reduce but do not completely inhibit cholesterol synthesis. Five agents are currently available in the United States: lovastatin, pravastatin, simvastatin, fluvastatin, and atorvastatin. These agents are highly effective in lowering LDL-cholesterol concentrations, the primary target of lipid-lowering therapy in most patients.[1] Data from primary and secondary prevention studies in nontransplant hyperlipidemic patients treated with these agents have shown reductions in cardiac mortality (33% and 42% risk reduction respectively).[25,26]

Several investigators have reported success with lovastatin in cardiac allograft recipients with elevated lipid levels (see Table 9.2). Lovastatin doses of 10-20 mg/day for more than 1.5 months reduced total cholesterol by 21-29% and LDL-cholesterol by 25-32%. Experience with pravastatin has been similar to that of lovastatin (see Table 9.2). Pravastatin

doses of 10-40 mg/day for more than 1.5 months reduced total cholesterol by 11-21% and LDL-cholesterol by 15-42%.

In a primary prevention study involving pravastatin, 97 heart transplant recipients were randomly assigned within two weeks after transplant to receive pravastatin 40 mg/day (n = 47) or no lipid-lowering therapy (n = 50).[33] Cholesterol levels at 3, 6, 9, and 12 months after transplantation were consistently lower in the pravastatin group (mean 193 mg/dl versus 248 mg/dl, p < .001). Clinically severe rejection leading to hemodynamic compromise within 1 year was less in the pravastatin group which translated into a one-year survival benefit of 94% compared to 78% in the control group (p = 0.02). Transplant CAD (diagnosed angiographically or at autopsy and as measured by intracoronary ultrasound, was less in the pravastatin group.) Natural killer cell cytotoxicity was assessed in a subgroup of 20 patients and was significantly lower in pravastatin patients. The investigators hypothesized favorable effects of pravastatin may reflect the importance of early reduction of cholesterol, a direct immunosuppressive effect, or both.

Several investigators have reported significant lipid reduction using simvastatin in cardiac transplant recipients. Simvastatin doses of 5-20 mg/day for more than 4 months reduced total cholesterol by 14-27% and LDL-cholesterol by 18-40% (Table 9.2).[34] In another primary prevention study, Wenke et al[35] randomly allocated 72 transplant patients immediately after transplant to simvastatin 5 to 15 mg/day (n = 35) or no simvastatin (n = 37).[34] After 4 years, the total cholesterol levels were 198 mg/dl and 228 mg/dl in the simvastatin and control groups, respectively (p = 0.03). Similar to the pravastatin study,[33] there was a benefit in four-year survival of 88.6% in the simvastatin group compared to 70% in the control group (p = 0.05). Death due to refractory rejection was 1 patient in the simvastatin group versus 5 patients in the control group (p = 0.1). Less experience is known with fluvastatin but limited results (see Table 9.2) appear similar to other agents in this class. No clinical results are available for atorvastatin at this time.

Table 9.1. Potential adverse effects and drug interactions in transplant recipients

Class of Lipid-Lowering Agent	Examples	Possible Adverse Effects	Comments
Bile Acid Resin	cholestyramine (Questran, Colestid)	may prevent absorption of fat-soluble vitamins; other poor compliance because of constipation and bloating	may inhibit absorption of cyclosporine and absorption of other fat-soluble drugs; space dose by 2 h with interacting drug; may increase triglyceride concentrations
Nicotinic Acid	nicotinic acid, and extended-release formulation	flushing, pruritus, increase in liver enzymes, increased uric acid concentrations, altered glucose tolerance, and exacerbation of peptic ulcer disease	concomitant cyclosporine prednisone use may exacerbate adverse effects
Fibric Acid Derivative	gemfibrozil (Lopid)	gallstones, myositis (especially in patients with decreased renal function), nausea, gastrointestinal uspet	increased risk of myositis concomitant HMG-CoA reductase inhibitors and immunosuppressive drugs
Antioxidant	probucol (Lorelco)	flatulence, loose stools, prolonged QT interval on electrocardiogram, decreased HDL	may interact with cyclosporine and cause fluctuation in cyclosporine concentrations
HMG-CoA Reductase Inhibitor	lovastatin (Mevacor) simvastatin (Zocor) pravastatin (Pravachol) fluvastatin (Lescol)	abdominal pain, flatulence, increase in transaminase concentration, myositis, sleep disturbances	may increase liver function enzymes; increased risk of myositis with high-dose HMG-CoA reductase inhibitors and/or fibric acid derivatives with cyclosporine

HDL = high density lipoprotein cholesterol; HMG-CoA = 3-hydroxy-3-methylglutaryl coenzyme A. (Modified from Ballantyne et al.[11])

There have been several reports in the literature of rhabdomyolysis in cyclosporine-treated heart transplant recipients receiving lovastatin.[28,36] These patients received high doses (80 mg/day) and/or concomitant therapy with gemfibrozil or niacin. The increased risk of myopathy with HMG-CoA reductase inhibitors in transplant recipients is most likely due to increased serum concentrations of HMG-CoA reductase inhibitors in patients receiving cyclosporine.[28,37,38] It is now generally accepted that low doses of

HMG-CoA reductase inhibitors can be used safely in patients treated with cyclosporine.

Bile Acid Sequestrants

Bile acid sequestrants (cholestyramine and colestipol) bind with bile acids, interrupting bile acid recirculation, and increase hepatic bile acid synthesis and LDL-receptor activity.[1] There are theoretical reasons to believe that these agents may interfere with absorption of lipid-soluble drugs, including cyclosporine.[11] In a small study of heart

Table 9.1. Potential adverse effects and drug interactions in transplant recipients

Author	Drug	Daily Dosage (mg)	N	F/U (mos)	TC	LDL	HDL	TG
Kuo (1989)[27]	lov	20-60	11	12	-29%*	-32%*	0%	-15%*
Kobashigawa (1990)[28]	lov	10-20	44	3	-26%	-25%*	-8%	-20%*
Ballantyne (1992)[11]	lov	20	15	13	-21%*	-31%*	+10%*	-9%
Kobashigawa (1992)[29]	pra	20-40	44	3	-19%*	-18%*	-14%	0%
Barbir (1991)[30]	sim	10	12	8	-38%*	-42%*	+18%*	-25%*
Vanhaecke (1994)[31]	sim	5-20	25	6	-27%*	-40%*	0%	-20%
Campana (1995)[32]	sim	10	20	4	-14%*	-21%*	+8%	-4%

F/U = follow-up; TC = total cholesterol; LDL = low-density lipoprotein cholesterol; HDL = high density lipoprotein cholesterol; TG = triglycerides; lov = lovastatin; pra = pravastatin; sim = simvastatin; flu = fluvastatin; * = p < 0.05

transplant recipients treated with cholestyramine, Keogh et al reported a 14% reduction in total cholesterol levels.[39] These investigators also monitored cyclosporine pharmacokinetics and found that the area under the whole blood cyclosporine concentration-time curve varied widely among the patients, ranging from a 23% decrease to a 55% increase from baseline, but cholestyramine did not affect cyclosporine A levels.

Fibric Acid Derivatives

Two fibrates are available in the United States: gemfibrozil and clofibrate.[1] Ballantyne et al reported the use of gemfibrozil as a single hypolipidemic agent in a small group of cardiac transplant recipients, resulting in marked reductions in triglycerides and small reductions in total and LDL-cholesterol levels.[11] In this group, gemfibrozil was well tolerated and was not observed to interfere with immunosuppressive therapy. Two newer fibric acid derivatives, bezafibrate and fenofibrate, are under study in Europe. Recent reports in heart transplant recipients have suggested possible nephrotoxicity with fenofibrate due to a cyclosporine interaction.[40] Hidalgo et al did not report nephrotoxicity with bezafibrate 400 mg/day which was as effective in cholesterol lowering compared with lovastatin 10 mg/day in cardiac transplant patients.[41]

Other Lipid Lowering Therapies

There have been a number of clinical trials examining the safety and efficacy of other lipid-lowering agents. Nicotinic acid reduces LDL and increases HDL-cholesterol levels and, unlike bile acid sequestrants, reduces triglyceride levels.[1] Clinical studies of this drug in transplantation have been lacking, mostly due to the many adverse effects (see Table 9.1). Probucol reduces both total cholesterol and LDL-cholesterol levels to a modest degree. It has been used in heart transplant patients and found to reduce both LDL- and HDL-cholesterol levels by 15%; thus, it did not produce a significant change in the ratio of LDL-cholesterol to HDL-cholesterol.[42] A 28% decrease in cyclosporine levels while receiving probucol was reported by the authors. Of note, apheresis in combination with diet and pravastatin 10 mg/day has demonstrated regression in transplant CAD. Larger trials are in progress for this lipid-lowering strategy.[43]

Summary

Elevated lipid levels have been reported in more than 60% of heart transplant recipients. Diet, genetic predisposition, and immunosuppressive agents have been identified as principal factors involved in elevated cholesterol levels in the transplant population.

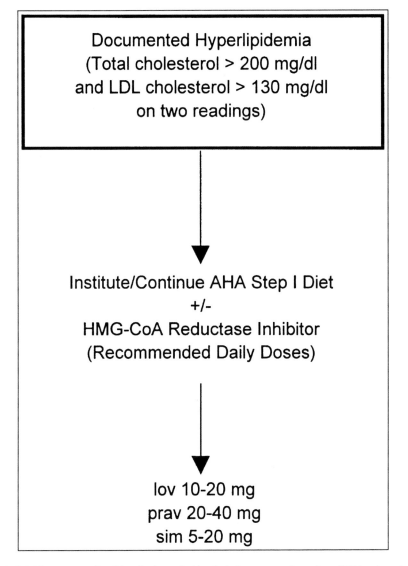

Fig. 9.2. Initial treatment algorithm for hyperlipidemia in heart transplantation. (AHA = American Heat Association, flu = fluvastatin, lov = lovastatin, pra = pravastatin, sim = simvastatin)

Elevated lipid levels have been correlated with an increased risk for cardiovascular disease in the graft vessels in heart transplant recipients. Unlike atherosclerotic disease in the nontransplant population, allograft vasculopathy progresses rapidly and can become clinically significant within 1 year.

Currently, no consensus exists regarding management. The role of the NCEP guidelines is controversial because they may not adequately reflect the need for aggressive lipid-lowering therapy following heart transplantation. Dietary therapy is the safest option, but has limited efficacy. It appears that the most effective agents to reduce cholesterol levels are the HMG-CoA reductase inhibitors; however, there is concern regarding associated myositis and rhabdomyolysis.

Based on the high cardiovascular risk of heart transplant patients, an initial treatment algorithm of diet and HMG-CoA reductase inhibitors is shown in Figure 9.2. For those transplant patients with (refractory to diet) hypertriglyceridemia (> 200 mg/dl), fibric acid derivatives appear to be well tolerated and an effective option. The safe and efficacious use of combination pharmacologic therapy awaits further study. Current research efforts are focusing on modulation of immunosuppressive therapy and the use of HMG-CoA reductase inhibitors. From autopsy studies demonstrating early lipid deposition in transplant coronary arteries and recent small clinical trials, it may be beneficial to initiate cholesterol lowering therapy with HMG-CoA reductase inhibitors early after heart transplant surgery.

References

1. Grundy SM. National Cholesterol Education Program. Second report of the Expert Panel on Detection, Evaluation, and Treatment of High Blood Cholesterol in Adults (Adult Treatment Panel II). Circulation 1994; 89:1329.
2. Miller LW. Allograft vascular disease: A disease not limited to hearts. J Heart Lung Transplant 1992;11 (3, Pt 2): S32.
3. Grady KL, Costanzo-Nordin MR, Herold LS et al. Obesity and hyperlipidemia after heart transplantation. J Heart Lung Transplant 1991; 10:449.
4. Taylor DO, Thompson JA, Hastillo A et al. Hyperlipidemia after clinical heart transplantation. J Heart Transplant 1989; 8:209.
5. Kobashigawa JA, Kasiske BL. Hyperlipidemia in solid organ transplantation. Transplantation 1997; 63:331-338.
6. de Groen PC. Cyclosporine, low density lipoprotein, and cholesterol. Mayo Clin Proc 1988; 63:1012.
7. Chan MK, Varghese Z, Moorhead JF. Lipid abnormalities in uremia, dialysis, and transplantation. Kidney Int 1981; 19:625.
8. Becker DM, Chamberlain B, Swank R et al. Relationship between corticosteroid exposure and plasma lipid levels in heart transplant recipients. Am J Med 1988; 85:632.
9. Ibels LS, Simons LA, King JO et al. Studies on the nature and causes of hyperlipidemia in uraemia, maintenance dialysis and renal transplantation. QJM 1975; 44:601.
10. Miller LW, Schlant RC, Kobashigawa J et al. 24th Bethesda Conference: Cardiac Trans-

plantation. Task Force 5: Complications. J Am Coll Cardiol 1993; 22:41.
11. Ballantyne CM, Radovancevic B, Farmer JA et al Hyperlipidemia after heart transplantation: Report of a 6-year experience with treatment recommendations. J Am Coll Cardiol 1992; 19:1315.
12. Kirk JK, Dupuis RE. Approaches to the treatment of hyperlipidemia in the solid organ transplant recipient. Ann Pharmacother 1995; 29:879.
13. Johnson MR. Transplant coronary disease: Nonimmunologic risk factors. J Heart Lung Transplant 1992; 11:S124.
14. Farmer JA, Ballantyne CM, Frazier OH et al. Lipoprotein (a) and apolipoprotein changes after cardiac transplantation. J Am Coll Cardiol 1991; 18:926.
15. Gao S-Z, Schroeder JA, Alderman EL et al. Prevalence of accelerated coronary artery disease in heart transplant survivors: Comparison of cyclosporine and azathioprine regimens. Circulation 1989; 80 (suppl 3): III-100.
16. O'Neill BJ, Pflugfelder PW, Singh NR et al. Frequency of angiographic detection and quantitative assessment of coronary arterial disease on and three years after cardiac transplantation. Am J Cardiol 1989; 63: 1221.
17. Hosenpud JD, Shipley GD, Wagner CR. Cardiac allograft vasculopathy: Current concepts, recent developments, and future directions. J Heart Lung Transplant 1992; 11:9.
18. McManus BM, Horley KJ, Wilson JE et al. Prominence of coronary arterial wall lipids in human heart allografts: Implications of pathogenesis of allograft arteriopathy. Am J Pathol 1995; 147:293.
19. Winters GL, Kendall TJ, Radio SJ et al. Posttransplant obesity and hyperlipidemia: Major predictors of severity of coronary arteriopathy in failed human heart allografts. J Heart Transplant 1990; 9:364.
20. Valantine HA. Role of lipids in allograft vascular disease: A multicenter study of intimal thickening detected by intravascular ultrasound. J Heart Lung Transplant 1995; 14:S234.
21. Pritzker MR, Lake KD, Reutzel TJ et al. Steroid-free maintenance immunotherapy: Minneapolis Heart Institute experience. J Heart Lung Transplant 1992; 11:415.
22. Renlund DG, Bristow MR, Crandall BG et al. Hypercholesterolemia after heart transplantation: Amelioration by corticosteroid-free maintenance immunosuppression. J Heart Lung Transplant 1989; 8:214.
23. Keogh A, Macdonald P, Harvison A et al. Initial steroid-free versus steroid-based maintenance therapy and steroid withdrawal af-

ter heart transplantation: Two views of the steroid question. J Heart Lung Transplant 1992; 11:421.

24. Kobashigawa JA, Stevenson LW, Brownfield EB et al. Corticosteroid weaning late after cardiac transplantation: Relation to HLA-DR mismatching and long-term metabolic benefits. J Heart Lung Transplant 1995; In press.

25. Shepherd J, Cobbe SM, Ford J et al. Prevention of coronary heart disease with pravastatin in men with hypercholesterolemia. N Engl J Med 1995; 333:1301.

26. Scandinavian Simvastatin Survival Study Group. Randomized trial of cholesterol lowering in 4444 patients with coronary heart disease: The Scandinavian Simvastatin Study (4S). Lancet 1994; 344:1383.

27. Kuo PC, Kirshenbaum JM, Gordon J et al. Lovastatin therapy for hypercholesterolemia in cardiac transplant recipients. Am J Cardiol 1989; 64:631.

28. Kobashigawa JA, Murphy FL, Stevenson LW et al. Low-dose lovastatin safely lowers cholesterol after cardiac transplantation. Circulation 1990; 82 (suppl 4):IV-281.

29. Kobashigawa JA, Brownfield ED, Stevenson LW et al. Effects of pravastatin for hypercholesterolemia in cardiac transplant recipients [Abstract]. J Am Coll Cardiol 1993; 21: 141A.

30. Babir M, Rose M, Kushwaha S et al. Low-dose simvastatin for the treatment of hyperlipidemia in recipients of cardiac transplantation. Int J Cardiology 1991; 33:241.

31. Vanhaecke J, van Cleemput J, van Lierde J et al. Safety and efficacy of low dose simvastatin in cardiac transplant recipients treated with cyclosporine. Transplantation 1994; 58:42.

32. Campana C, Iacona I, Regazzi MB et al. Efficacy and pharmacokinetics of simvastatin in heart transplant recipients. Ann Pharmacother 1995; 29:235.

33. Kobashigawa JA, Katznelson S, Laks H et al. Effect of pravastatin on outcomes after cardiac transplantation. N Engl J Med 1995; 333:621.

34. Wenke K, Thiery J, Meiser B et al. Long-term simvastatin therapy for hypercholesterolemia in heart transplant patients. Z Kardiol 1995; 84:130.

35. Wenke K, Meiser B, Thiery J et al. Simvastatin reduces graft vessel disease and mortality after heart transplantation: A four-year randomized trial. Circulation 1997; 96: 1398-1402.

36. East C, Alivizatos PA, Grundy SM et al. Rhabdomyolysis in patients receiving lovastatin after cardiac transplantation. N Engl J Med 1988; 318:47.

37. Regazzi MB, Iacona I, Campana IC et al. Altered disposition of pravastatin following concomitant drug therapy with cyclosporin A in transplant recipients. Transplant Proc 1993; 25:2732.

38. Arnadottir M, Eriksson L-O, Thysell H et al. Plasma concentration profiles of simvastatin 3-hydroxy-3-methylglutaryl coenzyme A reductase inhibitory activity in kidney transplant recipients with and without cyclosporin. Nephron 1993; 65:410.

39. Keogh A, Day R, Critchley L et al. The effect of food and cholestyramine on the absorption of cyclosporine in cardiac transplant recipients. Transplant Proc 1988; 20:27.

40. Boissonnat P, Salen P, Guidollet J et al. The long-term effects of the lipid-lowering agent fenofibrate in hyperlipidemic heart transplant recipients. Transplantation 1994; 58:245.

41. Hidalgo L, Zambrana JL, Blanco-Molina A et al. Lovastatin versus bezafibrate for hyperlipidemia treatment after heart transplantation. J Heart Lung Transplant 1995; 14:461.

42. Sundararajan B, Cooper DKC, Muchmore J et al. Interaction of cyclosporine and probucol in heart transplant patients. Transplant Proc 1991; 23:2028.

43. Park JW, Vermeltfoort M, Braun P et al. Regression of transplant coronary artery disease during chronic HELP therapy: A case study. Atherosclerosis 1995; 115:1.

CHAPTER 10

Immunosuppressive Drugs for the Prevention and Treatment of Transplant Coronary Artery Vasculopathy

Norman P. Briffa, C.R. Gregory and Randall E. Morris

Transplant coronary artery disease is the greatest limitation to long-term survival after heart transplantation.[1] The reported incidence of transplant coronary artery vasculopathy depends on the method of diagnosis. Using traditional coronary angiography, up to 40 percent of patients are found to have transplant coronary artery vasculopathy 5 years after transplantation[2] and it is said that for patients who have survived the first year after transplantation, the incidence of transplant coronary artery vasculopathy has not changed despite the introduction of cyclosporine A. Intravascular ultrasound suggests that the true figure is much higher with 60 percent of patients exhibiting changes at one year.[3]

Biology

Transplant coronary artery vasculopathy is characterized by the development of a neointimal layer in large and medium sized coronary arteries. This layer consists initially of mononuclear cells which are gradually replaced by smooth muscle cells. Later fibrosis leads to the eventual obliteration of these arteries. The main cause of transplant coronary artery vasculopathy is a chronic immune injury exacerbated by other factors such as reperfusion injury[4] CMV disease[5-7] and dyslipidemias.[8]

Much has been learned about the biology of transplant coronary artery vasculopathy or chronic rejection from study of patients and from various animal models of heart, kidney and arterial transplantation. T-cell activation by histocompatibility and other antigens on endothelial cells leads to activation of these cells and of macrophages which play a central role in the process. Upregulation of both T cell and macrophage type cytokines (IFNγ, IL-6, IL-1, TNFα, IL-1β)[9] and chemokines (RANTES and MCP-1)[10,11] and iNOS has been demonstrated.[12,13] The net result of macrophage and endothelial activation is the production of various growth factors such as PDGF[14] IGF, EGF FGF[15] and TGFβ[16] by macrophage and endothelial cells themselves and by smooth muscle cells. These growth factors lead to a transformation of donor smooth muscle cells in the media of the coronary arteries to a secretory phenotype and migration of these cells to form a neointima. Endothelial cell activation with upregulation of MHC class II and adhesion molecules such as ICAM and VCAM[17,18] has been demonstrated in human subjects. Labarrere et al have also demonstrated a change in endothelial cells from an anticoagulant to a procoagulant phenotype in endothelial cells.[19-22]

Transplant-Associated Coronary Artery Vasculopathy, edited by Marlene L. Rose.
©2001 Eurekah.com.

Studies of Chronic Rejection

There are three different types of studies which provide information on the effect of immunosuppressive drugs in chronic rejection.

1. In vitro studies of the effect of immunosuppressive drugs on smooth muscle cell proliferation.
2. Preclinical studies. There are several small animal models of chronic rejection-heart,[23] kidney[24] and aortic transplants[25] in rats, heart transplants in mice,[26] and heart[27] and aortic transplants[28] in rabbits. Large animal models of chronic rejection are less common. Madsen et al described a model of heterotopic heart transplants in inbred miniswine using IVUS[29] to monitor neointimal formation. We have developed a non-human primate model of chronic rejection also using IVUS and 3-D reconstruction to measure neointimal formation in an aortic transplant.[30]
3. Clinical Studies. These case-control studies are retrospective and use death, pathology of coronary arteries from explanted hearts, coronary angiography and more recently intravascular ultrasound as endpoints.

Mechanisms

Immunosuppressive drugs may prevent allograft coronary artery disease by different mechanisms (Fig. 10.1).

Early Inhibition of T Cell Activation

Suppression of early events in T cell activation by blocking of costimulation is known to inhibit graft vascular disease in rat models of heart and kidney transplantation.[31,32]

Inhibition of Smooth Muscle Proliferation

The formation of a neointima layer depends on migration and proliferation of smooth muscle cells. This has been quantified in the rat aortic transplant model.[33]

Inhibition of Antibodies Directed Against Antigens on Coronary Arteries

The importance of antibodies directed against MHC and other antigens on endothelial cells to the etiology of allograft coronary artery disease has been demonstrated in both preclinical and clinical studies. Deposits of IgM and complement are frequently found on arteries affected by graft vascular disease.[12,13] We have shown a strong correlation between IgM exposure and neointimal area in hamster to rats aortic transplants which we have used as a model of chronic xenogeneic vasculopathy[34] (Fig. 10.2). Russell et al produced coronary atherosclerosis in a transplanted heart in SCID mice with repeated injections of antidonor serum.[35]

Data from 240 recipients of heart transplants showed that patients who developed anti-HLA antibodies within the first three months and whose levels remained high at the end of the first year had a worse prognosis and were at high risk of developing transplant coronary artery vasculopathy.[36] In an anecdotal report, Zales et al found deposits of immunoglobulin and C3 in endomyocardial biopsies of children who went on to develop transplant coronary artery vasculopathy after heart transplantation.[37] Peptide specific antiendothelial antibodies were found by the Harefield group in 15 out of 21 patients with allograft coronary artery disease and in only out of 20 patients with no coronary disease. These antibodies were predominantly IgM and were directed against vimentin a fibrillary protein present in mesenchymal cells.[38,39] The same group has found an association between antiendothelial IgM detected by flow cytometry and changes in the antithrombin III component of the heparin-mediated anticoagulant pathway[40]—changes which are known to be risk factors for the development of transplant coronary artery vasculopathy. In contrast Hosenpud and his group could not find an association between antidonor antibodies and transplant coronary artery vasculopathy.[41]

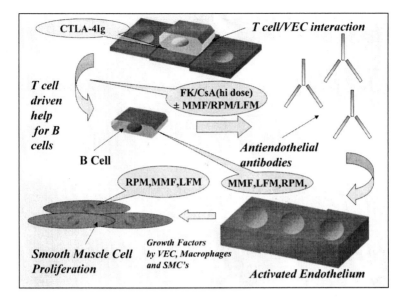

Fig. 10.1. Simplified schema of the immunological sequence of events leading to chronic transplant vasculopathy and sites of inhibition by different immunosuppressive agents known to inhibit the process. CsA–cyclosporine, FK–FK506, MMF–mycophenolate mofetil, RPM–rapamycin, LFM–leflunomide, VEC–vascular endothelial cells, SMCs–smooth muscle cells.

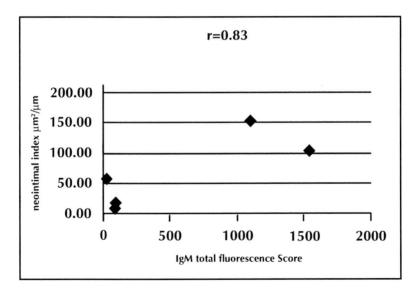

Fig. 10.2. Association between xenogeneic IgM exposure and neointimal area of hamster to rat aortic transplant harvested after 8 weeks.

Inhibition of Acute Myocardial Rejection

Although it is clear that chronic rejection requires a degree of histoincompatibility between donor and recipient, the nature of the relationship between acute and chronic rejection is unclear. T-cell activation by alloantigens results in acute rejection which in itself may lead to chronic rejection. Alternatively T-cell activation may lead independently to both acute and chronic rejection processes (Fig. 10.3.).

Evidence from preclinical studies in heart kidney and aortic transplant models, suggest that a certain amount of acute rejection is required for chronic rejection to develop.[42,23,43] Libby's group from Boston has demonstrated in transplanted rabbit hearts that smooth muscle cells are activated and start to proliferate during acute rejection.[44,45] In addition, simulating episodes of acute rejection by cyclosporine withdrawal accelerated the development of graft vascular disease in the same model.[27]

The evidence from clinical studies is less clear. Most studies looking at immunological risk factors for the development of chronic rejection in renal transplantation suggest that acute rejection is a risk factor, especially when severe and when it occurs late i.e., after 12 months.[46-48] However in Finland, where diversity of histocompatibility antigens is not as diverse as elsewhere and acute rejection is less common, chronic rejection has been seen in the absence of acute rejection.[49] In clinical studies of heart transplantation, some papers suggest acute rejection is a risk factor for the development of allograft coronary artery disease whilst others do not. Both the Stanford and Papworth heart transplant groups have not found acute rejection of any grade to be a risk factor for the development of transplant

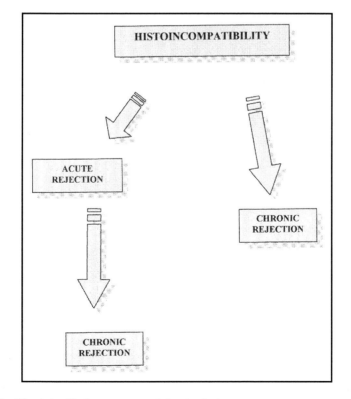

Fig. 10.3. Possible relationships between acute and chronic rejection.

coronary artery vasculopathy.[50,51] Uretsky in 1987[52] and Radonvancevic in 1990[53] found that patients who had experienced two or more rejection episodes had increased risk of developing transplant coronary artery vasculopathy. Narrod et al in 1989[54] and Schutz in 1990[55] found, as in renal transplantation that higher rejection scores and rejection after one year were risk factors for the development of transplant coronary artery vasculopathy. The pediatric heart transplant group at Loma Linda have found that acute rejection between 3 and 12 months after heart transplantation was a significant risk factor for the development of transplant coronary artery vasculopathy.[56] Kobashigawa et al have demonstrated that patients with higher cumulative rejection scores were more likely to develop a significant neointima on IVUS at one year.[57]

A group from the Netherlands has found that sustained elevation of TH1 cytokine production in endomyocardial biopsies taken after 1 year is associated with the development of transplant coronary artery vasculopathy.[40]

Early effective control of acute rejection and steroid withdrawal may lead indirectly to the prevention of transplant coronary artery vasculopathy. Opelz writing for the Collaborative Transplant Study has found that patients maintained after the first year on a steroid free immunosuppressive regimen after renal or heart transplantation live longer.[58] In renal transplantation however, this survival advantage occurred irrespective of whether the patient received any treatment for acute rejection in the first year.[59] It is not clear from these reviews, whether the excess deaths of patients maintained on steroids were related to chronic rejection or to cardiovascular causes.

Azathioprine

Azathioprine is the pro-drug for the purine antagonist mercaptopurine. Its immunosuppressive mode of action cannot however be explained by antagonism of purine biosynthesis alone. Although newer drugs are likely to take over its role, a few clinical studies have shown that its use is associated with

significantly less chronic rejection, less allograft coronary artery disease and better survival after both kidney and heart transplantation.[60-62]

Cyclosporine

Cyclosporine—microemulsion formulation (MEF) (Neoral) is the third generation derivative of cyclosporine. It is a newly formulated version of cyclosporine designed to increase the bioavailability of cyclosporine in the small bowel, which is where cyclosporine is absorbed.

The pharmacokinetics of cyclosporine MEF administered to human volunteers and stable renal heart and lung transplant patients, including patients with cystic fibrosis[63,64] differs from cyclosporine-SGC (soft gel capsule – Sandimmune) in the following ways:[65]

1. Shorter time to maximum blood level (T max)
2. A higher maximum blood level (C max)
3. A higher AUC (area under the time/ concentration curve)
4. Lower intrasubject variability for T max, C max, minimum blood level, AUC, and percentage peak-trough fluctuation.

Cyclosporine MEF shows a better correlation between trough concentrations and AUC. The majority of in vitro, preclinical, and clinical studies of chronic rejection and cyclosporine deal with the SGC formulation rather than with the MEF.

In Vitro Studies

Cyclosporine inhibits smooth muscle cell proliferation at near toxic concentrations.[66]

Cultured endothelial cells synthesize endothelin in the presence of cyclosporine.[67] Increased endothelin expression has been demonstrated in a rat cardiac allograft model of chronic vascular rejection[68] and endothelin-1 peptide expression has been demonstrated in patients with transplant coronary artery disease.[40] Endothelin is known to be a powerful mitogen and increases smooth muscle cell proliferation. However although cyclosporine increased endothelin concentrations fivefold in an endothelial cell conditioned medium, there was a decrease in smooth muscle cell proliferation when exposed to this medium.[69] The authors suggest that other unidentified endothelial cell derived factors inhibit smooth muscle cell proliferation and abolish the mitogenic effect of endothelin.

Preclinical Studies

The effect of cyclosporine on graft vascular disease in animal studies depends predominantly on the dose administered. Other factors which have an influence include the organ used, the species used, and the presence or absence of other immunosuppressive drugs. Higher doses of cyclosporine are more effective at inhibiting neointimal formation. Rats treated with 2.5, 5, and 10 mg/kg of cyclosporine had a 62%, 74% and a 97% reduction in intima after an aortic transplant.[70] Trough cyclosporine blood levels in the animals on 10 mg/kg exceeded 1000 ng/ml. In a separate study, 5 mg of cyclosporine MEF was more effective than the same dose of cyclosporine SGF at inhibiting neointimal formation in a rat model of aortic transplantation. WF rat recipients of heterotopic DA hearts were treated with triple immunosuppressive therapy consisting of methylprednisolone, azathioprine and three doses of cyclosporine—5, 10 and 20 mg/kg/day.[71] Low dose cyclosporine was associated with severe intimal thickening. The intermediate dose significantly inhibited formation of a neointima whilst 20 mg/kg of cyclosporine inhibited all vascular changes. In this study, there was an inverse correlation between mean cyclosporine blood levels and mean intimal thickness. Expression of VCAM, ICAM and MHC class II on epicardial and intramyocardial arteries was partially or completely inhibited in the animals on 10 and 20 mg/kg/day of cyclosporine. Lewis recipients of Fisher kidneys treated with 15 mg/kg on alternate days had longer survival, diminished proteinuria and lower plasma creatinine than untreated animals.[72] Late cyclosporine treatment (10 mg/kg) between 15 and 28 days was also effective at

preventing graft vascular disease in Lewis hearts transplanted into Fisher rats infected with CMV.[73]

Cyclosporine at a dose of 10 mg/kg/day suppressed transplant arteriosclerosis in the aorta allografted cholesterol clamped rabbit.[28] The same dose completely inhibited acute and chronic rejection and expression of ICAM and VCAM in transplanted rabbit hearts.[74]

Doses of cyclosporine (5 mg or less/kg) which produce levels in rats which we would accept as therapeutic and are effective for controlling acute rejection have either no effect or frequently have a deleterious effect on neointimal proliferation in models of chronic rejection.[75] Possible explanations for this phenomenon

1. In Lewis recipients of Fisher kidneys treated with 1.5mg/kg of cyclosporine, there is important up regulation of transforming growth factor β, endothelin, and HSP-70 as compared with control animals.[10]
2. Indirect T-cell allorecognition which is a cyclosporine resistant pathway for T-cell help for antibody production to donor MHC antigens may be important in endothelial cell antigen presentation and chronic rejection.[76]

Clinical Studies

Reports of the effect of cyclosporine on chronic rejection in both heart and kidney transplantation have changed over the past 10 years. Early reports[77,78] suggested that the half-life of the first cadaveric kidney transplants has not changed since the use of cyclosporine. In 1989 the Stanford group demonstrated that improved cyclosporine immunosuppression does not decrease the time related prevalence of transplant coronary artery disease.[79]

Anecdotal reports have implicated cyclosporine as a cause of accelerated endothelialitis.[6] However microscopic examination of heart sections from nonheart transplant patients on cyclosporine revealed no abnormalities.[80]

Recent reports suggest that cyclosporine therapy may have an impact on chronic rejection and allograft coronary artery disease. The most recent ISHLT registry suggests that half-life for heart transplants improved from 5.4 years in the 1980-1985 period to 8.7 years in the 1986-1990 period.[81] Vanrenterghem et al reported an increase of renal allograft half-life from 16.6-23.04 years when maintenance therapy was changed from azathioprine with high doses of steroids to cyclosporine and low dose steroids.[82] In a prospective randomized study, the frequency of clinically defined chronic rejection after four years was 25% for the patients on azathioprine and steroids and 9% for the patients on triple therapy.[60] Conversely in a separate study from the Netherlands patients were randomly assigned 3 months after renal transplantation to groups continuing to receive cyclosporine or changing to azathioprine. Eight years after transplantation survival was 75.3% in the cyclosporine group and 85.9% in the azathioprine group (p = 0.14).[83]

Cyclosporine dose and trough levels have been related to the development of chronic rejection in liver, kidney and heart transplantation.

Soin et al studied patients who had received a liver transplant over a four year period. Liver grafts in patients maintained on median cyclosporine levels (whole blood, trough level) of more than 175 mcg/L in the first 28 days posttransplant had a significantly lower incidence of chronic rejection (2 out of 49 vs.22 out of 97-p = 0.002).[84]

In a multivariate analysis of 587 kidney-alone transplants performed over a five year period, patients on less than 5 mg/kg of cyclosporine at one year were more likely to develop chronic rejection (p = 0.007).[85] Valantine et al[86] analyzed the results from 225 heart transplant patients who had survived one year after transplant and who fulfilled the defined inclusion criteria. The actuarial 5-year survival in patients whose average cyclosporine dose was less than 3mg/kg/day was 60% as compared with 77% in patients whose average dose was more than 3 mg/kg/day. The probability of freedom from CAD did not differ significantly in the two groups. However, prevalence of death

from CAD was significantly higher in patients whose dose was less than 3 mg/kg/day. In a case-control study of heart transplant patients, cyclosporine dosages were statistically lower in patients with chronic rejection in the first three postoperative years, whereas trough levels were significantly lower in the first six postoperative months in this same group of patients.[87]

Miller et al and the Sandoz/CVIS investigators[88] examined the effect of cyclosporine dose and trough blood levels on various indices of intimal thickening as measured by IVUS. There was a significant difference in intimal area between patients on more than 5 mg/kg/day and those on less than 3 mg at the end of the third year (p = 0.04). There was a greater increase in intimal index over one year in patients whose average trough cyclosporine level was less than 200 when compared with patients with levels greater than 400 (p = 0.016).

Area under the concentration-time kinetic curve (AUC) is a much better estimate of drug exposure than cyclosporine trough levels. Kahan and his group in Texas[89] have used a pharmacokinetic strategy for cyclosporine administration for more than 10 years. They stress that variability in bioavailability of cyclosporine is an important risk factor for the development of chronic rejection after renal transplantation. In a recent paper, Freimark et al[90] have shown that quilty lesions in cardiac allografts are related to reduced endocardial levels of cyclosporine. Variability of cyclosporine concentrations within a graft may also be a risk factor for the development of neointimal lesions within coronary arteries.

Tacrolimus

Tacrolimus (USAN for FK506) was discovered by Kino and Gotoh[91] of Fujisawa pharmaceuticals in 1984 during a program designed to discover and develop an alternative immunosuppressant to cyclosporine. It is a macrocyclic lactone derived from the actinomycete, *Streptomyces Tsukubaensis*.

Mechanisms of Action and Pharmacodynamics

Tacrolimus suppresses the immune system by similar mechanisms as cyclosporine.[92-95] It binds in the cytoplasm with FK binding proteins of which FKBP 12 is believed to be the isoform most responsible for participating in immunosuppressive reactions. FKBPs are the equivalent of cyclophilin, the cyclosporine binding protein. Tacrolimus-FKBP complexes associate with calcium dependant calcineurin-calmodulin complexes. Calcineurin is a serine threonine phosphatase that binds to NFAT. The combined NFAT (nuclear factor of activated T cells)/calcineurin complex migrates into the nucleus where it acts as a transcription factor for the activation of the promoter region of genes for various cytokines. The tacrolimus-FKBP12 complex inactivates the enzymatic activity of calcineurin thus ultimately inhibiting the transcription of cytokines namely—IL-2, IL-3, IL-4, IL-5, IFN-γ, TNF-α and GM-CSF. The limited amounts of calcineurin in immune cells as compared with nonimmune cells and the fact that calcineurin is critical to T cell activation accounts for the relative sensitivity of lymphocytes to tacrolimus which distributes and binds to FKBPs in all cells.

Tacrolimus is 50-100 times more potent than cyclosporine in vitro.[65] This is due to differences in partition coefficients and also due to increased binding affinity of tacrolimus to FKBP 12. The proliferative responses of T cells to alloantigens, plant mitogens and anti-CD2 and CD3 antibodies are inhibited by tacrolimus. Direct cytotoxic cell killing and stimulation of T cells via the calcium independent CD28 pathway are both resistant to tacrolimus.

Pharmacokinetics

The peak concentration of infused tacrolimus declines rapidly initially due to distribution.[96] It then slows down over the next 24 hours after reaching distribution equilibrium. Absorption after oral administration is poor with peak blood levels occurring 4 hours after intake. The oral bioavailability and T max of tacrolimus are highly variable and range from 5-67% and 0.5-8 hours. The low and variable bioavailability of tacrolimus is caused by its transformation in the gut wall by cytochrome P-450 (CyP 450) 3A4 enzymes and the counter transport of parent drug and metabolites by p-glycoprotein (Pgp). Unlike cyclosporine, tacrolimus is absorbed in a completely bile-independent manner. Trough blood levels (Cmin) correlate well with area under the concentration-time curve and with toxicity and in kidney transplantation, with efficacy.[96] Assessment of blood levels is critical to the use of tacrolimus because of its narrow therapeutic index.

Tacrolimus levels are assayed using an enzyme linked immunosorbant assay or a microparticle enzyme immuno assay of blood. These assays cannot distinguish between multiple metabolites of differing efficacy and toxicity. The whole blood/plasma ratio varies between 10:1 and 30:1 and like cyclosporine is temperature dependent.

Tacrolimus undergoes further biotransformation by CyP450 and other isoenzymes in the liver, producing metabolites that are excreted in bile.

In Vitro Studies

Like cyclosporine, tacrolimus induces expression of mRNA for endothelin in vascular endothelial cells.[97,98] Levels required to do this are however much higher than those that are clinically relevant.

Preclinical Studies

There are fewer studies of chronic rejection with tacrolimus when compared with cyclosporine and the results of these, all heart transplant studies in rats, are conflicting. In one study, rats receiving 0.25 mg/kg/day had far less graft vascular disease when compared to animals on 0.1 mg/kg/day.[99] In another study, heart allografts and isografts in recipient rats on 2 mg/kg/day developed severe graft vascular disease[100,101]

Clinical

Tacrolimus has been extensively studied in liver transplantation. In a randomized open trial involving eight European centers,[102] a tacrolimus-based immunosuppressive regimen was compared to a cyclosporine-based regimen. At the end of 2 years 1.5% of patients on tacrolimus developed chronic rejection as opposed to 5.3% on cyclosporine (p = 0.032). Gjertson et al[103] have analyzed the UNOS Kidney Transplant Registry and have shown that the half-life of kidney transplants in patients discharged on tacrolimus is 14.4 years whilst those discharged on cyclosporine or others is 8-9 years(p = 0.04).

The Pittsburgh group[104] have published the intermediate-term results of patients on tacrolimus after heart transplantation and compared them with their experience with cyclosporine. This was not a randomized study. Patients receiving tacrolimus had a lower risk of hypertension and of developing refractory rejection. At the most recent follow-up, 48% of patients on tacrolimus were free of steroids as compared to 17% of those on cyclosporine. At the end of 4 years, 82% of patients on tacrolimus and 73% of those on cyclosporine were free from allograft coronary artery disease (p = 0.07). In a prospective randomized open label study[40] comparing tacrolimus to cyclosporine in heart transplantation, more patients on cyclosporine developed new onset hypertension (79% vs.52%, p = 0.04) and more patients on cyclosporine required therapy for hypercholesterolemia (65% vs.28%, p = 0.02). There was however no difference in survival or rates of rejection.

Sera from patients who were enrolled in a prospective study[105] comparing tacrolimus with cyclosporine after heart transplantation were analyzed for the presence of anti-endothelial antibodies. Samples from patients taking tacrolimus contain significantly less antibodies as assessed by both Western blotting and by flow cytometry.

Mycophenolate Mofetil

Mycophenolate mofetil (MMF) (mycophenolic acid morpholinoethylester-RS-61443) is a pro-drug that when hydrolyzed by liver esterases produces the active metabolite mycophenolic acid (MPA).

Mycophenolic acid (MPA) was initially derived from cultures of the *Penicillium* species by Gosio in 1896[106] and purified in 1913. Its antibacterial and antifungal activities were recognized in the 1940s. It was not until the 1980s[107] that Nelson, Eugui and Allison of Syntex considered MPA for use as an immunosuppressant in the United States as part of their search for selective immunosuppressive agents in the treatment of autoimmune disease. A morpholinoethylester of MPA was selected from a number of derivatives on the basis of its structure, its ability to inhibit lymphocyte proliferation in vitro, its ability to inhibit antibody synthesis in mice and its greater bioavailability when compared to MPA. In 1987, Morris and colleagues at Stanford University[108] decided to evaluate MPA for use in transplantation.

Pharmacokinetics and Mechanism of Action

MPA is a noncompetitive reversible inhibitor of IMPDH (inosine monophosphate dehydrogenase). This NAD-dependant enzyme is the rate limiting enzyme in the de novo pathway for purine biosynthesis. The drug inhibits the type II isoform of the enzyme[95] more potently than the type I in lymphocytes.

Resting lymphocytes rely on the salvage pathway for purine biosynthesis and on both de novo and salvage pathways when activated. Therefore MPA suppresses T- and B- cell activation more potently than other resting cells or other cells for which de novo purine biosynthesis is not essential for proliferation.

A diminished supply of guanine nucleotides results in decreased DNA synthesis, allosteric feedback inhibition of purine and pyrimidine biosynthesis, inhibition of glycosylation of adhesion molecules,[109] decreased endothelial cell inducible nitric oxide synthase, increased apoptosis and decreased cyclin dependant kinase activity resulting in G0/G1 arrest.

Pharmacokinetics

The bioavailability of MPA is only 43% that of the ester, which is highly soluble at the lower pH of the upper GI tract and is absorbed more rapidly. The liver is the primary location for esterase mediated hydrolysis of mycophenolate mofetil into MPA. The liver is also the site of conversion of MPA to its primary metabolite, mycophenolate glucuronide. This inactive metabolite is excreted in the bile. Some of it is reconverted to MPA by gut glucronidases and undergoes enterohepatic recirculation. This recirculation causes secondary peaks to appear in the plasma 6-12 hours post dose. The high concentration of drug in the gut may account for the gastrointestinal side effects.

MPA does not extensively distribute into cellular fractions of the blood and has a blood to plasma ratio of 0.6. The mean apparent half-life of MPA is 18 hours. Clinical trials in renal transplant patients have shown immunosuppressive effects at doses starting at 2-3 g per day.

Preclinical Studies

Mycophenolate mofetil has been shown to be effective in the prevention of graft vascular disease in both rat models of aortic[110-113] and renal[114,115] transplantation. Both the adventitial inflammatory component and neointimal proliferation were inhibited by mycophenolate in the aortic transplant animals. In the renal transplant study,[114] allospecific IgM and IgG responses were absent in the treated group. Mycophenolate, on its own and when given with rapamycin was able to inhibit neointimal formation after balloon injury.[116] In a study of heterotopic primate cardiac xenografts,[117] mycophenolate mofetil was more effective than azathioprine when combined with cyclosporine and steroids in preventing graft vascular disease.

Clinical Studies

In a pooled efficacy analysis of three large randomized, double blind, clinical studies of renal transplantation[118] two different doses of mycophenolate, 2 and 3g were compared with azathioprine in a cyclosporine-based regimen. At one year, the results were as follows:

1. Graft survival was 90.4%, 89.2% and 87.6% in the MMF 2g, MMF 3g and azathioprine groups respectively. This difference did not reach statistical significance.
2. Rejection episodes were 19.8%, 16.5%, and 40.8% in the same three groups with relative risk of 0.46 for the MMF 2g group when compared to the azathioprine group and 0.38 for the MMF 3g group.
3. Renal function was consistently better for both MMF groups at 3, 6 and 12 months.

The use of cytolytic agents was analyzed as an indicator of severity of rejection; 19.7%, 8.8%, and 4.9% of patients in the azathioprine, mycophenolate 2 g, and mycophenolate 3 g groups received these agents respectively.

In a study of heart transplant patients (Costanzo MR: Proceedings ASTP 1997, 87), mycophenolate was compared to azathioprine in a cyclosporine-based protocol. At one year following surgery, patients receiving mycophenolate had a 0.33 mm^2 increase in coronary artery luminal area as measured by IVUS, while patients receiving azathioprine had a 0.81 mm^2 decrease in luminal area.

Sirolimus (Rapamycin)

Sirolimus (USAN for rapamycin) is another microbial natural product and is produced by the actinomycaete *Streptomyces hygroscopicus* isolated from Easter Island (Rapanui to its natives) soil samples. It emerged from an antifungal drug discovery program in the mid-1970s directed by Sehgal[119] at Ayerst Research in Montreal, Canada. The antifungal properties of the drug were not pursued when it became apparent that the drug caused involution of lymphoid tissue. Martel subsequently demonstrated that rapamycin suppresses experimental allergic encephalomyelitis and passive cutaneous anaphylaxis in the rat.[119] Structural analysis of rapamycin revealed a macrocyclic lactone. It was not until the newly discovered structure of tacrolimus was found to be remarkably similar to that of sirolimus that groups at Stanford (Morris) and Cambridge (Calne)[102,120] uncovered the potential of the drug as an immunosuppressant.

Medicinal Chemistry

Sirolimus shares with tacrolimus its origin from a Streptomyces species and its structure. Both contain the same tricarbonyl region consisting of an amide, a ketone and a hemiketal. The sirolimus ring contains in addition a triene segment. This difference in structure accounts for the lower stability of sirolimus in aqueous solution. Like tacrolimus, sirolimus is hydrophobic.

Pharmacodynamics and Mechanism of Action

Sirolimus enters cells easily because of its lipohilicity. In the cell, the section of the ring that is identical to tacrolimus binds to cytosolic FKBP. Although this complex is necessary for the biological action of sirolimus, the target of the complex is not yet known. In yeast cells, proteins called targets of rapamycin (TOR) (sirolimus effector protein or SER in mammalian cells)[121] have been identified, which may be the target of the FKBP-sirolimus complex. The effects of interaction between this complex and its target include inhibition of protein synthesis by inhibition of a kinase (p70 S6 kinase). This kinase normally acts on S6 ribosomal protein. Sirolimus is also known to inhibit kinase activity of CDk2/cyclinE complex in yeast cells. This would prevent the cell from progressing from G1 to S phase. This is unlike cyclosporine or tacrolimus, which inhibit cell cycle progression at the G0 to G1 stage.

As it interferes with events at a later stage than cyclosporine or tacrolimus, sirolimus is less efficient at inhibiting cytokine synthesis.

It does however inhibit pathways that are resistant to cyclosporine or tacrolimus—namely calcium-independent activation induced by exogenous cytokines or stimulation of the CD28 pathway.

Immunoglobulin synthesis and antibody-dependant cellular cytotoxicity are also inhibited by sirolimus albeit at a much higher concentration than that required to inhibit T-cell activation.

Pharmacokinetics

Like cyclosporine and tacrolimus, sirolimus is transformed in the gut wall and liver by cytochrome CyP 450 and counter transported in the gut lumen by the multidrug resistance transporter, Pgp. This accounts for its low bioavailability and high pharmacokinetic variability.

Levels of sirolimus in the blood are best measured by HPLC. Whole blood concentrations of sirolimus metabolites exceed those of the parent drug. These metabolites have shown immunosuppressive activity in vitro.

The drug is extensively bound to cells in the blood, much more so than cyclosporine or tacrolimus; 97% of labeled drug is contained within red and white blood cells. There have been few PK and toxicity studies in humans. Peak concentrations are reached within 2 hours of oral dosing in healthy volunteers and recipients of renal transplants.[122] Cmax and AUC correlate well with dose, apart from the lower dose of 3 mg/m^2/day where levels in renal transplant patients were higher than expected. Elimination half-life varied between 43.8 and 86.5 hours in renal transplant patients.

In Vitro Studies

Rapamycin has been shown to effectively inhibit both rat and human vascular smooth muscle proliferation.[66,123,126] It achieves this effect by reducing p33 (cdk2) kinase activity.[124] Rapamycin has also been demonstrated to inhibit porcine smooth muscle cell migration.[125]

Preclinical Studies

Continued treatment with rapamycin at clinically relevant doses prevents transplant vasculopathy in rat models of heart[126] and arterial transplantation.[124-129] In these models, it has been shown to effectively inhibit expression of Th1 cytokines and of various growth factors such as PDGF, FGF and TGF, whilst preserving expression of Th2 cytokines.[130]

Clinical Studies

Recently published phase II studies in renal transplant patients suggest the drug is able to decrease acute rejection rates from 40 to less than 10% amongst patients taking full dose cyclosporine.[131] This improvement is achieved with a nonsignificant increase in infectious complications. The authors of the study suggest that the drug may mitigate the need for long-term steroid therapy. The efficacy of rapamycin may be diminished by the adverse effect of the drug on cholesterol levels.[132]

Leflunomide/ Malononitrilamides

Leflunomide is a new immunomodulatory drug that is effective in experimental models of autoimmune diseases and in allo or xenotransplantation.[133] In a phase II clinical trial, leflunomide showed high tolerability and efficacy in patients with advanced rheumatoid arthritis. The immunomodulatory activity of leflunomide is attributed to its primary metabolite A77 1726, which is a malononitrilamide. The in vitro and in vivo mechanisms of action of this class of compounds are not defined completely. Several malononitrilamide analogues and A77 1726 inhibit T- and B-cell proliferation,[134,135] suppress immunoglobulin production and interfere with cell adhesion. Although no central molecular mechanism of action has been proposed to explain all the effects of the malononitrilamides, the inhibition of de novo pyrimidine biosynthesis[133,40] and of cytokine- and growth factor receptor-associated tyrosine kinase activity are leading hypotheses for the effects of A77 1726 on T- and B-cell proliferation and function.

In Vitro Studies

Leflunomide has been shown to inhibit proliferation of human rat and murine smooth muscle cell proliferation.[66,136] This proliferation can be reversed by the addition of uridine to the medium[137] suggesting that inhibition of de novo pyrimidine biosynthesis is the mechanism by which leflunomide has this effect.

Preclinical Studies

Leflunomide combined with cyclosporine has completely prevented organ rejection in two very difficult animal models across major histocompatibility barriers: kidney transplantation between mongrel dogs[138] and functional whole limb transplantation[139] between Brown Norway rat donors and Lewis recipients.

The efficacy of leflunomide in prevention of allograft and xenograft vasculopathy has been demonstrated in rat heart[140,141]and arterial transplant[129,142] models. In combination with cyclosporine, leflunomide completely prevented production of xenogeneic antibodies and neointimal changes in a xenogeneic aortic transplant model.[34] Leflunomide is effective at not only preventing chronic vasculopathy but was able to reverse[140] the arterial changes when therapy was started late in both allogeneic and xenogeneic[141] models of heterotopic heart transplantation.

Clinical Studies

Malononitrilamide analogues of A77 1726 are being evaluated for immunosuppressive efficacy in preclinical models of transplantation. If these analogues show efficacies and therapeutic indexes that are similar to leflunomide in these models and in phase I trials, the preclinical and phase I data will be used to select the analogues for phase II trials in organ transplant recipients.

Future

Therapies that are at the very early stage of preclinical development and that may have an impact on the development of graft vascular disease include

1. inhibition of the CD28-B7 T cell costimulatory activation pathway by the fusion protein CTLA-4Ig.[31,32] This has been shown to effectively inhibit acute and chronic rejection in rat cardiac and renal models.
2. Gene therapy using antisense oligonucleotides to cd-kinases involved in the cell cycle or to i-NOS, an enzyme expressed by activated macrophages, which are centrally involved in the process of chronic rejection.

Conclusion

The effect of cyclosporine on allograft coronary artery disease has not been as dramatic as its effect on acute rejection and early survival. A number of new immunosuppressive agents are currently being evaluated in heart transplantation. Cyclosporine-MEF by its pharmacokinetic profile, tacrolimus by its steroid-sparing effects, its inhibitory effects on T dependent antibodies directed against endothelial antigens and its relatively beneficial effects of blood pressure and lipid metabolism, mycophenolate and rapamycin by their effect on B cell function, their effect on smooth muscle cell proliferation, their effect on rates of acute rejection and their promising preclinical results, all have the potential of influencing the process of chronic rejection and allograft coronary artery disease. There is also great promise for leflunomide and its analogues by virtue of their anti-T and B cell effects, their effects on smooth muscle proliferation and their efficacy in preclinical models, when these drugs reach the clinic.

Avoidance of late acute rejection, which according to the latest registry figures,[81] is still an important cause of mortality is crucial if one is to have an impact on the incidence of transplant coronary vasculopathy.

Guidelines are required for the planning of studies of chronic rejection in heart transplantation. Should future studies be looking at primary or secondary prevention[143] and should we substitute IVUS or endomyocardial biopsy markers[144] as endpoints for the more traditional ones of pathology and coronary angiography?

References

1. Hosenpud JD, Shipley GD, Wagner CR. Cardiac allograft vasculopathy: Current concepts, recent developments, and future directions. J Heart Lung Transplant 1992; 11(1 Pt 1):9-23.
2. Olivari MT, Kubo SH, Braunlin EA et al. Five-year experience with triple-drug immunosuppressive therapy in cardiac transplantation. Circulation 1990; 82(5 Suppl): IV276-80.
3. Yeung AC, Davis SF, Hauptman PJ et al. Incidence and progression of transplant coronary artery disease over 1 year: Results of a multicenter trial with use of intravascular ultrasound. Multicenter Intravascular Ultrasound Transplant Study Group. Journal of Heart & Lung Transplantation 1995; 14(6 Pt 2):S215-20.
4. The International Society for Heart and Lung Transplantation 16th annual meeting and scientific sessions. New York, New York, March 15-18, 1996. Abstracts. J Heart Lung Transplant 1996; 15(1 Pt 2):S1-114.
5. Normann SJ, Salomon DR, Leelachaikul P et al. Acute vascular rejection of the coronary arteries in human heart transplantation: Pathology and correlations with immunosuppression and cytomegalovirus infection. Journal of Heart & Lung Transplantation 1991; 10(5 Pt 1):674-87.
6. Paavonen T, Mennander A, Lautenschlager I et al. Endothelialitis and accelerated arteriosclerosis in human heart transplant coronaries. J Heart Lung Transplant 1993; 12(1 Pt 1):117-22.
7. Gao SZ, Hunt SA, Schroeder JS et al. Early development of accelerated graft coronary artery disease: Risk factors and course. Journal of the American College of Cardiology 1996; 28(3):673-9.
8. Kemna MS, Valantine HA, Hunt SA et al. Metabolic risk factors for atherosclerosis in heart transplant recipients. Am Heart J 1994; 128(1):68-72.
9. Russell ME, Wallace AF, Hancock WW et al. Upregulation of cytokines associated with macrophage activation in the Lewis-to-F344 rat transplantation model of chronic cardiac rejection. Transplantation 1995; 59(4):572-8.
10. Nadeau KC, Azuma H, Tilney NL. Sequential cytokine dynamics in chronic rejection of rat renal allografts: Roles for cytokines RANTES and MCP-1. Proceedings of the National Academy of Sciences of the United States of America 1995; 92(19):8729-33.

11. Karnovsky MJ, Russell ME, Hancock W et al. Chronic rejection in experimental cardiac transplantation in a rat model. Clin Transplant 1994; 8(3 Pt 2):308-12.

12. Hancock WH, Whitley WD, Tullius SG et al. Cytokines, adhesion molecules, and the pathogenesis of chronic rejection of rat renal allografts. Transplantation 1993; 56(3): 643-50.

13. Hancock WW, Shi C, Picard MH et al. LEW-to-F344 carotid artery allografts: Analysis of a rat model of posttransplant vascular injury involving cell-mediated and humoral responses. Transplantation 1995; 60(12):1565-72.

14. Higgy N, Davidoff A, Benediktsson H et al. Platelet-derived growth factor receptor expression in chronic rejection of cardiac and renal grafts in the rat. Transplantation Proceedings 1991; 23(1 Pt 1):609-10.

15. Hayry P, Myllarniemi M, Aavik E et al. Chronic rejection: Potential new sites of therapy. Transplantation Proceedings 1996; 28(6):3225-6.

16. Waltenberger J, Miyazono K, Funa K et al. Transforming growth factor-beta and organ transplantation. Transplantation Proceedings 1993; 25(2):2038-40.

17. Salomon RN, Hughes CC, Schoen FJ et al. Human coronary transplantation-associated arteriosclerosis. Evidence for a chronic immune reaction to activated graft endothelial cells. American Journal of Pathology 1991; 138(4):791-8.

18. Labarrere CA, Pitts D, Nelson DR et al. Coronary artery disease in cardiac allografts: Association with arteriolar endothelial HLA-DR and ICAM-1 antigens. Transplantation Proceedings 1995; 27(3):1939-40.

19. Labarrere CA, Pitts D, Nelson DR et al. Vascular tissue plasminogen activator and the development of coronary artery disease in heart-transplant recipients. New England Journal of Medicine 1995; 333(17):1111-6.

20. Labarrere CA, Pitts D, Nelson DR et al. Coronary artery disease in cardiac allografts: Association with depleted arteriolar tissue plasminogen activator. Transplantation Proceedings 1995; 27(3):1941-3.

21. Faulk WP, Labarrere CA, Nelson DR et al. Coronary artery disease in cardiac allografts: Association with arterial antithrombin. Transplantation Proceedings 1995; 27(3): 1944-6.

22. Labarrere CA, Pitts D, Halbrook H et al. Tissue plasminogen activator, plasminogen activator inhibitor-1, and fibrin as indexes of clinical course in cardiac allograft recipients. An immunocytochemical study. Circulation 1994; 89(4):1599-608.

23. Adams DH, Tilney NL, Collins JJ, Jr. et al. Experimental graft arteriosclerosis. I. The Lewis-to-F-344 allograft model. Transplantation 1992; 53(5):1115-9.

24. Yilmaz S, Taskinen E, Paavonen T et al. Chronic rejection of rat renal allograft. I. Histological differentiation between chronic rejection and cyclosporin nephrotoxicity. Transplant International 1992; 5(2):85-95.

25. Halttunen J, Partanen T, Leszczynski D et al. Rat aortic allografts: A model for chronic vascular rejection. Transplantation Proceedings 1990; 22(1):125.

26. Russell PS, Chase CM, Winn HJ et al. Coronary atherosclerosis in transplanted mouse hearts. I. Time course and immunogenetic and immunopathological considerations. American Journal of Pathology 1994; 144(2):260-74.

27. Nakagawa T, Sukhova GK, Rabkin E et al. Acute rejection accelerates graft coronary disease in transplanted rabbit hearts. Circulation 1995; 92(4):987-93.

28. Andersen HO, Madsen G, Nordestgaard BG et al. Cyclosporin suppresses transplant arteriosclerosis in the aorta-allografted, cholesterol-clamped rabbit. Suppression preceded by decrease in arterial lipoprotein permeability. Arterioscler Thromb 1994; 14(6):944-50.

29. Madsen JC, Sachs DH, Fallon JT et al. Cardiac allograft vasculopathy in partially inbred miniature swine. I. Time course, pathology, and dependence on immune mechanisms. Journal of Thoracic & Cardiovascular Surgery 1996; 111(6):1230-9.

30. Progressive Graft Vascular Disease by Serial Intravascular Ultrasound (IVUS) in a novel model of combined aortic auto and allograft transplantation in nonhuman primates. International Society for Heart and Lung Transplantation Eighteenth Annual Meeting and Scientific Sessions.; 1998; Chicago.

31. Chandraker A, Russell ME, Glysing-Jensen T et al. T-cell costimulatory blockade in experimental chronic cardiac allograft rejection: Effects of cyclosporine and donor antigen. Transplantation 1997; 63(8):1053-8.

32. Azuma H, Chandraker A, Nadeau K et al. Blockade of T-cell costimulation prevents development of experimental chronic renal allograft rejection [see comments]. Proceedings of the National Academy of Sciences of the United States of America 1996; 93(22): 12439-44.

33. Lemstrom KB, Bruning JH, Bruggeman CA et al. Triple drug immunosuppression significantly reduces immune activation and allograft arteriosclerosis in cytomegalovirus-infected rat aortic allografts and induces early latency of viral infection. Am J Pathol 1994; 144(6):1334-47.

34. Study of Intimal Thickening in Vascular Xenografts—Central Role of Xenoantibody and its suppression by Cyclosporine and Leflunomide used Alone and in Combination. International Conference on Xenotransplantation 97; 1997; Nantes, France.

35. Russell PS, Chase CM, Winn HJ et al. Coronary atherosclerosis in transplanted mouse hearts. II. Importance of humoral immunity. J Immunol 1994; 152(10): 5135-41.

36. Rose EA, Pepino P, Barr ML et al. Relation of HLA antibodies and graft atherosclerosis in human cardiac allograft recipients. J Heart Lung Transplant 1992; 11(3 Pt 2):S120-3.

37. Zales VR, Crawford S, Backer CL et al. Spectrum of humoral rejection after pediatric heart transplantation. Journal of Heart & Lung Transplantation 1993; 12(4):563-71; discussion 572.

38. Dunn MJ, Crisp SJ, Rose ML et al. Anti-endothelial antibodies and coronary artery disease after cardiac transplantation. Lancet 1992; 339(8809):1566-70.

39. Wheeler CH, Collins A, Dunn MJ et al. Characterization of endothelial antigens associated with transplant-associated coronary artery disease. Journal of Heart & Lung Transplantation 1995; 14(6 Pt 2):S188-97.

40. Faulk WP, Rose ML, Meroni PL et al. Antibodies to endothelial cells can indentify myocardial damage and predict development of coronary artery, disease in patients with transplanted hearts. Human Immunol. 1999; 60:826-32.

41. Hosenpud JD, Everett JP, Morris TE et al. Cardiac allograft vasculopathy. Association with cell-mediated but not humoral alloimmunity to donor-specific vascular endothelium. Circulation 1995; 92(2):205-11.

42. Lemstrom KB, Raisanen-Sokolowski AK, Hayry PJ et al. Triple drug immunosuppression significantly reduces aortic allograft arteriosclerosis in the rat. Arteriosclerosis, Thrombosis & Vascular Biology 1996; 16(4):553-64.

43. Yilmaz S, Hayry P. The impact of acute episodes of rejection on the generation of chronic rejection in rat renal allografts. Transplantation 1993; 56(5):1153-6.

44. Program Issue, The International Society for Heart and Lung Transplantation. Thirteenth Annual Meeting and Scientific Sessions. The Journal of Heart and Lung Transplantation 1993; 12(1):S96.

45. Tanaka H, Sukhova GK, Swanson SJ et al. Endothelial and smooth muscle cells express leukocyte adhesion molecules heterogeneously during acute rejection of rabbit cardiac allografts. Am J Pathol 1994; 144(5): 938-51.

46. Massy ZA, Guijarro C, Wiederkehr MR et al. Chronic renal allograft rejection: Immunologic and nonimmunologic risk factors. Kidney International 1996; 49(2):518-24.

47. Hong JH, Sumrani N, Delaney V et al. Causes of late renal allograft failure in the cyclosporin era. Nephron 1992; 62(3):272-9.

48. Basadonna GP, Matas AJ, Gillingham KJ et al. Early versus late acute renal allograft rejection: Impact on chronic rejection. Transplantation 1993; 55(5):993-5.

49. Hayry P, Isoniemi H, Yilmaz S et al. Chronic allograft rejection. Immunological Reviews 1993; 134:33-81.

50. Gao SZ, Schroeder JS, Hunt SA et al. Influence of graft rejection on incidence of accelerated graft coronary artery disease: A new approach to analysis. Journal of Heart & Lung Transplantation 1993; 12(6 Pt 1):1029-35.

51. Stovin PG, Sharples LD, Schofield PM et al. Lack of association between endomyocardial evidence of rejection in the first six months and the later development of transplant-related coronary artery disease. Journal of Heart & Lung Transplantation 1993; 12(1 Pt 1):110-6.

52. Uretsky BF, Murali S, Reddy PS et al. Development of coronary artery disease in cardiac transplant patients receiving immunosuppressive therapy with cyclosporine and prednisone. Circulation 1987; 76(4):827-34.

53. Radovancevic B, Poindexter S, Birovljev S et al. Risk factors for development of accelerated coronary artery disease in cardiac transplant recipients. European Journal of Cardio-Thoracic Surgery 1990; 4(6):309-12; discussion 313.

54. Narrod J, Kormos R, Armitage J et al. Acute rejection and coronary artery disease in long-term survivors of heart transplantation. Journal of Heart Transplantation 1989; 8(5):418-20; discussion 420-1.

55. Schutz A, Kemkes BM, Kugler C et al. The influence of rejection episodes on the development of coronary artery disease after heart transplantation. European Journal of Cardio-Thoracic Surgery 1990; 4(6):300-7; discussion 308.

56. Bailey LL, Zuppan CW, Chinnock RE et al. Graft vasculopathy among recipients of heart transplantation during the first 12 years of life. The Pediatric Heart Transplant Group. Transplantation Proceedings 1995; 27(3): 1921-5.

57. Kobashigawa JA, Miller L, Yeung A et al. Does acute rejection correlate with the development of transplant coronary artery disease? A multicenter study using intravascular ultrasound. Sandoz/CVIS Investigators.

Journal of Heart & Lung Transplantation 1995; 14(6 Pt 2):S221-6.

58. Opelz G. Multicenter evaluation of immunosuppressive regimens in heart transplantation. The Collaborative Transplant Study. Transplantation Proceedings 1997; 29(1-2): 617-9.

59. Opelz G. Effect of the maintenance immunosuppressive drug regimen on kidney transplant outcome. Transplantation 1994; 58(4): 443-6.

60. Isoniemi H, Ahonen J, Krogerus L et al. Chronic rejection of renal allografts with four immunosuppressive regimens. Transplantation Proceedings 1992; 24(6):2716-7.

61. Beckingham IJ, Dennis MJ, Innes A et al. Prolonged renal allograft survival in chronic rejection by conversion to triple therapy. Transplantation Proceedings 1993; 25(2): 2100.

62. Addonizio LJ, Hsu DT, Douglas JF et al. Decreasing incidence of coronary disease in pediatric cardiac transplant recipients using increased immunosuppression. Circulation 1993; 88(5 Pt 2):II224-9.

63. Girault D, Haloun A, Viard L et al. Sandimmune neoral improves the bioavailability of cyclosporin A and decreases inter-individual variations in patients affected with cystic fibrosis. Transplant Proc 1995; 27(4):2488-90.

64. Mikhail G, Eadon H, Leaver N et al. An investigation of the pharmacokinetics, toxicity, and clinical efficacy of neoral cyclosporin in cystic fibrosis patients. Transplant Proc 1997; 29:599-601.

65. Morris RE. Mechanisms of action of new immunosuppressive drugs. Ther Drug Monit 1995; 17(6):564-9.

66. Mohacsi PJ, Tuller D, Hulliger B et al. Different inhibitory effects of immunosuppressive drugs on human and rat aortic smooth muscle and endothelial cell proliferation stimulated by platelet-derived growth factor or endothelial cell growth factor [see comments]. Journal of Heart & Lung Transplantation 1997; 16(5):484-92.

67. Bunchman TE, Brookshire CA. Cyclosporine-induced synthesis of endothelin by cultured human endothelial cells. Journal of Clinical Investigation 1991; 88(1):310-4.

68. Forbes RD, Cernacek P, Zheng S et al. Increased endothelin expression in a rat cardiac allograft model of chronic vascular rejection. Transplantation 1996; 61(5):791-7.

69. Leszczynski D, Zhao Y, Yeagley TJ et al. Direct and endothelial cell-mediated effect of cyclosporin A on the proliferation of rat smooth muscle cells in vitro. American Journal of Pathology 1993; 142(1):149-55.

70. Stoltenberg RL, Geraghty J, Steele DM et al. Inhibition of intimal hyperplasia in rat aortic allografts with cyclosporine. Transplantation 1995; 60(9):993-8.

71. Koskinen PK, Lemstrom KB, Hayry PJ. How cyclosporine modifies histological and molecular events in the vascular wall during chronic rejection of rat cardiac allografts. American Journal of Pathology 1995; 146(4): 972-80.

72. Paul LC, Chea R, Davidoff A et al. Efficacy of long-term cyclosporine treatment to prevent chronic renal allograft rejection in the rat. Transplantation Proceedings 1994; 26(5):2567-8.

73. Handa N, Hatanaka M, Baumgartner WA et al. Late cyclosporine treatment ameliorates established coronary graft disease in rat allografts. Transplantation 1993; 56(3):535-40.

74. Molossi S, Clausell N, Sett S et al. ICAM-1 and VCAM-1 expression in accelerated cardiac allograft arteriopathy and myocardial rejection are influenced differently by cyclosporine A and tumour necrosis factor-alpha blockade. Journal of Pathology 1995; 176(2): 175-82.

75. Mennander A, Tiisala S, Paavonen T et al. Chronic rejection of rat aortic allograft. II. Administration of cyclosporin induces accelerated allograft arteriosclerosis. Transplant International 1991; 4(3):173-9.

76. Sawyer GJ, Dalchau R, Fabre JW. Indirect T cell allorecognition: A cyclosporin A resistant pathway for T cell help for antibody production to donor MHC antigens. Transplant Immunology 1993; 1(1):77-81.

77. Cook D. Clinical Transplants. Los Angeles: UCLA Tissue Typing Laboratory, 1987:277.

78. Gjertson G. Clinical Transplants. Los Angeles: UCLA Tissue Typing Laboratory, 1991: 225.

79. Gao SZ, Schroeder JS, Alderman EL et al. Prevalence of accelerated coronary artery disease in heart transplant survivors. Comparison of cyclosporine and azathioprine regimens. Circulation 1989; 80(5 Pt 2):III100-5.

80. Barone JH, Fishbein MC, Czer LS et al. Absence of endocardial lymphoid infiltrates (Quilty lesions) in nonheart transplant recipients treated with cyclosporine. Journal of Heart & Lung Transplantation 1997; 16(6): 600-3.

81. Hosenpud JD, Bennett LE, Keck BM et al. The Registry of the International Society for Heart and Lung Transplantation: Fourteenth official report—1997. Journal of Heart & Lung Transplantation 1997; 16(7):691-712.

82. Vanrenterghem Y, Peeters J. Impact of cyclosporine on chronic rejection and graft vasculopathy. Transplantation Proceedings 1994; 26(5):2560-3.

83. Hollander AA, van Saase JL, Kootte AM et al. Beneficial effects of conversion from cyclosporin to azathioprine after kidney transplantation [see comments]. Lancet 1995; 345(8950):610-4.

84. Soin AS, Rasmussen A, Jamieson NV et al. CsA levels in the early posttransplant period—predictive of chronic rejection in liver transplantation? Transplantation 1995; 59(8):1119-23.

85. Almond PS, Matas A, Gillingham K et al. Risk factors for chronic rejection in renal allograft recipients. Transplantation 1993; 55(4):752-6; discussion 756-7.

86. Valantine H, Hunt S, Gamberg P et al. Impact of cyclosporine dose on long-term outcome after heart transplantation. Transplantation Proceedings 1994; 26(5):2710-2.

87. Mamprin F, Gamba A, Fiocchi R et al. Low dosage of cyclosporine and increased occurrence of chronic rejection in heart-transplanted patients. Transplantation Proceedings 1994; 26(5):2581-2.

88. Miller L, Kobashigawa J, Valantine H et al. The impact of cyclosporine dose and level on the development and progression of allograft coronary disease. Sandoz/CVIS Investigators. Journal of Heart & Lung Transplantation 1995; 14(6 Pt 2):S227-34.

89. Kahan BD, Welsh M, Rutzky LP. Challenges in cyclosporine therapy: The role of therapeutic monitoring by area under the curve monitoring. Therapeutic Drug Monitoring 1995; 17(6):621-4.

90. Freimark D, Czer LS, Aleksic I et al. Pathogenesis of Quilty lesion in cardiac allografts: Relationship to reduced endocardial cyclosporine A. Journal of Heart & Lung Transplantation 1995; 14(6 Pt 1):1197-203.

91. Gotoh T, Nakahara K, Nishiura T et al. Studies on a new immunoactive peptide, FK-156. II. Fermentation, extraction and chemical and biological characterization. Journal of Antibiotics 1982; 35(10):1286-92.

92. Sigal N, Dumont F. Cyclosporin A, FK 506 and Rapamycin: Pharmacologic Probes of lymphocyte signal transduction. Annu Rev Immunol 1992; 10:519-560.

93. Schreiber S, Crabtree G. The Mechanism of Action of Cyclosporin A and FK 506. Immunol Today 1992; 13:136-142.

94. Wiederrecht G, Lam E, Hung S et al. The mechanism of action of FK506 and cyclosporine A. Ann N Y Acad Sci 1993; 696: 9-19.

95. Brazelton TR, Morris RE. Molecular mechanisms of action of new xenobiotic immunosuppressive drugs: Tacrolimus (FK506), sirolimus (rapamycin), mycophenolate mofetil and leflunomide. Curr Opin Immunol 1996; 8(5):710-20.

96. Hausen B, E.Morris R. Review of Immunosuppression for Lung Transplantation—Novel Drugs, New Uses for Conventional Immunosuppressants, and Alternative Strategies. Clinics in Chest Medicine 1997; 18(2):353-366.

97. Takeda Y, Miyamori I, Wu P et al. Effect of FK 506 on the expression of endothelin receptor mRNA in the vasculature. Journal of Cardiovascular Pharmacology 1995; 26(Suppl 3):S290-2.

98. Takeda Y, Yoneda T, Ito Y et al. Stimulation of endothelin mRNA and secretion in human endothelial cells by FK 506. Journal of Cardiovascular Pharmacology 1993; 22(Suppl 8):S310-2.

99. Hisatomi K, Isomura T, Ohashi M et al. Effect of dose of cyclosporine or FK506 and antithrombotic agents on cardiac allograft vascular disease in heterotopically transplanted hearts in rats. Journal of Heart & Lung Transplantation 1995; 14(1 Pt 1): 113-8.

100. Meiser B, Reichart B. Graft vessel disease: The impact of immunosuppression and possible treatment strategies. Immunological Reviews 1993; 134:99-116.

101. Morris R, Meiser B. Identification of a new pharmacologic Action for an old compound. Med Sci Res 1989; 17:609-610.

102. Williams R, Neuhaus P, Bismuth H et al. Two-year data from the European multicentre tacrolimus (FK506) liver study. Transplant International 1996; 9(Suppl 1): S144-50.

103. Gjertson DW, Cecka JM, Terasaki PI. The relative effects of FK506 and cyclosporine on short- and long-term kidney graft survival [see comments]. Transplantation 1995; 60(12):1384-8.

104. Pham SM, Kormos RL, Hattler BG et al. A prospective trial of tacrolimus (FK 506) in clinical heart transplantation: Intermediate-term results. Journal of Thoracic & Cardiovascular Surgery 1996; 111(4):764-72.

105. Jurcevic S, Dunn MJ, Crisp S et al. A new enzyme-linked immunosorbent assay to measure anti-endothelial antibodies after cardiac transplantation demostrates greater inhibition of antibody formation by tacrolimus compared to cyclosporine. Transplantation 1998; 65:1197-1202.

106. Gosio B. Ricerche Batteriologiche e chimiche sulle alterazioni del mais. Revista di Igiene e Sanita Pubblica Ann 1896; 7:825-868.

107. Nelson PH, Eugui E, Wang CC et al. Synthesis and immunosuppressive activity of some side-chain variants of mycophenolic acid. Journal of Medicinal Chemistry 1990; 33(2):833-8.

108. Morris RE, Hoyt EG, Eugui E et al. Prolongation of rat heart allograft survival by RS-61443. Surg Forum 1989; 40:337-338.
109. Laurent AF, Dumont S, Poindron P et al. Mycophenolic acid suppresses protein N-linked glycosylation in human monocytes and their adhesion to endothelial cells and to some substrates. Exp Hematol 1996; 24(1):59-67.
110. Morris RE, Hoyt EG, Murphy MP et al. Mycophenolic acid morpholinoethylester (RS-61443) is a new immunosuppressant that prevents and halts heart allograft rejection by selective inhibition of T- and B-cell purine synthesis. Transplantation Proceedings 1990; 22(4):1659-62.
111. Steele DM, Hullett DA, Bechstein WO et al. Effects of immunosuppressive therapy on the rat aortic allograft model. Transplantation Proceedings 1993; 25(1 Pt 1):754-5.
112. Raisanen-Sokolowski A, Vuoristo P, Myllarniemi M et al. Mycophenolate mofetil (MMF, RS-61443) inhibits inflammation and smooth muscle cell proliferation in rat aortic allografts. Transplant Immunology 1995; 3(4):342-51.
113. Raisanen-Sokolowski A, Aho P, Myllarniemi M et al. Inhibition of early chronic rejection in rat aortic allografts by mycophenolate mofetil (RS61443). Transplantation Proceedings 1995; 27(1):435.
114. Azuma H, Binder J, Heemann U et al. Effects of RS61443 on functional and morphological changes in chronically rejecting rat kidney allografts. Transplantation 1995; 59(4):460-6.
115. Azuma H, Binder J, Heemann U et al. Effect of RS61443 on chronic rejection of rat kidney allografts. Transplantation Proceedings 1995; 27(1):436-7.
116. Gregory CR, Huang X, Pratt RE et al. Treatment with rapamycin and mycophenolic acid reduces arterial intimal thickening produced by mechanical injury and allows endothelial replacement. Transplantation 1995; 59(5):655-61.
117. O'Hair D, McManus RP, Komorowski R. Inhibition of chronic vascular rejection in primate cardiac xenografts using mycophenolate mofetil. Annals of Thoracic Surgery 1994; 58(5):1311-5.
118. Halloran P, Mathew T, Tomlanovich S et al. Mycophenolate mofetil in renal allograft recipients: A pooled efficacy analysis of three randomized, double-blind, clinical studies in prevention of rejection. The International Mycophenolate Mofetil Renal Transplant Study Groups. Transplantation 1997; 63(1):39-47.
119. Morris RE. New Immunosuppressive Drugs. Philadelphia London Toronto Montreal Sydney Tokyo: W.B.Saunders Company, 1995.
120. Calne R, Collier D, Lim S et al. Rapamycin for immunosuppression in Organ Allografting. Lancet 1989; 2:227.
121. Sabatini D, Erdjument-Bromage H, Lui M et al. RAFT-1: A mammalian protein that binds to FKBP12 in a rapamycin dependant fashion and is homologous to yeast cells. Cell 1994; 78:35-43.
122. Brattstrom C, Tyden G, Sawe J et al. A randomized, double-blind, placebo-controlled study to determine safety, tolerance, and preliminary pharmacokinetics of ascending single doses of orally administered sirolimus (rapamycin) in stable renal transplant recipients. Transplant Proc 1996; 28(2):985-6.
123. Cao W, Mohacsi P, Shorthouse R et al. Effects of rapamycin on growth factor-stimulated vascular smooth muscle cell DNA synthesis. Inhibition of basic fibroblast growth factor and platelet-derived growth factor action and antagonism of rapamycin by FK506. Transplantation 1995; 59(3):390-5.
124. Marx SO, Jayaraman T, Go LO et al. Rapamycin-FKBP inhibits cell cycle regulators of proliferation in vascular smooth muscle cells. Circulation Research 1995; 76(3):412-7.
125. Poon M, Marx SO, Gallo R et al. Rapamycin inhibits vascular smooth muscle cell migration. Journal of Clinical Investigation 1996; 98(10):2277-83.
126. Schmid C, Heemann U, Azuma H et al. Rapamycin inhibits transplant vasculopathy in long-surviving rat heart allografts. Transplantation 1995; 60(7):729-33.
127. Belitsky P, Gulanikar A, He G et al. Effect of immunosuppression on chronic rejection in the rat aortic allograft model. Transplantation Proceedings 1993; 25(1 Pt 2):935.
128. Gregory CR, Huie P, Billingham ME et al. Rapamycin inhibits arterial intimal thickening caused by both alloimmune and mechanical injury. Its effect on cellular, growth factor, and cytokine response in injured vessels. Transplantation 1993; 55(6):1409-18.
129. Morris RE, Huang X, Gregory CR et al. Studies in experimental models of chronic rejection: Use of rapamycin (sirolimus) and isoxazole derivatives (leflunomide and its analogue) for the suppression of graft vascular disease and obliterative bronchiolitis. Transplantation Proceedings 1995; 27(3):2068-9.
130. Wasowska B, Wieder KJ, Hancock WW et al. Cytokine and alloantibody networks in long term cardiac allografts in rat recipients treated with rapamycin. J Immunol 1996; 156(1):395-404.

131. Kahan B. Sirolimus: A New Agent for Clinical Renal Transplantation. Transplant Proceedings 1997; 29(1/2):48-50.

132. Murgia MG, Jordan S, Kahan BD. The side effect profile of sirolimus: A phase I study in quiescent cyclosporine-prednisone-treated renal transplant patients. Kidney International 1996; 49(1):209-16.

133. Silva Junior HT, Morris RE. Leflunomide and malononitrilamides. American Journal of the Medical Sciences 1997; 313(5):289-301.

134. Swan SK, Crary GS, Guijarro C et al. Immunosuppressive effects of leflunomide in experimental chronic vascular rejection. Transplantation 1995; 60(8):887-90.

135. Siemasko KF, Chong AS, Williams JW et al. Regulation of B cell function by the immunosuppressive agent leflunomide. Transplantation 1996; 61(4):635-42.

136. Nair RV, Cao W, Morris RE. Inhibition of smooth muscle cell proliferation in vitro by leflunomide, a new immunosuppressant, is antagonized by uridine. Immunology Letters 1995; 48(2):77-80.

137. Nair RV, Cao W, Morris RE. The antiproliferative effect of leflunomide on vascular smooth muscle cells in vitro is mediated by selective inhibition of pyrimidine biosynthesis. Transplantation Proceedings 1996; 28(6):3081.

138. Lirtzman RA, Gregory CR, Levitski RE et al. Combined immunosuppression with leflunomide and cyclosporine prevents MLR-mismatched renal allograft rejection in a mongrel canine model. Transplantation Proceedings 1996; 28(2):945-7.

139. Yeh LS, Gregory CR, Griffey SM et al. Combination leflunomide and cyclosporine prevents rejection of functional whole limb allografts in the rat. Transplantation 1997; 64(6):919-22.

140. Xiao F, Chong A, Shen J et al. Pharmacologically induced regression of chronic transplant rejection. Transplantaton 1995; 60(10):1065-72.

141. Lin Y, Vandeputte M, Waer M. Effect of leflunomide and cyclosporine on the occurrence of chronic xenograft lesions. Kidney International—Supplement 1995; 52:S23-8.

142. MacDonald AS, Sabr K, MacAuley MA et al. Effects of leflunomide and cyclosporine on aortic allograft chronic rejection in the rat. Transplantation Proceedings 1994; 26(6):3244-5.

143. Hunsicker LG, Bennett LE. Design of trials of methods to reduce late renal allograft loss: The price of success. Kidney International—Supplement 1995; 52:S120-3.

144. Labarrere CA, Nelson DR, Faulk WP. Endothelial activation and development of coronary artery disease in transplanted human hearts [see comments]. JAMA 1997; 278(14):1169-75.

Index

A

Adhesion molecules 171-172, 190, 198
Angiography 7, 11-12, 17, 76, 91, 94, 97,
    144, 146, 190, 191, 201
Antibodies 173, 183, 191, 196, 197, 201
Antigen presenting cells 71, 75
Antithrombin 191
Apoptosis 40, 49, 56, 57, 76, 129, 135, 175,
    198
Arrhythmia 6
Atherosclerosis 1-3, 7-16, 18, 26, 30, 31-32,
    35-39, 41, 44-45, 54-55, 58-60, 121,
    141-142, 154, 159, 181, 191
Atorvastatin 184
Azathioprine 182, 193, 194-195, 198-199

B

Bile acid 181-182, 184-186

C

Calcineurin 196
Calmodulin 196
CD4+ T cells 73-74, 78-79, 81, 84
CD40 52, 72, 78, 80, 82, 84
CD8+ T cells 72, 79, 82, 84, 123
Cholesterol 26-27, 29-31, 41-44, 46-50, 53
    56, 58-59, 61, 95, 100, 102, 106, 109,
    121-122, 124, 156, 181-187, 194, 200
Coagulation 8, 50-53, 58, 61, 90
Coronary spasm 6
Cyclosporine 174, 181-183, 185-186, 190,
    192-201
Cytokines 173, 175, 190, 196, 199, 200
Cytomegalovirus 166, 183

D

Dendritic cells 71
Denervation 4, 18
Diet 181, 183, 186

E

E-selectin 32-34, 36, 39, 47, 58, 59, 72, 78,
    171, 172

EDRF 48
Eicosanoids 45
Endothelial cell activation 190
Endothelial cells 167-168, 172, 175, 190,
    191-192, 194, 197
Endothelin 194-195, 197
Endothelium 166-168, 171-172, 175

F

Fatty streak 27, 28, 30
Fibric acid 184-186, 188
Fluvastatin 184-187

G

Growth factors 174, 190, 200

H

HMG-CoA reductase 181, 184-185, 187-188
Hyperlipidemia 181, 183, 187

I

ICAM-1 171-172, 174
Interferon g 40
Intravascular ultrasound 2, 7, 10-14, 90, 121

L

Leflunomide 192, 200, 201
LFA-3 72, 81
Lovastatin 184-187

M

Macrophages 171, 175, 190, 201
Matrix proteins 175
Monocytes 27-42, 44-47, 52, 56, 58, 61, 71,
    74, 76, 78, 81-82, 167-168, 175
Mycophenolate 192, 197-199, 201
Myocardial infarction 6, 10, 30, 33-34, 48,
    57, 59, 121

N

Nitric oxide 198

P

P-selectin 31-33, 36, 39, 45, 50, 52, 105
Platelets 16, 30, 33, 37-38, 47, 49, 51-53, 57,
        61, 124, 175
Pravastatin 184-187

Q

Quantitative angiography 10-11, 16

R

Rapamycin 192, 198-201
Rejection 1- 2, 8-9, 16, 71-74, 76-77, 80,
        82-84, 90-91, 95, 97-98, 100, 102,
        105-107, 109, 118, 120, 122-124, 130,
        136, 138, 141-145, 147, 150, 155,
        158-161, 166-167, 170-172, 174-176,
        183-184, 190-191, 193-201
Renal allografts 171
Rhabdomyolysis 185, 187

S

Simvastatin 184,-187
Sirolimus 199-200
Smooth muscle cells 16, 26-30, 34-43, 45-46,
        49, 50, 52-57, 59, 61, 71, 79, 81, 83,
        87-88, 93, 95, 98-99, 103, 106-108, 111,
        121, 129, 130, 132, 134-135, 137-138,
        147, 151, 154-156, 158, 160, 190-193

T

Tacrolimus 174, 196-197, 199-201
Tolerance 185
Triglycerides 183, 186

V

VCAM-1 31-32, 34-35, 47, 58, 59, 72, 78,
        80, 105, 122-124, 149, 163, 171-172